Argumentation in Chemistry Education
Research, Policy and Practice

Advances in Chemistry Education Series

Editor-in-chief:
Keith S. Taber, *University of Cambridge, UK*

Series editors:
Avi Hofstein, *Weizmann Institute of Science, Israel*
Vicente Talanquer, *University of Arizona, USA*
David Treagust, *Curtin University, Australia*

Titles in the Series:
1: Professional Development of Chemistry Teachers: Theory and Practice
2: Argumentation in Chemistry Education: Research, Policy and Practice

How to obtain future titles on publication:
A standing order plan is available for this series. A standing order will bring delivery of each new volume immediately on publication.

For further information please contact:
Book Sales Department, Royal Society of Chemistry, Thomas Graham House, Science Park, Milton Road, Cambridge, CB4 0WF, UK
Telephone: +44 (0)1223 420066, Fax: +44 (0)1223 420247,
Email: booksales@rsc.org
Visit our website at www.rsc.org/books

Argumentation in Chemistry Education
Research, Policy and Practice

Edited by

Sibel Erduran
University of Oxford, UK
Email: sibel.erduran@education.ox.ac.uk

ROYAL SOCIETY OF CHEMISTRY

Advances in Chemistry Education Series No. 2

Hardback ISBN: 978-1-78801-212-6
PDF ISBN: 978-1-78801-264-5
EPUB ISBN: 978-1-78801-579-0
Print ISSN: 2056-9335
Electronic ISSN: 2056-9343
Paperback ISBN: 978-1-83916-744-7

A catalogue record for this book is available from the British Library

The Royal Society of Chemistry is a charity, registered in England and Wales, Number 207890, and a company incorporated in England by Royal Charter (Registered No. RC000524), registered office: Burlington House, Piccadilly, London W1J 0BA, UK, Telephone: +44 (0) 20 7437 8656.

For further information see our web site at www.rsc.org

Printed in the United Kingdom by CPI Group (UK) Ltd, Croydon, CR0 4YY, UK

Preface

This book brings together scholars from around the world to consider the role of argumentation in chemistry education research, curriculum policy and practice. The authors, who are from the UK, USA, Spain, Ireland, Israel, Turkey and South Africa, converge on one fundamental question: how can argumentation help improve chemistry education? Argumentation is typically defined as the justification of knowledge claims with evidence. For example, what is the evidence for an exothermic reaction and how are claims about such reactions justified?

Argumentation has received much attention in the science education research community since the 1990s. Many science curricula around the world have included references to argumentation. Research has highlighted the importance of argumentation as a form of discourse that is important in the acquisition of scientific knowledge and the development of habits of mind in science. Argumentation plays a central role in the building of explanations, models and theories. Scientists construct arguments to relate the evidence they select to the claims they reach through use of warrants and backings.

Although argumentation has received much attention in the science education research community and has been advocated in the science curricula of many countries, its uptake within chemistry education is still fairly limited. This book is the first collection that focuses on argumentation in the context of chemistry education. The aim of the book is to contextualise argumentation in chemistry education by drawing on accounts from research, curriculum policy and practice. The overall purpose of the book is to contribute to knowledge on how argumentation can be infused in chemistry education in various senses: teaching strategies, learning resources, assessment, professional development of teachers as well as particular topics

Advances in Chemistry Education Series No. 2
Argumentation in Chemistry Education: Research, Policy and Practice
Edited by Sibel Erduran

Published by the Royal Society of Chemistry, www.rsc.org

(including organic and physical chemistry) and contexts (including the laboratory and the cultural environment).

The chapters are organised around three themes: overview of research, resources and strategies, and the context of argumentation in chemistry education. Where relevant, the chapters conclude with some practical examples and implications for teaching and learning summarised at the end of the chapter in a section called *Practical Digest*. This section provides some tangible materials and strategies for chemistry teachers, curriculum developers and teacher educators.

Sibel Erduran

Author Biographies

Mehmet Aydeniz is an associate professor of science education in the Department of Theory and Practice in Teacher Education at The University of Tennessee, Knoxville. His research centres on assessment of student learning in K-16 science classrooms. He pursues two lines of research. His first line of research focuses on helping students appropriate epistemic and social norms of science when engaged in scientific inquiry in the context of argumentation, modeling and computational thinking. His second line of research focuses on teacher learning. He studies how teachers develop understandings, knowledge and skills to effectively engage their students in scientific practices and with scientific content. Within that he focuses on pre-service science teachers' development of pedagogical content knowledge and skills to teach science through argumentation. Other interests include engineering education, computational thinking and STEM integration.

Renée S. Cole is a Professor of Chemistry at the University of Iowa. Her research focuses on issues related to how students learn chemistry and how that guides the design of instructional materials and teaching strategies as well on efforts related to faculty development. She is involved in a number of multi-disciplinary projects such as the *Increase the Impact Project*, which developed resources for PIs to improve the propagation of their innovations. She was named a 2018–2020 Collegiate Scholar by the University of Iowa in recognition of her outstanding scholarship, teaching, and service. She is a Fellow of the American Chemical Society (2015) and has served as a Councilor for the Division of Chemical Education, Chair of the Chemistry Education Research Committee, and as Program Chair for the Women Chemists Committee. She is also an Associate Editor for the *Journal of Chemical Education* and has been a co-editor for two books focusing on chemistry education research. She was awarded the prestigious Iowa Women

Advances in Chemistry Education Series No. 2
Argumentation in Chemistry Education: Research, Policy and Practice
Edited by Sibel Erduran

Published by the Royal Society of Chemistry, www.rsc.org

of Innovation Award for Academic Innovation & Leadership (2014), the University of Central Missouri College of Science & Technology Award for Excellence in Teaching (2010), and the Missouri Governor's Award for Excellence in Education (2009). She has over 45 publications, and over 100 international and national presentations.

Beatriz Crujeiras-Pérez is a lecturer in Science Education at the University of Santiago de Compostela, in Spain. Previously she has worked as a researcher in Chemistry. Currently, she is engaged in pre-service teacher development, teaching Chemistry and Physics Education in primary and secondary education. Her doctoral dissertation focused on high school students' engagement in scientific practices in the chemistry laboratory. Her research interests focus on students and teacher learning, including inquiry-based teaching and learning, scientific and epistemic practices and critical thinking. She is the principal investigator on a research project funded by the Spanish Ministry of Economy, Industry and Competitiveness called Epis-pract, that seeks to analyse the influence of epistemic knowledge in the development of scientific practices (inquiry, modeling and argumentation).

Alison Cullinane is a postdoctoral researcher in the Department of Education at the University of Oxford on Project Calibrate, a Wellcome Trust funded project (grant number 209659/Z/17/Z). This project examines methods of assessing the practical aspects of science at GCSE level. Prior to taking up this position, Alison was a research officer and PhD student at the National STEM education research centre, EPI*STEM based in the University of Limerick (UL). She has contributed to the teaching of an introductory biochemistry course and science pedagogy courses to both primary and secondary pre-service teachers at UL, NUI Galway and Mary Immaculate College Limerick, Ireland. Her research interests include assessment and assessment design, practical science, nature of science and teacher education.

Sibel Erduran is a Professor of Science Education and a Fellow of St Cross College at University of Oxford, UK. She also holds a Visiting Professorship position at Zhejiang Normal University, China. Previously she held a Distinguished Chair Professor position at National Taiwan Normal as well as Visiting Professorships at Kristianstad University, Sweden and Bogazici University, Turkey. She is an Editor of *International Journal of Science Education,* Section Editor for *Science Education* and an elected member of the Executive Board of European Science Education Research Association. Her higher education was completed in the USA (PhD Vanderbilt, MSc Cornell, BA Northwestern), and she was employed at University of Pittsburgh, University of Bristol and King's College London. Her research interests focus on the inclusion of epistemic practices of science in science education, particularly in the context of chemistry. Her work on argumentation has received international recognition through awards from NARST and EASE, and received funding from a range of institutions including the Wellcome Trust, Spencer Foundation and Science Foundation Ireland. Her co-authored book (Erduran and Kaya) entitled *Transforming Teacher Education through the Epistemic Core of Chemistry: Empirical Evidence and Practical Strategies* has recently been published by Springer.

Bryan Henderson received his PhD from Stanford University in Science Education. His research pursues two crucial objectives: (1) given the substantial empirical evidence for the importance of our prior thinking in the construction of new thinking, learners need to be provided spaces where they feel safe to share their thinking at whatever stage their ideas might be in; and (2) how we exchange ideas can vary in sophistication, and hence, supports are necessary for students to articulate their thinking and the sharing of those ideas in an increasingly critical, evidence-based fashion. Dr Henderson is interested is in the utilization of educational technology to facilitate critical, peer-to-peer science learning. His classroom-based research on critical speaking and listening intersects with his psychometric development of assessments that gauge how students learn science through evidence-based argumentation. As the director of the *Braincandy* project (www.braincandy.org), Dr Henderson has developed a cloud-based technology that affords students the safety of participating in classroom activities anonymously, and then makes discrepancies in anonymous student thinking transparent to the entire classroom through visualization tools. In turn, these differences in thinking set the stage for authentic, peer-to-peer argumentation as students seek to overcome uncertainty in the pursuit of classroom consensus. Dr Henderson is currently an Assistant Professor of Learning Sciences at Arizona State University, where he is a recipient of the *ASU Centennial Professorship* for outstanding teaching, leadership, and service. In addition to a PhD from Stanford, Dr Henderson also possesses three Bachelor's degrees in Physics, Astronomy, and Philosophy (with distinction) from the University of Washington and two Master's degrees in Physics and Education from Portland State University. He is a Principal Investigator on a 3-million-dollar collaborative research grant between ASU and UC Berkeley, funded by the National Science Foundation (NSF).

Avi Hofstein carried out his work for almost 50 years in the context of chemistry teaching and programs for advancing scientific literacy among high school students who do not opt to major in one of the sciences. One of his central concerns has been the design of learning environments and instructional strategies that foster the development of conceptual understanding and interest in chemistry. He investigated our understanding of the learning that takes place in these environments. In particular, Avi and his research group tried to design a format of inquiry labs that can promote the development of inquiry skills such as question asking, experiment design, and interpretation skills. The challenge has been to develop a format that is usable by many teachers (up-scaling), sustainable and effective. Special efforts were directed to the development of innovative alternative portfolio-based assessments that are essential for assessing and supporting the development of such complex skills. This work was accompanied by extensive research and development of professional development models. In a similar manner, Avi has carried out innovative work on introducing chemical industry into the main line of chemistry studies. These activities led to the establishment of a centre for chemical industry in Israel which has been a model for such centres in many locations worldwide. He has been one of the pioneers investigating out of school learning environments and has advocated the need to bridge the gap between the formal and informal settings. Ideas that have suggested and investigated many years ago are now being implemented all over the world as promising means for raising the interest and relevance of science studies and for the development of scientific literacy.

María Pilar Jiménez-Aleixandre is *Ad Honorem* professor of Science Education at the University of Santiago de Compostela, in Spain. Before completing her PhD about conceptual change on evolution, she was a high school teacher and served as science education coordinator for Spanish in-service teacher education. Her research focuses on argumentation, epistemic practices, critical thinking and socio scientific issues. She has published around 60 books and book chapters, among them the first volume about *Argumentation in Science Education* (Erduran & Jiménez-Aleixandre, 2008) co-edited with Sibel Erduran. She has also authored or co-authored about 70 papers in refereed journals in English, Spanish, French, Portuguese and Galician, among them some seminal work on argumentation in science education. She has participated in European Union funded projects about introducing argumentation and critical thinking in teacher education. She currently serves on the boards of *Science Education* (as co-editor of the Issues and Trends section), *Science & Education* and *Environmental Education Research*, among other journals. She is also an award-winning author of poetry and fiction, under the pen name of Marilar Aleixandre, and a member of the Royal Galician Academy or Real Academia Galega (RAG).

Dvora Katchevitch taught high school chemistry for the period of almost 40 years. In the last 20 years she was active in almost all the facets of chemistry teachers' professional development. During this period, she completed her PhD on argumentation in the chemistry laboratory, for which she was awarded a prestigious award. Currently, Dr Katchevitch is the head of the National Center for Chemistry Teachers at the Weizmann Institute of Science, which focuses on professional development of chemistry teachers in Israel. A variety of activities for chemistry teachers are conducted in the framework of the centre, *e.g.*, workshops for professional learning communities (PLCs), or courses for developing learning materials that will advance chemistry teaching and learning in Israel. In addition, Dr Katchevitch was one of the authors of several textbooks in chemistry, science and technology education, and she is also the editor of a local journal for chemistry teachers.

Rachel Mamlok-Naaman studied chemistry and chemistry education. She is employed in the chemistry group at the Department of Science Teaching, the Weizmann Institute of Science, where she served both as the head of the National Center for Chemistry Teachers until December 2015, and as the coordinator of the chemistry group at the Department of Science Teaching (until June 2016). In addition, she was involved as a package leader in European projects. Mamlok-Naaman is also the coordinator of a special MSc program for chemistry teachers, in a project of Professional Learning Communities (PLC), in ARTIST, a project in the framework of Erasmus (as an external evaluator). Her publications focus on topics related to teachers' professional development, and to students' learning, *e.g.*, development, implementation, and evaluation of new curricular materials; research on students' perceptions of chemistry concepts; inquiry-type teaching and learning; relevance in chemistry education; the nature of science, and education for sustainable development (ESD), for which she has been selected as a 2018 Awardee for the ACS-CEI Award for Incorporation of Sustainability into the Chemistry Curriculum. For her work on chemistry teachers' professional development in Israel, she received the 2016 Maxine Singer Prize for outstanding scientists at the Weizmann Institute.

Alena Moon is completing her postdoctoral research fellowship in Chemistry at the University of Michigan and will begin as assistant professor of chemistry at University of Nebraska-Lincoln in Fall 2018. She received her PhD in chemistry education from Purdue University. Her graduate research involved analysing the impact of curricular materials on students' causal reasoning used in classroom argumentation. She is currently investigating the effect of Writing-to-Learn across introductory STEM and developing novel ways of analysing the quality of students' scientific reasoning evidenced by writing. Results from this work has been published in *Science Education, Journal of Research in Science Teaching, Journal of Chemical Education*, and *Chemistry Education Research and Practice.*

Audrey Msimanga is a Senior Lecturer in Science Education and the Academic Head of Postgraduate at Wits School of Education in the University of the Witwatersrand in Johannesburg, South Africa. Her interest is understanding teaching and learning of science in contexts of socio-economic and socio-cultural diversity. At the micro level Dr Msimanga explores the role and dynamics of classroom interaction in the teaching and learning of science, specifically how science teachers and students talk; how talk helps students make sense of science; what talk reveals about student scientific reasoning; the role of silence and language in science learning.

Brighton Mudadigwa is a final year PhD Student at the University of the Witwatersrand in Johannesburg, South Africa and a Physical Science teacher for 22 years. Brighton's research interest is in teaching for conceptual understanding with a focus on pedagogical link-making and social construction of knowledge. He is involved in professional development of chemistry teachers specifically to explore teaching approaches for learner conceptual understanding.

Anne O'Dwyer is a lecturer of Science Education in the Department of STEM Education at Mary Immaculate College, Limerick, Ireland, where she lectures on undergraduate and postgraduate modules, to both pre-service and in-service elementary teachers. Anne was a postdoctoral researcher in EPI-STEM, a National STEM education research centre based in University of Limerick. Here she ran science educational research at all levels and also taught on Chemistry and Chemistry pedagogy modules. Anne's doctoral research focused on Chemistry Education and specialised in developing authentic and innovative curriculum materials for teaching Organic Chemistry at second level from which she has published findings. (O'Dwyer & Childs, 2014, 2015).

Jonathan Osborne holds the Kamalachari Endowed Chair in Science Education at the Graduate School of Education, Stanford University (2009-). He started his career teaching high school physics and then moved to teacher training and research at King's College where he was appointed the Chair in Science Education in 2003. He was President of the US National Association for Research in Science Teaching (2006–7) and has won the Association's award for the best research publication in the *Journal of Research in Science Teaching* twice (2003 and 2004) and the Distinguished Contribution to Science Education Award in 2018. He was a member of the US National Academies Panel that produced the *Framework for K-12 Science Education*. He also chaired the expert group for the science assessments conducted by the OECD PISA Program in 2015 when science was the primary focus. Currently he is PI on the SNAP program to develop assessments for the Next Generation Science Standards in California. His research interests are in the role of argumentation in science and improving the teaching of literacy in science.

Diana Ng Yee Ping is a doctoral student at the Oxford University Centre for Educational Assessment. Her doctoral thesis examined the construct validity of a scientific reasoning test for primary school children in Singapore. The thesis addressed the need to develop valid and reliable instruments for testing reasoning in science learning. Her academic portfolio included conferences, seminar presentations, poster sessions, publications, abstract and manuscript reviews, as well as educational outreach efforts. She was a former teacher and examiner of national examinations in Singapore. During her teaching career, she received numerous awards including the country's highest honor for outstanding educators in 2008 – the President's Award for Teachers. In 2012, she received her Master of Education (Educational and Psychological Measurement and Evaluation) degree from Nanyang Technological University in Singapore. For the degree, she received the certificate of commendation for the 2012 Singapore Teacher's Union Gold Medal as one of the top three graduates of the course.

Aybuke Pabuccu is an Assistant Professor of Chemistry Education at Canakkale Onsekiz Mart University, Turkey. From 2008, she teaches undergraduate and graduate courses such as, General Chemistry; Organic Chemistry; History of Science, Education of Environmental Protection; Research Project in Chemistry Education; Teaching Practice; Connecting Chemistry Education with Other Disciplines. She has been a Visiting Scholar at Bristol University, UK and University of Illinois at Urbana–Champaign, USA. She served as an elected member to the International Committee of NARST. She received her PhD, master's and bachelor's degrees in secondary science and mathematics education from Middle East Technical University at Turkey. She has been a chemistry teacher in a high school in Ankara, Turkey. Her research focuses on the nature of science and the epistemic practices of science. She is the co-writer of three books and a book chapter on chemistry education, the one in Turkish language: Bonding Chemistry and Argument-Teaching and Learning Argumentation through Chemistry Stories (2012), 5E Learning Cycle Laboratory Instruction Improving Understanding of Acid Based Concepts (2012), Kimya ve Argumantasyon (2012) and Promoting argumentation in the context of chemistry stories (Chapter in the book named Relevant chemistry education from theory to practice, 2015).

Courtney Stanford is a postdoctoral researcher at the Virginia Commonwealth University and will begin as assistant professor of chemistry at Ball State University in Fall 2018. She earned an M.S. in organic chemistry from the University of Connecticut and a PhD degree in chemistry education from the University of Iowa. Her current research has focused on designing resources to assist in the identification, development, and assessment of workplace skills in STEM classrooms, and investigating the connections between information processing and symbolic representations used in organic chemistry. As part of her graduate work she examined the influences of instructor facilitation and course materials on student argumentation, and the propagation of STEM educational innovations.

Marcy H. Towns is a Professor of Chemistry and Director of General Chemistry at Purdue University. In 2017 she received both the ACS Award for Achievement in Research for the Teaching and Learning of Chemistry and the most prestigious award the ACS offers for excellence in teaching, the James Flack Norris Award for Outstanding Achievement in the Teaching of Chemistry. She is a Fellow of the American Association for the Advancement (AAAS) 2009, and a Fellow of the American Chemical Society (ACS) 2012.

She received the Society of College Science Teachers and National Science Teachers Association 2015 Outstanding Undergraduate Science Teaching Award. She has won Purdue University's most prestigious honors for teaching including The Class of 1922 Outstanding Innovation in Helping Students Learn Award (2015) and the Charles B. Murphy Outstanding Undergraduate Teaching Award (2013). She also received the chemistry department's most prestigious honor for teaching, the Arthur B. Kelly Award in 2013. She was the American Chemical Society's (ACS) Division of Chemical Education Chair in 2015 and served on the ACS Examinations Institute Board of Trustees for 9 years. She has over 80 publications, over 1600 citations, and over 100 international and national presentations. She is an Associate Editor for the *Journal of Chemical Education*, focusing on manuscripts pertaining to chemistry education research.

Abha Vaishampayan is a former physics teacher who is pursuing her PhD in Science Education at The Pennsylvania State University. Her dissertation study focuses on understanding students' sensemaking practices in astronomy. Abha has been working on developing a middle-school astronomy curriculum in collaboration with the Harvard-Smithsonian Center for Astrophysics, which facilitates learning in multimodal ways. She is also involved in developing a teacher preparation program to adapt Ambitious Science Teaching that promote discourse-rich practices for secondary science teaching. Abha is a recipient of the Vincent N. and Lois W. Lunetta Fellowship in Science Education (2015–16) and the Dean's Graduate Assistantship A ward for engaged scholarship and research in Science Education.

Carla Zembal-Saul is a science education scholar and science teacher educator. She holds the Kahn Endowed Professorship in STEM Education at The Pennsylvania State University. Her work is situated in school–university–community partnerships in the United States and abroad. Zembal-Saul's research investigates instructional practices and tools that support preservice and in-service teachers in engaging children productively in scientific discourse and practices, with an emphasis on argumentation and constructing evidence-based explanations. She is deeply invested in practitioner inquiry and video analysis of teaching as mechanisms for advancing teacher learning and development across the professional continuum. In addition to contributing to the research community, Zembal-Saul is committed to collaborating with teachers, bridging research and practice, and co-authoring publications with practitioners. She was recognized as a National Science Teachers Association Fellow in 2015, and she served on the *National Academies of Sciences, Engineering and Medicine: Board on Science Education* consensus committee that authored the report, *Science Teachers' Learning: Enhancing Opportunities, Creating Supporting Contexts* (2015).

Acknowledgements

I would like to acknowledge and thank

- Connor Sheppard for his support and informative guidance in the production process;
- Michelle Carey for her help in setting up the project and for enabling continued support;
- Keith Taber for his initiative to make the book idea possible;
- Royal Society of Chemistry for recognising the significance of argumentation for chemistry education.

Advances in Chemistry Education Series No. 2
Argumentation in Chemistry Education: Research, Policy and Practice
Edited by Sibel Erduran

Published by the Royal Society of Chemistry, www.rsc.org

Dedication

I dedicate this book to Richard Duschl, my doctoral supervisor and career mentor who exposed me to argumentation studies in the 1990s. His progressive vision for science education continues to inspire me to this day.

Advances in Chemistry Education Series No. 2
Argumentation in Chemistry Education: Research, Policy and Practice
Edited by Sibel Erduran

Published by the Royal Society of Chemistry, www.rsc.org

Contents

Advances in Chemistry Education Series No. 2
Argumentation in Chemistry Education: Research, Policy and Practice
Edited by Sibel Erduran

Published by the Royal Society of Chemistry, www.rsc.org

CHAPTER 1

Argumentation in Chemistry Education: An Overview

SIBEL ERDURAN

University of Oxford, Oxford, UK
Email: sibel.erduran@education.ox.ac.uk

1.1 Introduction

Many chemistry lessons include activities that promote a sense of awe and wonder in students. Consider, for instance, the demonstration where a solution of ammonia is poured into three beakers which contain (unknown to the students) small amounts of phenolphthalein, lead nitrate and copper(II) sulfate solutions. The beakers' contents turn red, milky white and deep blue respectively. Pouring the contents of the beakers into acid reverses the changes, to give a colourless solution.[1] The changes in colour are impressive. The activity is likely to enthuse and engage the students but what does it communicate about chemistry? Does the observation of these colour changes count as doing 'chemistry'? What is 'chemical' about this demonstration? What chemistry do students learn by observing such a demonstration? From the standpoint of students who have not been introduced to the background information on the chemicals involved, without a language to explain why the colour changes, the observation practically amounts to magic!

Next, suppose that the teacher explains to the students that the solutions in the beakers are phenolphthalein, lead nitrate and copper sulfate. Phenolphthalein turns red, the lead nitrate forms a milky white precipitate of lead(II) hydroxide and the copper sulfate forms the deep blue

Advances in Chemistry Education Series No. 2
Argumentation in Chemistry Education: Research, Policy and Practice
Edited by Sibel Erduran

Published by the Royal Society of Chemistry, www.rsc.org

$[Cu(NH_3)_4(H_2O)_2]^{2+}$. Furthermore, the teacher explains, the colour changes because the following reactions are taking place.

$$Pb(NO_3)_2(aq) + 2NH_3(aq) + 2H_2O(l) \rightarrow Pb(OH)_2(s) + 2NH_4NO_3(aq)$$

$$[Cu(H_2O)_6]^{2+}(aq) + 4NH_3(aq) \rightarrow [Cu(NH_3)_4(H_2O)_2]^{2+}(aq) + 4H_2O(l)$$

In the subsequent step, the reactions are reversed in acid as follows:

$$Pb(OH)_2(s) + 2HNO_3(aq) \rightarrow Pb(NO_3)_2(aq) + 2H_2O(l)$$

$$[Cu(NH)(HO)]^{2+}(aq) + 4H + (aq) + 4HO(l) \rightarrow [Cu(HO)]^{2+}(aq) + 4NH\ +(aq)$$

The teacher uses the formulae and equations to explain the chemical reactions that account for the changes in colour. Students might ask questions about particular aspects of the equations that might be confusing to them and eventually the class settles on an understanding of how the colour change is a result of the chemical reactions represented in the equations. Let us examine how the teacher and the students engage in this demonstration if the demonstration proceeds as described above. From the onset, the teacher already has the background knowledge including knowledge of the chemical formulae and equations that help interpret the changes in colour. The students do not. From their point of view, this is an aesthetic experience. The teacher then tells the students what the chemical composition of the liquids are and what accounts for the colour changes. When the teacher explains what is happening through chemical terminology and conventions like chemical equations, he or she makes a *claim* about what is in the beakers and why the colour changes.

Obviously, the chemical formulae and equations do not appear in the demonstration itself. They are in the mind of the teacher. They are abstract notions that chemists have produced as part of a language to explain chemical phenomena. They are representations that are institutionalised in the professional community of chemists. The language gives chemists a means to interpret phenomena but the phenomena themselves at the observation level do not provide any direct clues about the chemicals represented in the symbolism. The symbolism itself has been produced over centuries and it needs to be learned. It is not accessible through direct experience with chemical phenomena. If the observer does not have the chemical language, the entire experience has no basis in chemistry. In the worst case scenario, the activity is mere entertainment. In the best case scenario in many lessons, it is about a dogmatic expression of a series of claims superimposed onto some visually stimulating phenomena. As observers, the students have to take the word of the teacher that the provided explanations and the reasoning are true, making the teacher the owner of indisputable knowledge. In terms of the dynamics of interaction, the teacher disseminates the knowledge and expects it to be believed. The students assume the position of having little knowledge, ready to accept the claims being made by the teacher.

1.2 Infusing Argumentation in Teaching and Learning

At this point, we can ask: are the students actually engaged in chemistry in this episode? What makes a demonstration scientific as opposed to magical? What are some key features of science that school science should include such that students experience authentic science? A significant aspect of science, including chemistry, is its reliance on the justification of claims with evidence. Without evidence, science could not operate. Indeed, the reliance on evidence is a central defining feature of science.[2] Can we identify claims and evidence in the hypothetical scenario about ammonia and colour changes? We can certainly attribute the teacher's assignment of the chemicals to each test tube's content as a claim but what can we make of the evidence provided? It is as if further claims are being made about what the chemical formulae are and how the equations underlying the demonstration explain the observations. From the standpoint of the observing students, the entire demonstration is a set of claims, some of which are treated as evidence. Indeed, the information that is being presented to explain the observations is symbolic and abstract, and does not have any direct link to the observed colours, for instance. Could not any other formula, for instance, account for a similar colour change? How would the students be able to differentiate another set of chemicals corresponding to the observations or not? Of course, is not always easy for students to access the evidence in the first place. How are the students supposed to guess what is in the beakers or the liquid that is being added to the beakers? Much of chemistry actually relies on many similar macroscopic properties (*e.g.* colourless liquids) that, unless one has the chemistry language to define and reason with, are impossible to decipher unless advanced chemical testing such as high-performance liquid chormatography or gas chromatography is conducted.

Such testing would be inconceivable for the purposes of school chemistry to determine the component of every single chemical that the students might be expected to use. Indeed this would provide a major distraction to the pedagogical goals that the teacher might have for learning about a particular chemical phenomenon. Can we, then, at least engage students in some modes of thinking that resemble evidence-based reasoning even if they still need to take the teacher's word for many aspects of what they are exposed to? In the hypothetical example of the ammonia demonstration, how can students be put into a role where they are more empowered to reason with some form of evidence to reach some conclusions themselves rather than being told about these conclusions as claims made by the teacher? Suppose that in this scenario, the students were given the main formulae on separate pieces of paper and they were tasked to figure out how they can put them together to account for the observations from the demonstration. The students can get together in groups, research each formula and reason what goes with what in order to produce the end colour. Furthermore, they can be expected to justify why they think so.

Suppose also that in a classroom, some students end up generating wrong chemical equations or attributing the wrong equation to the particular observation. The diversity of 'claims' about the chemical equations can be a rich ground for discussion between groups in order to provide some reasons for why one equation is appropriate and also why another is not. Being put into a position to justify your own claim, and to refute an opposing claim, would at least engage the students in a mode of thinking that is typical of scientists. Alternatively, the tasks could be 'biased' in a way so as to actually get the students to refute a wrong claim. A wrong equation could be presented in a list of correct ones and students are asked to choose and justify which equation they agree with and which they do not, providing their reasons for either case. Without such potential strategies to engage students in discussions about evidence and justifications, it is difficult to imagine how students could adopt a scientific mindset. Watching the teacher and believing in what he or she claims to be the case is possibly the most unscientific role that students could be expected to play in a science lesson.

In science education research, the work surrounding students' as well as teachers' reasoning with evidence, justifications and claims is referred to as 'argumentation'. There is now a substantial body of work in this area as evidenced by content reviews of key journals[3] and the presence of books dedicated to this theme.[4] In its simplest definition, an argument consists of a claim justified with evidence, and argumentation refers to the process of constructing arguments.[4] The intention of this book is to contextualise argumentation in chemistry education by drawing on accounts from research, curriculum policy and practice. Although there is vast amount of work on argumentation in science education at the present, there are few studies in the context of argumentation in chemistry education.[5] Hence the overall purpose of the book is to contribute to knowledge on how argumentation can be infused in chemistry education in various senses: the curriculum, teaching strategies, learning resources, assessment, professional development of teachers as well as particular topics (including organic and physical chemistry) and contexts (including the laboratory and social issues and the cultural environment). In the following sections, a brief review of the main rationale for including argumentation in chemistry education is presented, along with a synopsis of how argumentation is represented in school curricula.

1.3 Curriculum Context of Argumentation

There are at least five potential contributions of argumentation to science education.[4] Argumentation can support: (a) the access to the cognitive and metacognitive processes (*i.e.* evidence-based reasoning) that characterise expert performance and model such processes for students; (b) the development of communicative competences and particularly critical thinking; (c) the achievement of scientific literacy and empowerment of students to engage in ways of talking in science; (d) the enculturation into the practices of the scientific culture and the development of epistemic criteria for knowledge evaluation; (e) the development of reasoning, particularly the choice of theories or positions based on rational criteria.

Many curricula from around the world, from Chile to Taiwan have recognised the contributions that argumentation can make to science teaching and learning.[5] For example, since the 1990s seminal curriculum standards in the United States have been advocating the inclusion not only of scientific knowledge but also how we get to construct scientific knowledge.[6,7] In England and Wales, there have been a series of science curricula that emphasised argumentation-related themes, for example through the *Ideas and Evidence*[8] and *How Science Works*[9] components of the National Science Curriculum. The latest national science curriculum for GCSE is organised around four main themes that are applied to each subject area of biology, chemistry and physics: (a) working scientifically, (b) experimental skills and strategies, (c) analysis and evaluation, and (d) vocabulary, units, symbols and nomenclature. The document includes learning objectives such as "*being objective, evaluating data in terms of accuracy, precision, repeatability and reproducibility and identifying potential sources of random and systematic error*" and "*presenting reasoned explanations, including relating data to hypotheses.*"[10] Such references emphasise the epistemic aspects of argumentation that concern the generation and evaluation of evidence. There are other references in the curriculum documents that stress the importance of debate and communication of arguments, for instance as suggested by the learning objective of "*...recognising the importance of peer review of results and of communication of results to a range of audiences.*"[10]

Despite such curriculum reform rhetoric, the effective implementation of argumentation in everyday chemistry classrooms remains a challenge. How can learning objectives on argumentation be transformed for teaching and learning purposes? What teaching strategies need to be in place to ensure that students can engage in argumentation effectively? What tools including technology can be used to support the teaching and learning of argumentation? How can argumentation be linked to cross-curricular subjects? How can it be assessed? How can teachers be supported in their professional development to infuse argumentation into their lessons? What does argumentation look like in different contexts such as the laboratory or a classroom with limited resources? This book is intended to address such questions by providing some insight into the theoretical underpinnings of argumentation in chemistry education and to offer some practical guidelines for teachers, curriculum developers and teacher educators.

1.4 Overview of the Book

The chapters in this book are organised around three themes: overview of research, resources and strategies and finally, the context of argumentation in chemistry education. Most chapters conclude with some practical examples for teaching and learning summarised at the end of the chapter.

These concluding sections are meant to distil ideas presented in the chapter for practical use and to provide some concrete suggestions to teachers. Sometimes these might include actual lesson resources that teachers could use (Chapters 2, 4 and 10) including examples of online tools (Chapter 5). Some chapters illustrate the richness of argumentation discourse at secondary (Chapter 12) and tertiary (Chapter 11) levels of education. Links are made to related interdisciplinary concepts such as STEM (science, technology, engineering and mathematics) (Chapter 3) and socio-scientific issues (Chapter 9) as well as particular learning contexts such as the laboratory (Chapter 8). As a relatively unfamiliar strategy, argumentation may present challenges for chemistry teachers. In this respect, teacher education including pre-service and in-service teacher education becomes critical for successful enactment of argumentation in lessons. Hence, one chapter tackles the issue of teacher education providing some insight into how teacher education practices need to be reformed to infuse argumentation (Chapter 7). In a similar fashion, assessment of argumentation skills or understanding of argumentation demands new perspectives and strategies, and some examples are presented (Chapter 6).

An important framework for the definition of argumentation has been Toulmin's Argument Pattern (TAP) which has been used extensively in research in science education.[11] In Chapter 2, Aydeniz provides a summary and a critique of argumentation studies conducted in chemical education and discusses implications for practice and future research. A critical part of this chapter is the discussion on the nature of chemical knowledge and how to promote philosophy of chemistry in argumentation-based teaching and learning.[12] This chapter also provides a sample argumentation task that can be adopted and used by high school chemistry teachers, professors of introductory chemistry courses and pre-service teacher educators.

Considering argumentation concerns issues of language, knowledge construction and social interaction, it is, by definition, an interdisciplinary concept. Hence, in Chapter 3, Crujeiras-Pérez and Jiménez-Aleixandre target the issue of interdisciplinarity in addressing the role of argumentation in chemistry education. The authors discuss the characterisation of interdisciplinary learning, understood as the capacity to integrate knowledge and modes of thinking from two or more disciplines to produce a cognitive advancement in ways that would have been impossible or unlikely through single disciplinary means. The benefits and challenges for implementing interdisciplinary learning are addressed. The chapter includes examples of interdisciplinary chemistry tasks set in real-life contexts that require participants' engagement in a range of argumentation practices such as drawing evidence-based conclusions or evaluating arguments.

Following on from the initial chapters that provide the foundation of the book through reviews of curriculum policy, the role of argumentation in teaching and learning, and interdisciplinarity, the next three chapters focus more closely on teaching and learning resources and strategies. In Chapter 4,

Cullinane and O'Dwyer provide some example resources and strategies useful for supporting the teaching and learning of argumentation in the secondary chemistry classroom from both authors' personal experiences, as well as from a review of current literature. The authors describe how strategies could be operationalised at the level of the classroom. Selected strategies include concept cartoons, evidence-based reasoning, and scenario analysis on social media. Some ideas and suggested resources are drawn from existing sources and some have been developed for the chapter and therefore are a unique contribution to the area of chemistry argumentation. The resources further include the examination of social media platforms to address plastic pollution and energy sources which contextualise chemistry in order to enthuse students. The authors examine chemistry concepts and socio-scientific issues and how they can be situated in relation to argumentation in secondary chemistry education.

In Chapter 5, Henderson and Osborne explore how technology and evidence-based argumentation are both central to the work of the professional scientific community. The authors state that the use of technology to promote the learning of science through argumentation is a recent educational practice, and they provide some examples of how technology can support chemistry students' engagement in argumentation, as well as how technology can aid analysis of student argumentation by both educational researchers and chemistry instructors. They present several contemporary projects that developed practical, next-generation technology to support argumentation. These advances have the potential to scaffold and improve the quality of the argumentation process in the chemistry classroom.

Assessment plays a central role in educational practice. As such, argumentation as a novel strategy for chemistry education creates new demands on how teachers' and students' understanding of and engagement in argumentation can be assessed. In Chapter 6, Ng cautions that although recent revisions to the primary and secondary science curricula of some countries incorporated argumentation as part of the learning outcomes, there are still few guidelines on the assessment of these abilities. The chapter reviews existing learning progression research on scientific argumentation and proposes a model for the design and construction of test items.

The inclusion of argumentation in chemistry education has implications for teachers' professional learning. In Chapter 7, Zembal-Saul and Vaishampayan illustrate the demanding nature of supporting teacher development in argumentation by using conceptual, epistemic, and social lenses. In the context of current reform efforts in science education, the authors argue that participation in scientific discourse and practices is a necessary condition for learning disciplinary core ideas. The chapter concludes with what the authors believe to be one of the most pressing considerations in moving forward with the continuous professional development of teachers: that of equity in opportunity and access to high quality science learning experiences for every student.

The next five chapters situate argumentation in context: in the laboratory, in socio-scientific issues, in organic and physical chemistry and in low-income countries. The purpose of this section of the book is to highlight what issues emerge in infusing argumentation in different contexts be it the setting or the topic. The assumption made in this section of the book is that there may be some aspects of chemistry that affords particular opportunities and constraints to the implementation of argumentation.[7] In Chapter 8, Hofstein, Katchevitch and Mamlok-Naaman focus on the processes through which students construct arguments in the chemistry laboratory while conducting different types of inquiry experiments. The authors have carried out studies with students based on experiments that differed in level of complexity. It was found that the more complex experiments served as a better platform for developing arguments. Moreover, the authors identified factors that support the teaching and learning of arguments in the laboratory context, namely the role of asking questions and strategies for dealing with unexpected results from experiments.

In Chapter 9, Christodolou and Grace discuss socio-scientific issues-based education and the role that argumentation plays in the context of chemistry education. They discuss the importance of formulating and developing socio-scientific reasoning and the engagement of young people in active citizenship. The authors present a pedagogical approach called "socio-scientific inquiry based-learning" (SSIBL) which incorporates socio-scientific reasoning and citizenship education within a wider framework of responsible research and innovation. They consider the SSIBL framework in the context of "Green & Sustainable Chemistry" as an example of how socio-scientific argumentation can be contextualised in chemistry education.

Pabuccu problematises in Chapter 10 the difficulties that university students encounter in learning organic chemistry. She presents studies on argumentation in organic chemistry education. Both instructors and students might perceive organic chemistry as a subject that does not offer much to debate but rather demands much memorisation. The author illustrates example activities to support the incorporation of argumentation in organic chemistry lessons. Unlike regular organic chemistry activities, these activities are not only aimed at helping students understand organic chemistry concepts, but also to develop their ability to reason with evidence and justifications. The chapter also includes the author's own reflections as a teacher educator on teaching organic chemistry through argumentation.

In Chapter 11, Towns, Cole, Moon and Stanford investigate the use of argumentation in physical chemistry. The authors highlight how they combined argumentation analysis with methodological frameworks from chemistry, the chemical thinking learning progression and Johnstone's triangle, to provide further insight into the nature of argumentation in physical chemistry classrooms. They also adapted a framework from undergraduate mathematics education research to illustrate the discursive moves that instructors make in facilitating argumentation in their classes. Combining these methodological approaches, the authors offer a novel way to examine effective argumentation strategies.

Finally, in Chapter 12 Msimanga and Mudadigwa make the case that argumentation is a high level form of engagement which is difficult to establish, support or sustain, particularly in constrained teaching and learning contexts that prevail in many low-income contexts. The authors use examples drawn from the South African chemistry education context, which is as diverse as the country's general population in terms of social, economic, cultural and linguistic differences. This diversity is partly responsible for the persistent disparate learner experiences in spite of huge government efforts towards equitable provision of education. Some of the challenges include resourcing of schools and inadequate infrastructure. For various social and political reasons science is taught and learned in English in spite of general low proficiency in learners. Using vignettes from real classroom interactions the authors illustrate how argumentation plays out in some chemistry classrooms in the context of a low-income country.

Overall, the chapters in this book build on a vision for chemistry education that brings to the foreground the use of evidence and justifications in making claims, the underpinning components of argumentation. It is through engagement in argumentation that students will develop scientific habits of mind so that while they enjoy the awe and wonder of chemistry, they can also question what they are observing and adopt a critical attitude towards the claims being made about macro-level phenomena and how they are explained at micro- and symbolic levels. Expecting students to believe in claims, even if they are teachers' claims, is not good enough for a science lesson because at its core this approach not scientific. It is dogmatic. We need to move from the landscape of magic and aesthetic experience coupled with dogmatic claim making towards a more scientifically authentic engagement between students and teachers, as limited as this may be due to difficulties in accessing chemical evidence fully. Research on argumentation provides some clues as to how to implement argumentation in chemistry lessons, and this book is intended to provide some background knowledge and practical guidelines.

References

1. Nuffield Foundation and the Royal Society of Chemistry, 2015.
2. R. Giere, *Understanding Scientific Reasoning*, Holt, Rinehart and Winston, Fort Worth, TX, 3rd edn, 1991.
3. S. Erduran, Y. Ozdem and J. Y. Park, *Int. J. STEM Educ.*, 2015, **2**, 5.
4. S. Erduran and M. P. Jimenez-Aleixandre, *Argumentation in Science Education: Perspectives from Classroom-Based Research*, Springer, 2007.
5. E. Kaya, *Int. J. Sci. Educ.*, 2013, **35**(7), 1139–1158.
6. American Association for the Advancement of Science, *Benchmarks for Science Literacy*, New York, 1993.

7. National Research Council, *National Science Education Standards*, National Academy Press, Washington, DC, 1996.
8. Department for Education and Skills and Qualifications and Curriculum Authority, Science: The National Curriculum for England, HMSO, 2004.
9. Qualifications and Curriculum Authority, *How science works*, 2007.
10. Department for Education Science Programmes of Study: Key Stage 4, *National Curriculum in England*, HMO, London, 2014.
11. S. Toulmin, *The Uses of Argument*, Cambridge University Press, 1958.
12. S. Erduran, *Found. Chem.*, 2007, **9**(3), 247–263.

CHAPTER 2

Teaching and Learning Chemistry through Argumentation

MEHMET AYDENIZ

The University of Tennessee, Department of Theory and Practice in Teacher Education, A 408 Jane & Bailey Education Complex, 1126 Volunteer Blvd, Knoxville 37996, USA
Email: maydeniz@utk.edu

2.1 Teaching and Learning Chemistry through Argumentation

2.1.1 Introduction

Chemistry is a fundamental component of the science curriculum starting from elementary education all the way to the graduate school. Students learn about atoms, molecules, mixtures, solutions in elementary grades, and then learn about classification, chemical composition of materials, and chemical models in middle school, and finally about chemical reactions and energy in high school. When they come to university, they are guided to expand on their knowledge of chemistry acquired in high school, and learn about models, interactions, energy, analysis of chemical materials.[1] More, many students' success in other STEM courses ranging from physics to electrical engineering and medicine relies heavily on their knowledge of chemistry. Considering the abundance of chemical knowledge in university curricula,

Advances in Chemistry Education Series No. 2
Argumentation in Chemistry Education: Research, Policy and Practice
Edited by Sibel Erduran

Published by the Royal Society of Chemistry, www.rsc.org

and its importance for students' acquisition of knowledge and problem-solving skills in related subjects, considerable effort has been invested in helping students to develop a solid knowledge base and acquire robust problem-solving skills in chemistry courses. For instance, chemistry educators have used several reform-based instructional strategies such as inquiry, argumentation and group work to help students with their academic achievement in chemistry. This chapter focuses on the use of argumentation in chemistry education for the purposes of teaching and learning.

Historically, chemistry education has placed a heavy emphasis on students' conceptual understanding of core scientific ideas and acquisition of problem-solving skills through lectures, readings, problem solving and imitating scientific experiments in preparation for standardized assessments.[2] However, recent science education reform efforts expect students to learn science as practice, using epistemologies, methods and practices of scientists,[3–5] and develop and enact scientific identities.[6] It is believed that the ultimate goal of science education should be helping students to become legitimate participants of the scientific community.[3,6,7] This requires becoming proficient in using valid and reliable evidence to develop explanatory and persuasive arguments, developing, evaluating and critiquing explanatory or predictive models of natural phenomena, and productively engaging in argumentation using epistemic, conceptual and social norms of science.[2,3,8] Above all, recent science education reform documents call for students to assume greater epistemic agency and authority[2,9] with regard to learning, production, evaluation, justification and critique of knowledge.[2,3,10] Consequently argumentation, a prominent practice of scientists, has received significant and growing attention from science educators across different curricular contexts. This chapter focuses on chemistry educators' uptake of argumentation, provide a summary and critique of argumentation studies conducted in the field of chemical education, and discuss implications for practice and future argumentation studies in chemistry education.

In this chapter, I draw evidence from several bodies of literature to emphasize the intersection of chemistry and argumentation. First, I draw from philosophy of science literature to provide an overview of the nature of chemical knowledge and its practices. In doing so, I create a context for establishing the relationship between nature of knowledge and role of argumentation in advancements in our knowledge of chemistry. Second, I refer to learning sciences literature in order to situate the importance of argumentation for students' conceptual understanding of core scientific ideas and their acquisition of scientific inquiry and explanation skills. Third, I provide a review of argumentation literature in chemical education both at the K-12 and higher education levels. Then, I provide a discussion of key issues related to argumentation, challenges in curriculum development and assessment of student learning in the context of argumentation-based interventions. Finally, I provide an exemplary argumentation task that can be adopted and implemented by practitioners in their unique contexts.

2.1.2 Nature of Chemical Knowledge

Erduran and colleagues argue how we define chemical knowledge, and how much we know about the way chemical knowledge is constructed, has significant implications for how we teach chemistry and what the students learn in chemistry classrooms.[11,12] More precisely, they maintain that such an understanding is important because it has implications for "the ways in which we engage the students in activities and modes of thinking that characterize important aspects of chemistry" (p. 106 in ref. 11). This is consistent with the argument forwarded by modern science educators who argue that science curricula should reflect the nature of scientific inquiry and practices of the scientific community.[3,13–15] If in fact chemistry is a unique domain of science, with interest in unique questions and uses special tools and methods of investigations, K-16 chemistry curriculum should reflect those practices. This uniqueness of chemical knowledge and the practices of chemistry merit a discussion before I jump into my review and discussion of argumentation studies in the literature. So, answers to such questions as: what is chemistry? what is the nature of chemical knowledge? how do chemists study the nature? what types of tools they use? who benefits from the chemical knowledge? become critical.

Chemistry is a field of science that studies and attempts to understand, classify, manipulate and transform the structure and behaviors of chemicals.[11,16] In doing so, chemistry uses a special language, methods, way of thinking and making sense of nature. Chemistry as a field primarily uses a descriptive language, and uses experimentation to formulate knowledge about nature. Chemistry relies on models to simulate and describe the structure of chemicals and the interactions between different chemicals and their environments.[11,16–18] While chemistry is fundamentally considered to be a purely experimental field,[17] the theoretical arguments have also made significant contributions to the advancement of the field.[12] However, physics has primarily dominated the field of philosophy of science, leaving chemistry little room to act as a theoretical and philosophical field. This has changed, however, in recent years, because there is an explicit effort to establish chemistry as a field that is not totally divorced from philosophy.[11,12,19–24]

Abstractions, representations, and problem solving are key to the practices of chemists and thus to the advancement of chemical knowledge.[17] Francl states "Abstract objects are attractive in that they give chemists a common language, one that builds on the ability of such objects to encapsulate chemists' understanding of molecular structure and behaviour." (p. 428 in ref. 25). Further, "the use of abstract objects as the basis of language in chemistry invites an insidious insistence on the primacy of chemical intuition, as captured by these objects" (p. 428 in ref. 25). Because the abstract concepts are descriptions of our imagination of what may be possible, rather than what is being perceived,[11] at best abstraction aids our ability to approximate an atom or a molecule's structure and behaviors. As noted by Francl, "Notions such as bond order and formal charge serve as proxies for a constellation of reactions and structures of interest" (p. 427 in ref. 25). Chemists use inscriptions to talk about this abstract nature of chemical knowledge.[16,17] This abstract nature of chemical knowledge, importance of empirical evidence, hypothesis testing, relationships and modeling lends itself to argumentation. Therefore, argumentation should play a key role in learning and teaching of chemistry.

Another concept that is common to chemistry is models. Woody describes a model as "an approximation of a complete theoretical representation for a phenomenon." (ref. 26 as cited in ref. 11, p. 116). Models have played a key role in the development and communication of chemical knowledge (for an elaboration discussion on the topic see ref. 11 and 27) "Chemists use models to understand and explain the physical and chemical properties of matter" (p. 118 in ref. 11). They then use this understanding to make predictions about the interactions between different chemicals or between chemicals and their environments. Erduran and Duschl argue that modelling the structure and function of matter help chemists tackle many problems in their field.[11,28] One problem with the use of models in chemistry education is that "Chemical models have been presented to students as final versions of our knowledge of matter: copies of real molecules in contrast to approximate and tentative representations"[29,30] (as cited in ref. 11, p. 122). This perspective on use of models and modeling in chemistry education overlooks "the motivations, strategies and arguments underlying the development, evaluation and revision of chemical models" (p. 122 in ref. 11). This perspective leaves no room for the students to question, evaluate and critique the models – this limits the place of argumentation in teaching of chemistry.

Another relevant concept in the teaching and learning of chemistry is the notion of experimentation. Chemical experimentation involves modeling and predicting interactions between different atoms and/or molecules as precisely as possible to observe certain behaviors.[17] Chemists often repeat their experiments to achieve higher precision and to reach a level of confidence to make claims to knowledge. Therefore, they often engage in reflective activities about the design of their experiments, the safety and reliability of their instruments, the experimental data and their interpretations. In the process, they construct explanatory and predictive models and engage in persuasive and dialogical argumentation both to make sense of their models and to critique the utility of their models, the robustness of their data and the predictive power of their models. As noted by Erduran and Duschl, "Chemical experimentation, however, has rarely been translated in the educational environment as an activity through which models are developed, evaluated and revised.[11] Rather, experimentation is typically implemented as data collection and interpretation" (p. 123). However, "effective teaching and learning of chemistry requires" teachers who have knowledge of the models, arguments, methods and practices of the field they teach such that classroom discourse must "manifest 'what chemists do' and reason about" (p.129 in ref. 11). This type of knowledge and skills can potentially help teachers to problematize the models presented to the students and help them question the evidence, and critique models and arguments.[12] Erduran and Mugaloglu argue that chemistry curricula should reflect both the history and nature of chemical knowledge and point out the dangers of exposing students to a chemistry curriculum that is divorced from its philosophy.[12]

Despite importance of modeling, inquiry and argumentation in the progress of science, argumentation has rarely been associated with chemistry. This is partly because chemistry as a field has enjoyed being associated merely with "experimentation, logical reasoning and positivism." Chemists typically develop hypotheses, design experiments to test these hypotheses, collect and analyze data, and develop experimental models or theories about behaviors of chemicals. This practice has also guided how chemistry is taught in science classrooms. However, with increasing emphasis on students' engagement with scientific practices, and the social nature of learning, argumentation has been introduced to be an effective pedagogical approach for helping students to make sense of chemical phenomena, develop a durable and meaningful knowledge base of chemistry and engage in epistemic practices of science.

2.1.3 Argumentation Defined and Operationalized

Argumentation is often used to imply various meanings; while some use argumentation to persuade others of their opinions or claims to knowledge,[31,32] others use it to reach a consensus about critical decisions in a variety of contexts,[33] and yet others use argumentation to engage in self-reflection regarding their reasoning, evidence and warrants.[34,35] Argumentation can also be used for the purposes of "stimulating, fostering, and facilitating reflection" (p. 370 in ref. 34).

A scientific argument can be described as the justification of a viable claim backed with sufficient, relevant evidence and reasoning.[3,31,36] Argumentation is the process whereby the author of a claim to knowledge makes an attempt to persuade his/her audience of the validity of his/her evidence, using verbal reasoning and rebuttals or in a written format or using both methods. While some argue that argumentation must involve at least two cognitively or epistemologically confronting parties[32,37,38] others view argumentation as a form of self-reflection one goes through while refining, justifying his/her claims to knowledge.[34] Arguments are constructed and argumentation is advanced not always for the sole purpose of persuasion; they are often used to critique existing models, to identify limitations of counter arguments to reach a consensus about the validity and predictability of a model. Argumentation is an epistemic,[31,39] social, and motivational activity whose purpose is to justify or refute a claim to knowledge using professional tools of reasoning and valid and sufficient evidence.[40,41] No matter how we define argumentation, it has found a solid place in school curricula across grade bands. Similar to its definition in authentic scientific contexts, argumentation has been used in teaching and learning through different frameworks.[42]

Argumentation has been discussed in science education literature for its cognitive and social benefits for the learners. First, proponents of argumentation claim that arguments serve as tools for students to articulate their understanding, to engage in reflective reasoning, to make connections between their claims and evidence to propose, evaluate or critique a model or an explanation.[10,31] Second, they argue that argumentations serve as contexts in which students evaluate each other's ideas, provide opportunities to put their claims to knowledge to the test, and to receive feedback from the other members of the community.[3,31] The third function of argumentation is to help students recognize and engage in scientific practices, and appropriate the norms of authentic science and acquire scientific dispositions such as skepticism, criticism, open-mindedness.[43] Finally, proponents argue that argument construction and argumentation are both metacognitive, in that they both afford the opportunity for students to identify weaknesses in their claims, evidence, reasoning and explanations and opportunity to address these weaknesses, gaps in their evidence, reasoning and explanations.[31] Proponents argue that by engaging students in such practices, we expect the students to develop scientific identities. Here, identity is not limited to an interest in science and science careers but such dispositions as skepticism, critical thinking, creativity. Collectively, all of these learning activities contribute to students' epistemic, conceptual, motivational and social learning. It is because of these theoretical assumptions and empirical backing that argumentation has become a popular instructional approach in science education and more specifically in chemistry education.

Several review studies found that science educators approach argumentation from different perspectives and when they use argumentation as a context for their studies; they emphasize different aspects of argumentation. Some focus on students' use of evidence,[44] others focus on the impact of argumentation on students' learning outcomes[45,46] and yet others focus on the interactional patters/discourse during argument-based learning.[47]

In a thematic review study of argumentation conducted by Erduran, Ozdem and Park, the authors found that 116 articles focused on the linguistic aspect of argumentation, 99 focused on epistemic aspects of argumentation and 77 on both aspects of argumentation.[48] They defined epistemic aspects of argumentation as practices such as justification, use of evidence and warrants, or "wider epistemic practices such as 'inquiry' and 'explanation'" (p. 7). The linguistic aspects emphasize "discourse" and "discussion." They found that "A comparison between the epistemic aspects and linguistic aspect displays that researchers emphasized the linguistic aspect of argumentation more than the epistemic aspects." (p. 10). While science educators have extensively studied the role of argumentation students' acquisition of argumentation skills, the impact of argumentation on student learning and teachers' pedagogical knowledge and skills in argumentation, the number of studies conducted in chemistry classrooms are limited. Most argumentation studies in chemistry are limited to pre-service science teachers' pedagogical knowledge of argumentation, the quality of

Pre Service Teachers' (PSTs') arguments, or the impact of argumentation on students' conceptual understanding of chemical concepts. I provide a critical review of these studies in the next section.

2.1.3.1 Argumentation Studies in K-12 Education

While most argumentation studies in science education focus on non-chemistry subjects, a few science educators have conducted argumentation studies in the context of chemistry. Sampson and Clark designed a study to investigate whether students develop arguments of better quality when they work in groups than when they construct arguments individually, and whether students' engagement in group-based argumentation resulted in better learning outcomes than when the students constructed arguments individually.[49] The authors collected data from 168 high school chemistry students who were randomly assigned to collaborative or individual argumentation conditions. The authors formed a three-group structure for their research design. In order to generate the three comparison groups, they used students' individual scores on an ice-melting blocks problem. The authors put together the top five performing students in classrooms of 30, and four from classrooms of 28 to form the top performing group, then put the students performing at the next level into the middle group and the students with the lowest performance were put into the third group. This configuration ending up with 20 groups of three students with 60 individuals as the comparison group. The results showed that students who engaged in argument construction in group settings did not produce arguments of higher quality than those who worked alone. However, the results did show that students who engaged in argumentation in collaborative group settings performed better on mastery and transfer problems. The authors used a 3×2 ANOVA test to evaluate the differences in group achievement levels with regard to argumentation quality. The results of the ANOVA test indicated a significant main effect for the group, $F\ (2,\ 78) = 87.05$, $p < 0.001$, $\eta^2 = 0.69$. The results did not show a significant main effect for gender, $F\ (1,\ 78) = 1.9$, $p = 0.17$. The results showed that the triads ($M = 7.21$, $\mathrm{SD} = 1.64$) produced significantly better arguments than the bottom third group of individuals ($M = 4.75$, $\mathrm{SD} = 1.14$, $p < 0.001$, $d = 1.74$) and arguments that were equivalent in quality to those produced by the middle third individual group ($M = 6.75$, $\mathrm{SD} = 1.01$, $p = 0.54$). However, the top third group of individuals ($M = 9.39$, $\mathrm{SD} = 1.29$) produced arguments that were significantly better than the arguments produced by the triads ($p < 0.001$, $d = 1.48$).

McNeill investigated six science teachers' use of an 8-week chemistry curriculum designed explicitly to support students in "the construction of scientific arguments to explain phenomena in which students justify their claims using evidence and reasoning" (p. 1 in ref. 50). The three questions that the author intended to answer consisted of the following: What instructional practices do teachers engage in during the unit to support students in writing scientific explanations? How does student learning of scientific explanation vary by teacher? How does the support the teachers provided their students relate to any differential learning of scientific explanation? To characterize the nature of support that the teachers provided their students for constructing scientific argumentation, the author collected data from videotapes, teacher questionnaires, and to measure the impact such teacher support had on subsequent student learning the author used pre- and post-test data. The pre- and post-tests consisted of 15 multiple-choice items and four open-ended response questions. The open-ended questions required students to develop three scientific explanations for each of the three main content goals of instruction: substance and properties, chemical reactions, and conservation of mass.

McNeill (2008)[50] found that "the way teachers defined scientific argumentation, characterized teachers' support and influenced the other practices they used in their classrooms" (p. 1). More precisely, she found that when the teachers' definitions of scientific argumentation did not align with the "intended learning goal in the curriculum materials" the instruction deviated from the intended practice. Moreover, teachers' "simplification of this complex inquiry practice" negatively impacted students' learning gains "in terms of students' ability to write scientific arguments to explain phenomena using appropriate evidence and reasoning." (p. 1). More precisely, McNeill ran an analysis of covariance, ANCOVA, "with the teacher as the fixed factor, the pre-test score as the covariate, and the gain score as the outcome variable" (p. 255). She considered "students' total scientific explanation scores, claim scores, evidence scores, reasoning scores, and their multiple-choice scores" as the dependent variable. She found that on all five analyzes, the teacher effect was a significant predictor of students' explanation scores. The fact that these teachers had attended targeted professional development programs around argumentation shows that helping teachers to implement argumentation in the classroom is not an easy task. Simplifications of argumentation and scientific explanation may result in teachers' design of cognitively less-demanding tasks, which in turn could limit students' active and meaningful engagement with cognitively complex task of inquiry-based learning and problem solving. Collectively, these results echo with findings from other parallel studies which point out the importance of teachers' definition and understandings of argumentation for student learning through argumentation.[51–54]

Heng, Surif and Seng conducted a study with Malaysian secondary school students and explored students' argumentation skills in the context of chemistry.[55] The authors randomly assigned 120 students to either individual argumentation or group argumentation group conditions. Those who were assigned to group argumentation worked in groups of four. They administered an argumentation test called SAT, and based on the SAT, students were further divided into two groups, one where students had constructed arguments with valid chemistry concepts, and the other where students had constructed arguments that were irrelevant, illogical, or with invalid concepts. The SAT focused on three sub-concepts of acids and bases, namely neutralization, properties of acids and bases, and the strengths of acids and bases.[55] The authors presented each concept with a scenario and data, followed by six questions that corresponded with the six components of argumentation as described in the Toulmin's Argument Pattern (TAP) framework. The authors found that "students' mastery of scientific argumentation on the concepts of acids and bases was weak in both individual and group argumentations" (p. 512). More specifically, only less than 10% of arguments constructed by either group of students had valid chemistry concepts.

The results also showed that students who were assigned to the group argumentation condition developed a more complex argumentation scheme than those who were assigned to the individual condition. "More, students who were assigned to the group condition were able to argue with more scientific elements and showed their understanding of macro and sub-micro concepts." (p. 505). Finally, scientific argumentation schemes of students who were assigned to the group conditions, "consisted of a claim that was constructed based on the data and supported by warrants and backings" (p. 514) and students attended both the macroscopic and microscopic properties in their arguments. However, the scientific argumentation scheme of the students assigned to the individual conditions only "consisted of a claim which was based on single data and supported by warrant and backing at the macro level with misconceptions" (p. 514). These results show that if scaffolded carefully, and facilitated effectively, group-based learning environments can make a greater contribution to students' learning outcomes than the individual learning environments in the context of argumentation.

Collectively, these studies show that K-12 students struggle with developing complex evidence-based arguments but when supported through written scaffolds, or organizational scaffolds such as designing small group-based learning environments that encourage critique and questioning, students are able to develop complex arguments. One limitation of the argumentation studies in K-12 classrooms is that most interventions last for a relatively short period of time and students' limited experiences with argumentation may have a differential effect on the reported outcomes. The field would benefit from designing and studying the effects of long-term argumentation interventions. Such studies that rely on large sample size may help us to establish statistically robust causal inferences about the impact of argumentation on student learning outcomes.

2.1.3.2 *Argumentation Studies at the University Level*

Argumentation studies conducted at the university level are rare. However, a few chemistry educators have been able to explore university students' experiences with argumentation and the impact of argumentation on university students' learning outcomes. Kulatunga, Moog, and Lewis[56] used a modified version of Toulmin's argumentation scheme by Erduran, Simon and Osborne[57] to investigate the characteristics of student group argumentation in Peer-Led Guided Inquiry sessions in a General Chemistry course in the U.S. Peer-Led Guided Inquiry (PLGI) is a cooperative small group inquiry learning method in which students co-construct targeted chemistry concepts through specially designed paper-and-pencil activities. Weekly peer-led sessions with a focus on two small groups were recorded. They used Erduran and colleagues'[57] argumentation framework to measure both the quality and nature of the argumentation that took place in these groups. The results of their analysis of students' group argumentations revealed that students engaged in "co-constructed arguments, with more than one student providing evidence and reasoning" (p. 1207). While they found that students frequently "supported their answers with evidence and reasoning" they often failed to "elaborate on their reasoning or further validate their explanations" (p. 1207). The quality of students' arguments and the effectiveness of their argumentation skills may be enhanced by presenting students with more cognitively complex learning tasks.

Pabuccu and Erduran conducted a study with 89 undergraduate students in which they studied the impact of argumentation activity on students' conceptual understanding of conformational analysis of butane.[58] The study measured the impact of argumentation on: (a) pre-service teachers' understanding of conformations of alkanes; (b) the nature of the pre-service teachers' discourse; and (c) the quality of pre-service teachers' argumentation. The authors placed the students in groups of three to five, and "provided one written frame, which included a discussion question, a graph, evidence statements, and a table." The participants were asked to work together to solve the problems provided and develop an explanation to justify their answers. The problem asked the students to place a missing formula on the potential energy relationships among the various conformations of butane and constructing an argument to support their choices. The students were further provided evidence statements (some of which only served as distractors) related to types of strains and types of isomers to analyze and use for constructing their arguments. The researchers recorded and later analyzed students' discussions while attempting to form claims, and arguments in their groups. The results of their study showed that "high performing groups had multiple rebuttals in their argumentation and low performing groups had problems in evaluating the credibility of evidence." The authors point out the importance of learners' abilities to evaluate evidence and their spatial skills in quality of learners' arguments. This is expected considering the limited number of experiences students have with learning through argumentation in college courses. Students often do not know much about the mechanistic components and epistemic norms of a scientific argument, and short instructions aiming to help students become familiar with the epistemic norms and mechanistic components of an argument may not be of much help to the students. Therefore, studies aiming to measure the impact of argumentation on students' conceptual argumentation and study the quality of students' arguments should do a better job of training the participants about the epistemic norms and mechanistic aspects of a scientific argument.

Sanford, Moon, Towns, and Cole[59] investigated the impact of instructor facilitation of discourse on university students' argumentation skills and interactions between students in a physical chemistry course designed based on principles of Process Oriented, Guided Inquiry Learning (POGIL) learning environment. POGIL is an inquiry-based learning environment that encourages student collaboration. In this learning environment, students are expected to work in groups of three or four. "The POGIL activities are designed to lead students through a learning cycle of exploration, concept development, and application to help them construct new knowledge" (p. 1502).[59] The activities focus on core concepts with a premise of encouraging a deep understanding of the content and acquisition of higher-order thinking skills.[60] The authors recorded classroom conversations and collected data on student interactions in two iterations of the same course, with a slight scaffolding modification between the two iterations.

After analyzing data using Toulmin's argumentation framework,[61] they found that a small modification in scaffolding of argumentation sessions impacts upon the type of student interactions and nature of students' arguments. The authors report this slight modification encouraged students to take increased ownership over their learning, caused an "increase in the number of initial arguments, an increase in the number of warrants being generated overall, and more iterative arguments being generated" (p. 1511) by the students. For instance, the percentage of single student-constructed arguments increased from 15% in the first iteration to 32% in the second iteration of the course. Similarly, the number of claims, the amount of data, the number of alternative arguments and backings increased in the second iteration of the course.

Erduran and Villamanan[62] investigated engineering students' written arguments in the context of the thermodynamics principles involved in refrigeration. The students were provided with guiding questions to develop written arguments consistent with the components of the Toulmin's Argument Pattern[61] "following investigations on thermoelectric coolers" (p. 119). "The writing frame engaged the students in the recording, analysis and interpretation of experimental data including data from a simulation program" (p. 119). The students ($n = 131$) worked in groups of three or four producing 31 reports following their experimental investigations. "Each student group had to study the Peltier effect, a principle that they would use in their professional future designing devices to either heating or refrigeration appliances in the industry context" (p. 120). The authors used Toulmin's Argument Pattern[61] to analyze the quality of students' written arguments. They analyzed "concordance between experimental evidence and established conclusions" (p. 122). "Conclusions were made from three different points of view: (1) a theoretical view were inserted in a scientific context; (2) the experimental facts themselves; (3) a descriptive account" (p. 122).[62]

Results of this study showed that "very few students were able to accurately interpret the experimental data. Most texts used conclusions that were not derived from the evidence used" (p. 119). Only 35% of argued texts presented a concordance between facts and conclusions (p. 122) despite the fact that the authors had provided a writing scaffold (in the form of guiding questions) for the students to write evidence-based arguments. The results highlight the difficulties that engineering students face with the writing of scientific arguments and the need for prolonged engagement with argumentation and integration of more metacognitive scaffolds for students to effectively engage both with content and with argument construction.

Each of the studies presented above offer valuable insights into university students experiences with argumentation and highlight the unique challenges experienced by the students. They also offer valuable suggestions on how to form cognitively complex and socially functioning groups to help students maximize their learning during argumentation-based learning.

2.1.3.3 Argumentation Studies in Pre-service Teacher Education

Significant work has been done in the context of pre-service teacher (PST) education both for helping PSTs to develop pedagogical knowledge and skills to teach chemistry through argumentation and to identify the gaps in PST's knowledge of argumentation. Erduran[51] studied PSTs' adoption of argumentation teaching materials in the context of a project which "aimed to produce resources for teachers and pupils (grades 11–14) to facilitate the teaching and learning of ideas and evidence in science" (p. 1).[51] The university professors collaborated with school-based mentors to develop the instructional materials for the PSTs teachers to try in their teaching. The teaching materials produced placed "particular emphasis on the role of evidence" (p. 2) in justification of scientific explanations.[51] The students were expected "to evaluate statements and justify why a particular statement supports or does not support a theory, generate criteria for evaluating the statements, *e.g.* relevance of the evidence, and oppose the counter-theory by constructing counter arguments and providing an explanation for why the other theory is not plausible" (p. 3)[51] at the end of the each unit. The project team sought both verbal and written feedback about the resources from mentors and PSTs. They found that both mentor teachers and PSTs valued the resources and found them useful for promoting students' learning of chemical concepts through argumentation and engaging students in complex reasoning and justification of scientific explanations. This finding suggests the importance of teachers' access to curricular resources related to argumentation. However, I must note that access to curricular resources alone will not be sufficient for teachers to effectively implement argumentation in their classrooms. Teachers also need to become familiar with and engage in the epistemic practices of science around argumentation for effective implementation of argumentation in science classrooms.

Aydeniz and Ozdilek[63] developed a study to investigate the impact of engaging students in argument construction and evaluation on pre-service science teachers' self-efficacy to teach science through argumentation and explore the challenges they experienced while implementing argumentation lessons. They engaged 40 PSTs in argument construction, evaluation, and critique around chemical concepts, and then had them teach three argumentation lessons, conduct peer observation of teaching, and reflect on their teaching skills. The intervention lasted for 11 weeks and focused on participants' understanding of argumentation as a scientific practice and as a pedagogical tool. Data were collected through the Argumentation Self-Efficacy Scale and an open-ended questionnaire. The results of their study showed that the intervention had a significantly positive effect on PSTs' self-efficacy to teach chemistry through argumentation. Despite this self-reported positive self-efficacy by participants, they experienced significant challenges in guiding their students to construct scientific arguments and assessing the quality of arguments developed by their students.

Erduran, Ardac and Yakmaci-Guzel[64] studied pre-service Turkish chemistry teachers' use of argumentation in their classrooms. The participants consisted of 17 PSTs enrolled in a teacher education program in Turkey. The authors used the IDEAS pack in the training of chemistry PSTs. The training was integrated into a practicum methods course in chemistry and lasted over a six-week period. The authors investigated how PSTs designed argumentation lessons and supported argumentation in their classrooms after attending a series of training sessions. The materials used for professional development focused on pedagogical aspects of argumentation, and placed an emphasis on: "(1) how to introduce argument; (2) how to manage small group discussions; (3) how to teach argument; (4) what resources can be used to support argumentation by students; (5) how to evaluate arguments; and (6) how to model them for pupils." (p. 6).[64] The participants were given two weeks to prepare an argumentation lesson around a chemistry topic that is part of high school chemistry curriculum following the training provided. They then implemented their argumentation lessons. By analyzing multiple types of data, "teacher talk, student group talk, students' written work, teacher lesson plans, teacher interviews after training and teacher written responses to argument questions," the authors report that among other effective strategies all PSTs "used questioning, modeling and presentations as strategies to promote and support argumentation as well as explicit statements on the norms of argumentation" (p. 9).[64] The only observed difference between PSTs was "their use of meta-talk and the quality of the feedback provided to the students" (p. 10). These results show even the short-term training provided helped the course instructors achieve the intended pedagogical goals for the PSTs in the course. The authors warn, however, that while these results are promising we need to be mindful of school contexts and demands of curricula to determine if these PSTs will actually continue to implement these strategies in their future classrooms. More specifically, they argue that "without a shift in what gets assessed in terms of teaching and learning performances" it is unlikely that these PSTs will continue to teach science through argumentation and make argument construction skills the focus of their assessment. This suggests that while teaching PSTs about reform-based instructional strategies is important, when challenged by the demands of curricula or standardized tests, teachers may go back to business as usual and rely on lectures in delivering instruction. However, if we want teachers to engage students in types of reasoning and practices as scientists do, we need to help them develop sophisticated epistemologies.[65,66] Only teachers with sophisticated epistemologies related to science may be able to make classroom talk and assessments more accountable to the epistemic norms of scientific reasoning and discourse.[67,68] It follows that in addition to teaching their students about pedagogies of argumentation, PST educators should create contexts and opportunities for their students to understand the nature of chemical knowledge, how such knowledge is constructed, evaluated and validated.

2.1.4 Conclusion and Discussion

There has been a recent and persistent argument among science educators that science teaching should move away from lecturing and assessment of students' accumulation of canonical facts of science and focus more on student-centered approaches to instruction which engages students in construction, evaluation and critique of arguments and models.[32] This perspective is consistent with modern philosophical arguments around science, which state that scientific models and theories are subject to change, falsification and transformation.[69,70] This form of instruction requires the positioning of students as the source of epistemic and cognitive authority.[2] Consequently, this places an emphasis on collaborative nature of learning *i.e.* the structure and nature of group discourse, and the cognitive complexity of learning tasks. Argumentation discussions in science education have been situated within this framework of learning. Positioning students as the source of epistemic and cognitive authority, authors and contributors of shared knowledge, and emphasis on students' development of scientific identities in argument-based learning brings a host of issues and unique opportunities for the teacher, the students, curriculum developers and researchers. The review of studies in chemistry education allowed me to synthesize these challenges and reflect on opportunities offered by argumentation in the context of teaching and learning of chemistry.

2.1.4.1 Challenges with Learning to Argue and Arguing to Learn

Despite persistent calls to engage students in scientific practices such as argumentation and modeling, fostering students' productive engagement with scientific argumentation has proven difficult. Research has consistently reported that students often struggle with argumentation tasks[44,52,71] because argumentation requires them to propose viable claims, support their claims with sufficient and relevant evidence, engage in critical discourse by questioning both their own evidence, reasoning and claims, and that of other members of the learning community.[52] Others,[72–75] however, have claimed the opposite, suggesting that students bring with them a set of productive epistemic and cognitive resources that they draw upon when asked to construct explanations or arguments. Both to understand the sources of this difficulty and to provide scaffolding[76–79] necessary for students' effective engagement in argumentation scholars have conducted several studies in various contexts with diverse goals in mind.

I provided a summary of some of these studies in the context of chemistry education. Several themes emerged from my review of relevant studies in chemistry education. Some of these themes highlight what works for students' effective engagement with argumentation and argument construction practices, others focus on areas that still need further evidence and clarification, yet others focus on the source of challenges teachers face when attempting to teach chemistry through argumentation. These challenges related to student learning include: the difficulty students experience with tying evidence with claim to draw conclusions; failure to engage with content at a deeper level; and failing to effectively engage with one another using epistemic resources and exerting epistemic agency in group-based discussions. The challenges related to PSTs' pedagogical capacity to teach science through argumentation include lack of a sophisticated epistemology related to the way chemical knowledge gets constructed, evaluated and critiqued due to the experiment-driven nature of chemistry. The review of relevant studies also highlights the importance of mastery experiences and argumentation-based teaching resources in teachers' abilities to facilitate argumentation in the classroom. For instance, in Erduran's study[51] PSTs developed pedagogical knowledge and skills to implement argumentation by using instructional resources designed based on the principals of scientific argumentation. In Aydeniz and Ozdilek's study,[63] mastery experiences helped PSTs to develop the self-efficacy to teach science through argumentation. More, because of the practice component of the intervention, PSTs started to become aware of the potential challenges they may experience when teaching through argumentation. These findings suggest that teacher educators must work hard to create opportunities for both pre-service and in-service teachers to engage in authentic practices related to argumentation.

2.1.4.2 Promising Practices That Help Facilitate Student Learning in Argumentation

Designing argumentation tasks and facilitating argumentation in the classroom is a difficult endeavor. However, curriculum design projects and discourse studies in argumentation have provided invaluable perspectives on the aspects of learning tasks and characteristics of productive critical discourse. I will draw from several projects and studies to make some recommendations for designing cognitively challenging and meaningful learning tasks and facilitating productive argumentation sessions. The curriculum design projects that I draw from are the IDEAS project, and Argument-driven Inquiry (ADI) project.

2.1.4.2.1 Design Principles for Argumentation Tasks. First, any argumentation-based learning task should make the goal of the activity explicit to the students. Second, learning tasks should engage students in culturally relevant and academically important problems. Third, learning tasks should provide explicit scaffolding for students' cognitive engagement with the chemical ideas and social engagement with members of the learning community. Forth, learning tasks should engage students in epistemic questions about the nature of chemical knowledge. They should engage students in questions such as how are scientific ideas, explanations and models are constructed, critiqued and validated? What is the evidence behind claims to knowledge? Is the evidence provided relevant and sufficient? When I use evidence, can I develop an argument that will stand up to peer criticism? Learning tasks should engage students in a search for answers to these questions both individually and collectively with their peers. Finally, structures must be put in place for epistemic and cognitive division of labor so that all students are equally engaged with the construction, evaluation and critique of knowledge. Furthermore, teachers must create conditions, for all students to equally exercise their epistemic authority for making meaningful contributions to the shared knowledge building activities.

2.1.4.2.2 Design Principles for Group Argumentation. Group-based learning activities can be challenging because of the group dynamics, the purpose of grouping and the context in which the learning takes place. However, by keeping certain principles in mind, one can effectively facilitate argumentation-based learning groups. First, the instructor should explicitly communicate the goals of the group assignment to the students. If the goals are not clear, students will be confused about the expectations and are unlikely to effectively engage with argumentation. Second, the teacher should introduce students to the linguistic, mechanistic, epistemic and social aspects of argumentation. If you have ever taught in a high school or university classroom you would know that most group conversations are not generative; typically one student dominates the discussion and others confirm the accuracy of his/her claim to knowledge. This happens because most students are not familiar with the epistemic, social and linguistic norms of scientific arguments and argumentation. Students should be taught how to listen, how to critique, how to question, how to come to a consensus and how to build on other's ideas to extend the conversation. The teacher can model these discourse practices with their students before they engage with the activity. Third, teachers should use assessment strategies that will promote both positive interdependence and individual accountability. This is important for students to both work collaboratively and to assume epistemic agency to contribute to the construction, evaluation and critique of knowledge. For instance, the teacher can develop a protocol for students to self-assess their performance, and a peer assessment protocol where students evaluate the contributions of their group members to the construction and critique of arguments.

Practical Digest

A SAMPLE ARGUMENTATION TASK (GRADES 10–11)

PROBLEM and TASKS

Four different 2.0 liter flasks are filled with H_2, O_2, Ne and CO_2 gases respectively, at STP. The flasks are closed after being filled with the gases. After carefully analyzing the flasks filled with each of these gases, some students make the following claims about the behavior of the gas molecules in the flasks.

Claims

- Each flask has the same number of gas molecules.
- The velocity of the O_2 molecules in the flask is eight times bigger than that of the H_2 molecules.
- The velocity of the gas molecules is the same in each flask.
- The density of each gas is the same.
- There are eight times more O_2 molecules than H_2 molecules in the flask.

PART A. Your job is to confirm or refute each of these claims in your groups.

1. Justify or refute each of the following claims based on what we know about the gas laws and the molecular kinetic theory.
2. After discussing your answers with your group members develop a written explanation about each claim.
3. In your explanations, you need to support your explanation with evidence and warrants using the scientific principles about the gas laws. If computations are required for you to gather evidence, show your work and use your computations as part of your explanations.

PART B. If the temperature of each flask is doubled what can be said about the pressure inside the flasks? Please justify or refute each of these claims about gas pressures inside these flasks based on what we know about gas laws and kinetic molecular theory.

Claims

- The gas pressure inside each flask will stay the same.
- The gas pressure inside each flask will be equal.
- The gas pressure inside the flask with CO_2 will be four times larger than the gas pressure inside the flask with O_2 molecules.

1. After discussing your answers with your group members develop a written explanation either supporting or refuting each claim.
2. In your explanations, you need to support your explanation with evidence and warrants using the scientific principles about the gas laws. If computations are required for you to gather evidence, show your work and use your computations as part of your explanations.

PART C. Please answer the following questions

- What gas laws and/or theories did you use in answering these questions?
- Who contributed to the development of these theories or establishment of these laws?
- How were these theories or laws established within the scientific community? What experimental evidence or philosophical discussions contributed to the development of these theories?
- How did these theories evolve over time? What discussions or questions contributed to the evolution of these theories over time?

PART C questions are related to the nature of science. This can be assigned as homework and later used as a starting point for a discussion about the nature of chemical knowledge. This discussion can focus on the nature of chemical laws and/or theories related to the concepts under inquiry.

Implementing the Lesson: Facilitating Intra- and Intergroup Discussions. First, students will be placed in groups of three to read the problem and work together either to refute or justify the accuracy of the claims presented. After each group of students come up with their explanations for each claim, the teacher will facilitate an intergroup discussion to engage students in argumentation around claims provided. Second, the discourse language must be modeled by the teacher before students engage in explanation or argument construction, so they engage in a rich critical scientific discourse both within and across group discussions. It must be recognized that most students are not familiar with the scientific discourse. Therefore, teacher must familiarize their students with the epistemic, linguistic and social norms of argumentation and the mechanistic features of a scientific argument at the beginning of the learning activity to ensure students productive engagement with argumentation and development of scientific arguments. Finally, the teacher may adopt the Accountable Talk Stems (Learning Institute, 2003) for agreements, disagreements, clarifications, confirmations and elaborations. It should be noted that it may take some practice before the students appropriate the language of scientific argumentation and effectively navigate across different features of an argument to develop defendable evidence-based arguments.

The Accountable Talk Stems may be accessed at the following link: https://educationcloset.com/wp-content/uploads/2015/09/AccountableTalk-Stems.pdf

References

1. V. Talanquer, *Sci. Educ.*, 2013, **22**(7), 1757.
2. D. Stroupe, *Sci. Educ.*, 2014, **98**(3), 487.
3. R. Driver, P. Newton and J. Osborne, *Sci. Educ.*, 2000, **84**(3), 287.
4. National Research Council, *The National Science Education Standards*, National Academy Press, Washington, DC, 1996.
5. National Research Council, *A Framework for K-12 Science Education: Practices, Crosscutting Concepts, and Core Ideas*, The National Academies Press, Washington, DC, 2011.
6. G. S. Aikenhead, *Stud. Sci. Educ.*, 1996, **27**, 1.
7. R. Aikenhead and G. Aikenhead, *Sci. Educ.*, 1992, **76**, 559.
8. M. J. Ford, *Cognit. Instr.*, 2012, **30**, 207.
9. C. I. Damsa, P. A. Kirschner, J. E. B. Andriessen, G. Erkens and P. H. M. Sins, *J. Learn. Sci.*, 2010, **19**(2), 143.
10. M. J. Ford, *Sci. Educ.*, 2008, **92**(3), 404.
11. S. Erduran and R. A. Duschl, *Stud. Sci. Educ.*, 2004, **40**, 105.
12. S. Erduran and E. Z. Mugaloglu, *International Handbook of Research in History, Philosophy and Science Teaching*, ed. M. Matthews, Springer, Dordrecht, 2014, pp. 287–315.
13. J. T. Leach, R. Millar, J. Ryder and M. G. Sere, *Learn. Instr.*, 2000, **10**, 497.
14. J. Donnelly, A. Buchan, E. Jenkins, P. Laws and G. Welford, Investigations by Order: Policy, Curriculum and Science Teachers' Work Under the Education Reform Act, in *Studies in Education*, Nafferton, 1996.
15. G. J. Kelly, S. Druker and C. Chen, *Int. J. Sci. Educ.*, 1998, **20**(7), 849.
16. R. Hoffman, *The Same and Not the Same*, Columbia University Press, New York, 1995.
17. R. Hoffman, *Synthese*, 2007, **155**, 321.
18. N. Mbajiorgu and N. Reid, *Factors Influencing Curriculum Development in Chemistry: A Physical Sciences Practice Guide*, Higher Education Academy of Physical Sciences Centre, Hull, United Kingdom, 2006.
19. N. Bhushan and S. Rosenfeld, *Of Minds and Molecules*, ed. N. Bhushan and S. Rosenfeld, Oxford University Press, Oxford, 2000.
20. L. Mcintyre, *Found. Chem.*, 1999, **1**(1), 57.
21. E. Scerri, *Chem. Intell.*, 1997, **3**, 44.
22. J. Schummer, *Chemistry in the Philosophical Melting Pot*, ed. D. Sobczynska, P. Zeidler and E. Zielonacka-Lis, Peter Lang, Frankfurt am Main, 2004, 395–410.

23. R. L. Stein, *HYLE*, 2004, **10**, 5.
24. J. Van Brakel, *The Philosophy of Chemistry*, Leuven University Press, Leuven, 2000.
25. M. Francl, *Nat. Chem.*, 2002, **4**(6), 427.
26. A. Woody, *Proceedings of the Third International History, Philosophy, and Science Teaching Conference*, ed. F. Finley, D. Allchin, D. Rhees and S. Fifield, University of Minnesota, Minnesota, 1995, vol. 2, pp. 1295–1304.
27. R. Justi and J. K. Gilbert, *Chemical Education: Towards Research-Based Practice*, ed. J. K. Gilbert, O. de Jong, R. Justi, D. F. Treagust and J. H. van Driel, Springer, Netherlands, 2002, pp. 47–68.
28. S. Erduran and J. H. Hotchkiss, *J. Food Sci.*, 1995, **60**(5), 946.
29. K. Grosslight, C. Unger, E. Jay and C. Smith, *J. Res. Sci. Teach.*, 1991, **29**, 799.
30. M. A. Weck, Proceedings of the Third International History, Philosophy and Science Teaching Conference, University of Minnesota, Minnesota, 1995, **2**, 1286–1294.
31. M. P. Jiménez-Aleixandre, and S. Erduran, *Argumentation in Science Education. Perspectives from Classroom Based Research*, ed. S. Erduran and M. P. Jiménez-Aleixandre, Springer, Dordrecht, The Netherlands, 2007, pp. 3–27.
32. J. Osborne, *Science*, 2010, **328**(5977), 463.
33. L. K. Berland and V. R. Lee, *Int. J. Sci. Educ.*, 2012, **34**(12), 1857.
34. M. H. G. Hoffman, *Argumentation*, 2016, **30**(4), 365.
35. D. Kuhn, *Harvard Educ. Rev.*, 1992, **62**(2), 155.
36. L. K. Berland and K. L. McNeill, *Sci. Educ.*, 2010, **94**(5), 765.
37. J. Andriessen, M. Baker and D. Suthers, *Arguing to Learn-Confronting Cognitions in Computer-Supported Collaborative Learning Environments*, ed. J. Andriessen, M. Baker and D. Suthers, Springer, Dordrecht, 2003.
38. M. Kim and W.-M. Roth, *Pedagog. Int. J.*, 2014, **9**, 300.
39. C. Lumer, *Informal Logic*, 2005, **25**, 189.
40. A. I. Goldman, *J. Philos.*, 1994, **91**, 27.
41. A. I. Goldman, *Informal Logic*, 2003, **23**, 51.
42. J. B. Henderson, K. L. McNeill, M. González-Howard, K. Close and M. Evans, *J. Res. Sci. Teach.*, 2018, **55**(1), 5.
43. L. K. Berland and B. Reiser, *Sci. Educ.*, 2009, **93**, 26.
44. W. A. Sandoval and K. Millwood, *Cognit. Instr.*, 2005, **23**(1), 23.
45. M. Aydeniz, A. Pabuccu, S. Cetin and E. Kaya, *Int. J. Sci. Math. Educ.*, 2012, **10**(6), 1303.
46. V. Sampson, P. Enderle, J. Grooms and S. Witte, *Sci. Educ.*, 2013, **97**, 643.
47. M. Felton and D. Kuhn, *Discourse Processes*, 2001, **32**, 135.
48. S. Erduran, Y. Ozdem and J. Y. Park, *Int. J. STEM Educ.*, 2015, **2**(5), 1.
49. V. Sampson and D. B. Clark, *Sci. Educ.*, 2008, **92**, 447.
50. K. L. McNeill, *Sci. Educ.*, 2009, **93**(2), 233.
51. S. Erduran, *Sch. Sci. Rev.*, 2004, **87**(321), 1.
52. J. Osborne, S. Erduran and S. Simon, *J. Res. Sci. Teach.*, 2004, **41**(10), 994.

53. J. Osborne, S. Simon, A. Christodoulou, C. Howell-Richardson and K. Richardson, *J. Res. Sci. Teach.*, 2013, **50**, 315.
54. I. Tabak and E. Baumgartner, *Cognit. Instr.*, 2004, **22**(4), 393.
55. L. L. Heng, J. Surif and C. H. Seng, *Int. J. Sci. Educ.*, 2015, **37**(3), 505.
56. U. Kulatunga, R. S. Moog and J. E. Lewis, *J. Res. Sci. Teach.*, 2013, **50**(10), 1207.
57. S. Erduran, S. Simon and J. Osborne, *Sci. Educ.*, 2004, **88**, 915.
58. A. Pabuccu and S. Erduran, *Chem. Educ. Res. Pract.*, 2016, **17**, 523.
59. C. Sanford, A. Moon, M. Towns and R. Cole, *J. Chem. Educ.*, 2016, **93**, 1501.
60. R. S. Moog and J. N. Spencer, *Process Oriented Guided Inquiry Learning (POGIL)*, ed. R. S. Moog and J. N. Spencer, American Chemical Society, USA, 2008.
61. S. Toulmin, *The Uses of Argument*, Cambridge University Press, Cambridge, 1958.
62. S. Erduran and R. Villamanan, *Educ. Quím.*, 2009, **20**, 119.
63. M. Aydeniz and Z. Ozdilek, *Int. J. Sci. Math. Educ.*, 2016, **14**(7), 1255.
64. S. Erduran, D. Ardac and B. Yakmaci-Guzel, *Eurasia J. Math. Sci. Technol. Educ.*, 2006, **2**(2), 1.
65. S. Erduran, A. A. Bravo and R. M. Naamon, *Sci. Educ.*, 2007, **16**(9–10), 975.
66. R. S. Russ and M. J. Luna, *J. Res. Sci. Teach.*, 2013, **50**(3), 284.
67. M. D. Huling, PhD Thesis, University of South Florida, 2014.
68. R. S. Russ, J. E. Coffey, D. Hammer and P. Hutchison, *Sci. Educ.*, 2009, **93**(5), 875.
69. T. S. Kuhn, *The Structure of Scientific Revolutions*, Chicago, University of Chicago Press, 2nd edn, 1970.
70. K. Popper, *The Logic of Scientific Discovery*, Routledge, London, 2002 [1959].
71. M. P. Jiménez-Aleixandre, A. B. Rodríguez and R. A. Duschl, *Sci. Educ.*, 2000, **84**, 757.
72. D. Hammer, *Am. J. Phys.*, 2000, **68**(S1), 52.
73. D. M. Hammer and A. Elby, *J. Learn. Sci.*, 2003, **12**(1), 53.
74. E. B. Moje, T. Collazo, R. Carrillo and R. W. Marx, *J. Res. Sci. Teach.*, 2001, **38**(4), 469.
75. I. Tabak and B. Reiser, *Proceedings of the CSCL Conference*, ed. R. Hall, N. Miyake and N. Enyedy, 1997, pp. 289–298.
76. G. J. Kelly, J. Regev and W. Prothero, *Argumentation in Science Education: Perspectives from Classroom-based Research*, ed. S. Erduran and M. P. Jimenez-Aleixandre, Springer, Netherlands, 2008, pp. 137–158.
77. C. W. Keys, B. Hand, V. Prain and S. Collins, *J. Res. Sci. Teach.*, 1999, **36**(10), 1065.
78. D. Kuhn, J. Black, A. Keselman and D. Kaplan, *Cognit. Instr.*, 2000, **18**, 495.
79. K. L. McNeill, D. J. Lizotte, J. Krajcik and R. W. Marx, *J. Learn. Sci.*, 2006, **15**(2), 153.

CHAPTER 3

Interdisciplinarity and Argumentation in Chemistry Education

BEATRIZ CRUJEIRAS-PÉREZ* AND MARÍA PILAR JIMÉNEZ-ALEIXANDRE

Universidade de Santiago de Compostela, Departamento de Didácticas Aplicadas, Facultade de Ciencias da Educación. Avda. Xoan XXIII, 15782, Santiago de Compostela, Spain
*Email: beatriz.crujeiras@usc.es

3.1 Introduction

Chemistry is widely recognized as a difficult discipline for students, which, in many cases leads to negative attitudes or lack of motivation for learning it.[1,2] There are some problems recurring across studies, such as presenting contents as isolated facts, lack of transfer or application of chemical knowledge to everyday life, a lack of relevance as well as inadequate emphasis in how chemistry should be learned.[3–5] Chemical knowledge is necessary for people to make sense of their everyday lives and environment[6] since many complex problems in society, such as the energy crisis, climate change or food security, involve chemical knowledge. Therefore, all citizens should master some chemical understanding to help them to take part in public debates. In other words, all citizens should possess some understanding about chemistry so that they can make informed decisions affecting their everyday lives, which means being a chemical literate individual.[7]

Advances in Chemistry Education Series No. 2
Argumentation in Chemistry Education: Research, Policy and Practice
Edited by Sibel Erduran

Published by the Royal Society of Chemistry, www.rsc.org

Chemical literacy can be understood as a contribution from chemistry to scientific literacy,[8] considered as the ability to engage with science-related issues, and with the ideas of science, as a reflective citizen.[9] To do so, individuals need to develop 21st century skills such as problem solving, critical thinking, communication, collaboration, and self-management.[10]

A potential way of improving student motivation and interest in science and, in particular in chemistry, is to use social and real-life contexts and practical applications, which demands the use of interdisciplinary knowledge.[11] Although interdisciplinarity in science has been widely acknowledged in terms of integrating disciplines such as physics, chemistry and biology or even mathematics, there are other interdisciplinary approaches that involve, for instance, the incorporation of economics, history, and politics, among others, required to reflect and make decisions in the modern society.[12]

Argumentation contexts may be an appropriate environment for interdisciplinary learning, in particular when the focus is on socio-scientific issues (SSI), which by their nature require to draw on diverse domains and to integrate them. Argumentation is a central practice of science since one of science's attributes is the production of new knowledge about the natural world through objective argument and critique.[13] This practice is recognized also as one of the goals for scientific literacy, since it is relevant in the societal decisions citizens make in their everyday lives.[14]

This chapter discusses the characterization of interdisciplinary learning, the benefits and challenges for its implementation, as well as its connection to argumentation. Furthermore, we present examples of interdisciplinary chemistry tasks set in real-life contexts that require students' engagement in a range of argumentation practices such as drawing evidence-based conclusions or evaluating arguments.

3.2 Interdisciplinarity in Chemical Education

3.2.1 Interdisciplinary Thinking for Attaining Chemical Literacy

Interdisciplinary thinking is understood as the capacity to integrate knowledge and modes of thinking from two or more disciplines or established areas of expertise to produce a cognitive advancement in ways that would have been impossible or unlikely through single disciplinary means.[15] In this approach, students need to be able to comprehend and translate disciplinary languages, ways of knowing and methods, and then balance, synthesize and integrate them.[16] Although the term "interdisciplinary" is applied to a variety of curricular arrangements and it is frequently used interchangeably with several synonyms such as thematic teaching or integrated learning, they do not represent the same aspects[17] and there is a need for distinguishing among them. In an interdisciplinary approach the unique characteristics of each discipline are preserved, and distinctions among the various academic disciplines remain clear. However, in an integrated approach there is no clear distinction between or within the subjects and the curriculum is organized around relevant real-world problems. Thematic learning is akin to integration, but themes tend to be broader in focus.[18] Other approaches used to define the involvement of more than one discipline are multidisciplinary and transdisciplinary programs. The former draw ideas and methods from individual disciplines while acknowledging clear disciplinary boundaries, whereas the latter focuses on addressing solely the holistic problem or issue that involves multiple disciplinary knowledge.[19] According to these authors, the integrated learning approach is the most frequently used in educational contexts.

Interdisciplinary thinking is required in order to be prepared for the 21st century,[20] since many of the challenges will necessitate innovative solutions that have a basis in complex reasoning. Societies will demand a cadre of well-educated scientists to undertake the research and nurture the innovation that will be essential to meet the economic, social and environmental challenges that the world faces.[9] Hence a solid base of scientific literacy is necessary for all to become better citizens and discerning consumers.[21] Regardless of their career choice, citizens will face a range of situations involving science and technology knowledge: personal choices such as adoption of new environmentally-friendly renewable energy systems, public debates such as those around policy related to climate change, or toward ensuring food safety and security.[22,23] Although scientific literacy usually refers to specific knowledge, abilities, and values shared by the general public[24,25] there is not a unique definition. In this chapter, we draw from Roberts'[26] vision II of scientific literacy, understood as knowledge ability to face science-related situations that students are likely to encounter in everyday life, such as personal and public decisions about social problems. This vision can be translated to chemical education in terms of chemical literacy, which involves understanding key chemical ideas, a notion of what chemists do, essential skills, and chemical contexts.[27] In contrast to scientific literacy, there is a consensual definition of chemical literacy including four components:[28] chemical content knowledge, chemistry in context for explaining everyday situations, higher-order thinking skills and affective aspects. So, according to Shwartz *et al.*,[28] a chemically literate person is able to: (a) Acknowledge the importance of chemical knowledge in explaining everyday phenomena; (b) Use his or her understanding of chemistry in his or her daily life, as a consumer of new products and technologies, in decision-making, and in participating in social debates regarding chemistry-related issues; (c) Understand the relations between innovations in chemistry and sociological processes. All these operations, required to develop chemical literacy, demand an interdisciplinary understanding.

3.2.2 Interdisciplinary Teaching and Learning in Chemistry

Interdisciplinary education has a crucial role in science teaching, because it supplies new resources for the progress of science and technology.[29] According to St. Clair and Hough,[17] an interdisciplinary approach: (a) provides a positive classroom environment and enhances the learning process, and even achievement; (b) is a more holistic approach and therefore it supports students' development of a more holistic view of the world; (c) is more interactive and more global in content, preparing students better for citizenship in the 21st century; (d) improves problem-solving abilities by presenting multifaceted material and differing views; (e) promotes teacher collaboration.

The benefits of interdisciplinary approaches for learning are recognized in the literature; McComas and Wang[30] summarized them from four perspectives: (1) from a philosophical perspective, students can see science as a humanistic endeavour instead of belonging only to selected individuals; (2) from a psychological perspective, students are able to see the connections between single pieces of disciplines; (3) from a pedagogical perspective, integrated lessons can enhance teaching by promoting collaboration among peers; and (4) from a pragmatic perspective, an interdisciplinary approach promotes more interest in science and more understanding of science and its applications.

More specifically, an interdisciplinary learning approach: helps students create deep and holistic understanding about complex phenomena;[31] facilitates higher-order thinking;[32] leads students to a more meaningful learning experience, which enables them to reach higher levels of academic achievement;[33] promotes positive attitudes and motivation toward learning;[34] and improves reasoning abilities.[35]

In short, interdisciplinary learning enables students to understand the big picture of a given concept or problem with knowledge from multiple science disciplines.[36] Despite these benefits, interdisciplinary learning involves many challenges such as the lack of teachers' preparation to implement interdisciplinary curricula,[37] the absence of a robust cognitive model,[38] the lack of effective assessments[39] and the fact that it is more time consuming than standard approaches.[40]

Regarding students, interdisciplinary learning is more demanding because it requires knowing the interconnected concepts of the fields involved and synthesizing the knowledge from each discipline.[35] One of the main challenges for learning in interdisciplinary contexts is knowledge integration, due to students' tendency to isolate concepts as separate factual knowledge[36] and to the difficulties for connecting concepts across disciplines.[42] Interdisciplinary students must learn to interrogate multiple ways of knowing and the structure of knowledge itself. They must develop a reflective and explicit knowledge of how disciplines work.[42] Therefore, for interdisciplinary understanding to exist, some disciplinary knowledge is required, hence it is important to assess students' interdisciplinary understanding to distinguish whether an inability to solve a problem is due to the lack of the disciplinary building blocks, or to the inability to apply them together to solve interdisciplinary problems.[43]

According to Zhang and Shen,[39] a student's disciplinary foundation may help or hinder his or her interdisciplinary problem solving, which is influenced by the conceptualization of the problem elements and structure, the particular strategies of the discipline, and the attitudes toward interdisciplinary and disciplinary learning. In addition, interdisciplinary activities can reinforce disciplinary understanding, but to do so students need to differentiate knowledge from individual disciplines.[19]

In science, interdisciplinary problems require solvers that not only possess a relevant knowledge base but are also able to integrate knowledge and skills from different scientific areas.[39] Knowledge integration is recognized as one of the main challenges for successful interdisciplinary knowledge.[19] Shen, Liu and Sung[38] distinguished three operations involved in knowledge integration: (1) translation of scientific terms between disciplines so that people with different disciplinary backgrounds would be able to effectively communicate; (2) transfer of explanatory models, concepts and tools from one disciplinary context to another; and (3) transformation of a system typically considered in a given discipline into a new system applying explanatory models and concepts from one discipline.

Transfer is recognized as problematic in the science education community.[44,45] Students are able to apply scientific knowledge only to the original context in which the knowledge was first acquired. The problem of transfer can be effectively addressed if science instructions help students to articulate their full array of ideas, add new ideas that are normative and compelling, develop criteria to distinguish between these ideas, and form a more coherent and consistent view of scientific phenomena.[44]

Rennie, Venville and Wallace[46] proposed a theoretical framework for integrated learning comprising three lenses: (1) an integrated lens, that examines the ability to transfer ideas from one context to another, the application of understanding to practical contexts and students' general motivation; (2) a discipline-based lens, that focuses on traditional disciplinary conceptual knowledge; and (3) a sources-of-knowledge lens that examines how students access knowledge in order to make key learning decisions.

One of the roles of science teachers regarding interdisciplinary instruction is to support students in dealing with natural phenomena and associated real-world problems, which are not easily comprehensible or resolvable from a single disciplinary framework.[36] Students often develop fragmented ideas,[47] and therefore science instruction needs to help them to integrate ideas from distinctive sources in order to form coherent explanations and solve complex problems.[48,49]

Integration helps students to form deeper understandings, build connections among central concepts and become interested and motivated in school.[37,50] Although there is not a consensual description of integration, Beane[51] proposed four defining characteristics: (1) the curriculum is organized around problems and issues that are of personal significance in the real world; (2) the use of relevant knowledge in the context of the topic; (3) using knowledge to study an existing problem rather than for a test or grade level outcome; and (4) emphasis is placed on projects and activities with real application of knowledge and problem solving.

Linn, Slota, Terashima, Stone, and Madhok[52] proposed five processes of knowledge integration: (1) eliciting students' prior ideas to create relevant connections between new and already existing ideas in a learning context; (2) adding new ideas to explore the relationships among all of their existing and new ideas, in order to eventually form connections between them; (3) distinguishing productive ideas from unproductive ones to connect scientifically relevant and normative ideas; (4) sorting out the various connections among these ideas to develop a coherent understanding of the subject; and (5) developing criteria for the relationships between ideas.

In chemistry education, interdisciplinary research has focused on discipline integration for enhancing students' learning[35,39,53,54] and on interdisciplinary assessment.[38,43] Araújo, Morais and Paiva[53] explored the impact of poetry in pre-university students' motivation and learning about the periodic properties of the elements. The authors found some difficulties for interpreting the poem, however, they highlight the potential of poetry as a motivation strategy for engaging students in the construction of significant knowledge. Schaal *et al.*[35] carried out an interdisciplinary hypermedia assisted learning unit on hibernation and thermodynamics in secondary education. Their findings point to a successful interdisciplinary learning through the hypermedia-learning environment due to students' acquisition of higher-order domain-specific thinking. Zhang and Shen[39] investigated the strategies used by graduate students to integrate different scientific disciplinary knowledge to solve two interdisciplinary problems on the phenomenon of osmosis. Their findings point to the use of both disciplinary and interdisciplinary strategies, such as relying on mathematical formula, carrying out experiments (disciplinary) and neglecting one competing factor to simplify the situation and applying a 'mixed strategy' to reconcile the two competing factors (interdisciplinary). Kelly *et al.*[54] examined the methods undergraduate students use to solve ill-structured interdisciplinary problems such as the possibility to produce energy from the gas released by a herd of cows. Their findings point to satisfactory problem-solving skills except for monitoring and reflecting processes.

In the last years, some research about assessing interdisciplinary understanding has emerged. Shen *et al.*[38] examined college students' performances and difficulties in an interdisciplinary assessment about osmosis. Their findings demonstrate that the students developed conceptual understanding of osmosis over the years, but their interdisciplinary understanding was much limited compared to the growth of their understanding of individual scientific concepts such as physical pressure or water molecules. You *et al.*[43] assessed students' levels of disciplinary and interdisciplinary knowledge about carbon cycling. The empirical results indicated that the majority of students have mostly single disciplinary knowledge, biological knowledge and limited connections to chemistry.

In short, interdisciplinary education provides unique opportunities for meaningful learning. However, more attention needs to be paid to knowledge integration processes in order to achieve the purposes of this perspective. In the next section we discuss some approaches for knowledge integration.

3.3 Interdisciplinary Approaches for Knowledge Integration

Knowledge integration requires cohesive and active teaching and learning approaches that encompass real-world problem-based learning, so that it can be meaningful for students.[55] In science education, different approaches such as STEM (the acronym for Science, Technology, Engineering, Mathematics) learning, learning through SSI or learning through history and philosophy of science (HPS) and nature of science (NOS) are relevant environments to foster knowledge integration.

3.3.1 STEM Learning

Although there is not a unique definition for STEM learning,[56] it can be understood as the acquisition of knowledge and skills through experience and study, integrated though multiple lenses allowing for the appreciation of the encompassing complexity and cross-cutting ideas across the STEM disciplines as a whole.[57] This approach seeks to prepare students with the appropriate skills required in a rapidly changing society, which will need creative and competent STEM professionals.[58] In STEM education, students may develop cognitive skills while engaged in the study of specific STEM-related social or global situations.[59] There is a set of skills required for addressing the challenges of the 21st century, such as adaptability, complex communications, no routine problem solving, self-management, and systems thinking.[60] These skills, as well as knowledge and abilities, can be developed through STEM learning framed in scientific inquiry, technological innovation and mathematical computation.[59]

A related approach is STEAM learning, a balanced methodology that fosters Science, Technology, Engineering, Arts and Mathematics, defined by a transdisciplinary style of teaching encouraging a wide variety of knowledge and skill sets that promote problem solving.[61] It is based on the premise that STEM and Arts function better together than they do apart.[62] The goal of this approach is to prepare students to solve the world's pressing issues through innovation, creativity, critical thinking, effective communication, collaboration, and ultimately new knowledge.[40]

Although in the last years a number of studies regarding STEM and STEAM education have been conducted, there is still little research about how to better integrate knowledge from the disciplines or about what type of connections improve student learning.[40,41] Hence, one of the main challenges for educators is to elucidate how the disciplines can be effectively integrated, while ensuring the integrity of each one.[63]

3.3.2 Knowledge Integration Through HPS and NOS

HPS approaches are widely recognized as necessary to understand science and to develop scientific literacy.[64,65] HPS contributes substantially to the increase of the conceptual sophistication of science education by allowing for a deeper understanding of the concepts and the methods of science.[66] In particular, the history of science provides valuable background about the origins of the concepts and theories and the philosophy of science can contribute insights about the structure and growth of scientific knowledge through penetrating discussions of methodology and norms of inference.[67]

Teaching and learning with HPS has been widely acknowledged as beneficial for students' learning.[64] According to Matthews,[68] science teaching in high school must be stimulating, so that students will perceive the philosophical and cultural implications of scientific knowledge. However, this approach is not frequently applied in school science. Höttecke and Silva[64] analysed the constraints that interfere with effective implementation of HPS in school science teaching. These authors identified as main obstacles: (1) the institutional framework of science teaching, focusing on curriculum development; (2) lack of adequate HPS content in textbooks; (3) different epistemological beliefs about teaching each scientific subject, for instance physics and chemistry teachers tend to see science as "standing pure and separate from all involvements in society" compared to biology and environmental science teachers.[69] Other studies examining the application of HPS highlight teachers' insecurity in using the history of science in their classrooms[70] and teachers' complaints about the amount of extra time devoted to these aspects.[71]

In the last decades, a range of strategies and techniques have been used with the objective of including the HPS approach in science education. Narrative approaches for teaching HPS have yielded significant and interesting results.[72,73] The technique of role play is recommended, especially for enacting and reflecting scientific process in the classroom.[74] However, teachers usually lack teaching skills for supporting learning though this technique hence is not as frequently used as desired in the classrooms.[75]

Within HPS, or overlapping with it, NOS has been recognized as a critical issue to be considered for teaching and learning science.[64,76,77] NOS is of particular relevance to interdisciplinary approaches since, by its very nature, it is rooted in interdisciplinary fields. It has also a complex, dialectical nature, which may explain why NOS teaching is fraught with controversy. There are differences, for instance, between teaching NOS and teaching *about* it. The answers about how to introduce NOS in science teaching are diverse and even may involve formulating the question in different terms. For instance, Abd-El-Khalick[78] articulates it as the identification of crucial aspects of NOS that ought to be learned, and that may be used in order to design curricula and teach science. A different perspective has been proposed by Duschl,[79] with a focus on designing learning environments as epistemic communities of practice, rather than on what NOS content to teach. Duschl and Grandy[80] argue that both science and science education should be conceptualized in terms of cognitive, epistemic and social practices, as well as material and technological contexts, and therefore that "NOS learning occurs when students' engagement are situated in these practices" (p. 2012). Allchin,[81] and

Erduran and Dagher[82] have made substantial contributions to this second perspective. Allchin[81,83] emphasizes the nature of *whole* science, where whole means weaving together experimental/material, conceptual and social strands of doing science. He makes the case for including aspects such as the social interaction of scientists, the role of funding, motivation or cognitive biases. Erduran and Dagher's reconceptualization of what to teach about NOS seeks to coordinate the epistemic, cognitive and social aspects of science for the purposes of achieving a more inclusive portrayal of science in science teaching. Erduran and Dagher's ambitious proposal highlights the co-existence and interactions of cognitive, epistemic and social components. One of their original contributions is a heuristic approach, based on a visual benzene ring analogy, to infuse various aspects of the NOS into the curriculum. Our research programme is framed in this perspective related to epistemic communities of practice.

The implications of the HPS perspectives have been explored in the context of chemistry education. Kousathana, Demerouti and Tsaparlis[84] used acid-base models and the investigation of the role of history and philosophy in anticipation of students' misconceptions in acid–base equilibria. These authors highlight the need for a deep immersion in HPS so that students develop conceptual understanding of scientific progress and practice. Paixao and Cachapuz[85] developed an approach for teaching and learning mass conservation in chemical reactions, departing from the combustion reactions and their contemporary economic, environmental, social and political contexts. Erduran[86] developed a framework for applying HPS, in particular the concept of reduction in chemistry education, highlighting that reduction reinforces a deeper understanding of chemistry apart from being applied to chemical concepts such as chemical composition or bonding. Niaz[87] investigated the implementation of activities based on various controversial chemical episodes, such as Thompson's experiments for clarifying the nature of cathode rays or the Millikan-Ehnrenhaft controversy with respect to the determination of the elementary electrical charge. The findings of this study point to the potential of these activities for enhancing participants' understanding of the NOS through reflection and discussion. Guerra, Braga and Reiss[88] developed a project aimed to present science content using an historical philosophical approach focusing on the HPS from a social perspective, for instance Lavoisier and sciences in the Enlightenment. The results derived from the implementation showed that a social approach to the HPS stimulates interdisciplinary discussions in science classes, enabling students to reflect on the NOS.

Although the benefits of the HPS approach for learning are widely acknowledged, teachers need support for implementing it in their classrooms.

3.3.3 Knowledge Integration Through SSI

SSI are recognized as effective contexts for developing the knowledge and processes that contribute to attain scientific literacy.[89,90] They are relevant for students' lives, and they involve not only scientific content but also ethical, economic or cultural aspects.[91,92] It should be noted that SSI make part of science or, in other words, that the arguments about them always involve scientific notions. SSI may be defined as dilemmas or controversies grounded on scientific notions. Put into other words, in SSI science is involved in a social debate, often in a decision-making context[93] involving decisions that may be of a personal or of a socio-political nature.

Kolstø[94] proposes a framework to examine the contributions of SSI to students' empowerment and science literacy for citizenship. He argues that teaching through controversial SSI might support students in learning about science and what he terms *content-transcending knowledge*, in particular about four dimensions: science as a social process; limitations of science; values in science; and critical attitude. Thus, for instance, viewing science as socially constructed means acknowledging that both consensus and disagreements play a role in "science-in-the-making," in other words that disagreements do not always correspond to particular interests or to incompetence, and that they are necessary. On the other hand, about science limitations, he points out that data may be interpreted according to particular interests of different actors; the efforts of some companies to downplay the impact of human activity on global warming may be an example. Next, we present some examples of SSI teaching in chemistry education.

Ingo Eilks and colleagues at the University of Bremen have been carrying out a research program focusing on the relevance of chemistry education, combining action research, education for sustainability and SSI. For instance, Stolz, Witteck, Marks and Eilks[95] designed and implemented a lesson based on the SSI of doping in professional sports in chemistry education, oriented to prepare secondary school students to take part in decision-making. Eilks[96] reports about a socio-critical approach for teaching about biodiesel.

Eastwood, Schlegel and Cook[97] compared reasoning and perceptions of students participating in an interdisciplinary program integrating science with social aspects, taught through SSI, and students following a traditional program. It was found that SSI students showed higher levels of reasoning and tendencies toward incorporating different perspectives into their decision-making. On the other hand, they also acknowledged the importance of recognizing different perspectives and were more perceptive to controversy and complexity of science-related situations. The authors conclude that traditional university teaching provides few opportunities for critical evaluation of interdisciplinary problems.

Some of the challenges involved in teaching SSI in physics and chemistry are explored in Barrett and Nieswandt's[98] study on pre-service teachers' beliefs about teaching ethics through SSI. They identified four archetypes or subject-specific identities: model scientist/engineer; model individual; model teacher; and model citizen. Their findings suggest that, depending on their identity and on how they view the relationships between ethics and science education, the participants could experience teaching SSI as more or less challenging. Cole's[99] thesis examined why secondary chemistry teachers incorporated SSI into the curriculum and how their beliefs influenced the way in which they planned for and enacted SSI in their teaching. It was found that SSI-based lessons were used as a way to get rid of the stigma that chemistry is irrelevant, and that teachers who incorporated SSI-based lessons understood the complexities of chemistry and therefore believed it was important for students to also understand these complexities. As Kolstø *et al.*[100] pointed out, by definition, SSI involves scientific claims and arguments, in addition to the political, personal, or ethical questions of what action to choose. In the literature SSI have been identified as appropriate teaching contexts to learn argumentation, which is discussed in the next section.

3.4 Argumentation in Interdisciplinary Settings in Chemistry Education

In chemistry education, argumentation is conceived as one of the dimensions in the language of chemistry, as there are specific logical argumentation patterns in scientific arguing and writing.[101] The promotion of linguistic skills is also one of the key objectives of chemistry teaching. It is important for students to learn the language of chemistry with its technical terms, formulae and patterns of argumentation. The more limited students' skills are in understanding and using the language of chemistry, the more difficulties they will have in using teaching and learning materials in their future chemistry education and understanding and expressing scientific concepts.[19]

3.4.1 Argumentation as a Scientific Practice: Participating in Science's Epistemic Goals

Argumentation may be characterized in different ways, but in science education there is a consensus about two concurrent meanings: first, the evaluation of knowledge claims by articulating empirical evidence with relevant theory, and second persuasion of an audience.[102] In recent years argumentation has been framed in the notion of scientific practices, aligned with a view of learning science that involves students' participation in the epistemic goals of science.[79] These epistemic goals are components of epistemic cognition, which concerns how people acquire, understand, justify, change and use knowledge.[103] For Iordanou, Kendeou and Beker[104] argumentative reasoning encompasses on the one hand skills related to the construction and evaluation of arguments, and on the other hand meta-level knowledge, that is individuals' beliefs about knowledge and how it is constructed. Argumentation is one of the processes by which knowledge is achieved. Iordanou and colleagues point out that, according to the literature, more sophisticated epistemic beliefs relate to better argumentative reasoning.

Jiménez-Aleixandre and Crujeiras[105] suggested that a way to promote students' participation in the epistemic goals of science is placing scientific practices—modelling, argumentation and inquiry—at the centre of science teaching. It needs to be emphasized that a defining feature of scientific practices is activity. Students should be *engaged* in scientific practices, carrying out modelling, argumentation or investigation. However, opportunities for students to engage in argumentation are not frequent, and one of the factors involved may be teachers' notions of and practice of argumentation. Sampson and Blanchard[106] have examined secondary science teachers' views about engaging students in argumentation, as well as how these teachers evaluated alternative explanations and how they generated arguments to support a particular explanation. The results showed that teachers experienced similar problems as students do for engaging in argumentation, and that they mentioned a number of barriers to the integration of argumentation in science classrooms.

The problem of how to design classroom learning environments to foster the use of scientific evidence, and therefore of argumentation, has been addressed by McNeill and Berland[107] in relation to problems found in classroom practice. They have developed design heuristics to inform this design, aligned with the vision of science as a set of practices, about features of information: (a) phenomenon-based, consisting of empirical data; (b) transformable by students when they can manipulate it to find patterns; and (c) used dialogically.

3.4.2 Relevance of Argumentative Practices for Interdisciplinary Learning in Chemistry Education

Jiménez-Aleixandre and Erduran[102] outlined five intertwined potential contributions from the introduction of argumentation in science classrooms, supporting: (a) the access to cognitive and metacognitive processes characterizing expert performance; (b) critical thinking; (c) scientific literacy and talking science; (d) the development of epistemic criteria for knowledge evaluation; and (e) the enculturation into the practices of scientific culture. Similarly, Cavagnetto[108] stressed the importance of a steady increase in argument-based interventions to foster scientific literacy since it supports developing communication skills, metacognitive awareness and critical thinking. In this section we discuss chemistry education studies, in the context of interdisciplinary learning, which report about the impact of argumentative practices on a range of learning dimensions.

Infusion of argumentation teaching in teachers' education is the focus of several studies, because teachers' strategies are pivotal in introducing argumentation in science classrooms. An instance of interdisciplinary learning is Pabuccu and Erduran's[109] study with pre-service science teachers, which addresses four issues: understanding of conformation of alkanes, the nature of participants' discourse, the quality of their argumentation, and their spatial ability. The findings point to synergies among these issues, for instance high performing groups produced arguments of better quality, while low performing groups had problems in evaluating the credibility of evidence. Furthermore, it was found that spatial abilities play an important role in participants' engagement in argumentation, which is an original contribution. This study would lend support to the notion of how interdisciplinary learning benefits conceptual understanding. Studies by Kaya[110] and Aydeniz and Dogan[111] found that pre-service teachers learning through argumentation had a better understanding of chemical equilibrium than a control groups learning it through standard instruction. Cigdemoglu, Arslan and Cam[112] also reported how argumentation practices in teacher education contributed to chemical literacy on the domain of acids and bases.

In higher education, Becker *et al.*[113] examined field-dependant features of arguments, in particular epistemic criteria related to evidence within chemistry contexts, which they call socio-chemical norms. They found that what were considered as acceptable explanations relating to thermodynamics—heat, energy, enthalpy or entropy—were grounded in particulate-level evidence and justifications. In their inquiry approach, mathematical expressions and energy transfer were also used as normative types of explanations. As mentioned above, it has been proposed that argumentation would support the development of metacognitive skills. This is the focus of Bowen *et al.*'s[114] study on competencies related to thinking like a scientist, such as argumentation and metacognition. They found that an inquiry approach, where students designed the experiments, analysed data and generated and justified their own conclusions, enabled them to better understand the concepts and prompted the use of metacognitive skills. Kulatunga, Moog and Lewis[115] examined argumentation and participation patterns in general chemistry taught through an inquiry approach. An interesting finding from this study is that students were able to resolve wrong claims through argumentation, which is an indication of independent learning.

SSI offer appropriate contexts to engage high school students in argumentation. Jin, Mehl and Lan[116] used SSI related to energy consumption to address two dimensions: how content knowledge was incorporated into arguments about causal connections; and which discourse moves were used to interpret, refute or substantiate arguments. Karpudewan, Roth and Sinniah[117] reported about the impact of a green chemistry curriculum on high school students' understanding of acid–base concepts and argumentation skills. The students following the green curriculum exhibited better understanding and higher levels of argumentation than the control students.

These studies consider the impact of argumentative practices on several aspects of learning: content knowledge, metacognition, epistemic criteria, or causal reasoning. There are also studies that explore the impact of particular strategies on argumentation. Sampson and Clark[118] examined three questions about the impact of collaboration on argumentation in high school chemistry: whether groups craft better arguments than individuals; to what extent do individuals adopt arguments crafted by the group; and whether individuals working in groups were learning more than students working on their own. The findings indicate that students did not produce better arguments when working in groups; many of them, however, adopted at least some elements of their group's arguments. One interesting finding is that students from the collaborative condition demonstrated superior performance on the transfer problems.

As a summary, we may say that argumentation practices have positive impacts on a range of learning dimensions, from supporting better understanding of diverse chemistry concepts, to supporting transfer and metacognition.

3.5 Argument-based Interdisciplinary Learning in Chemistry Education: Examples from Classroom Studies

In this section we discuss instances from our work about argumentation and use of evidence in interdisciplinary settings in chemistry education. These studies make part of a research program framed in Duschl's[79] approach to the design of learning environments as epistemic communities of practice, and focusing on student's engagement in cognitive, epistemic and social practices.

3.5.1 Use of Interdisciplinary Evidence for Drawing a Conclusion about how to Prevent Browning in Apples

In this section, we describe the potential of an inquiry-based task framed in an everyday context for promoting interdisciplinary argumentation, in particular the use of interdisciplinary evidence for drawing a conclusion.[119] The task is set in the context of food industry and it consists of finding the best solution to prevent browning in sliced apples (Figure 3.1). To do so, the participants are required to design and implement a plan according to the requirements included in the handout of the task in order to draw a conclusion for preventing browning.

To solve the problem, the participants need to understand the browning reaction as an interdisciplinary issue involving chemical and biological

Freshfruit is a company that is planning to sell sliced fruit bags (for example apples) in high school snack bars, as an alternative to pastries and in order to promote healthy eating habits. The problem is that fruit changes colour when it is sliced to be packed. This fact could cause selling problems, since people would reject it because it would not look good.
What should we do to achieve preserving the fruit with the same aspect as freshly sliced fruit?

To solve the problem, you can use:

- Scientific or empiric knowledge about chemical reactions and living beings that lets you explain the browning of sliced fruits
- Material resources similar to those that Freshfruit company had:
  (The pH of the different materials is also indicated.)

Apples: pH = 3, 2	Orange juice: pH = 4
Tap water: pH = 7	Lemon juice: pH = 2, 3
Salt	Vinegar: pH = 3, contains: acetic acid and water
Sugar	Sodium bicarbonate: pH = 8, 4, antacid effect
Alimentary wrapping plastic	Other material that you consider necessary

To solve the problem,you have to prepare an experimental design in order to contrast your hypothesis. Once you get a resolution proposal you must justify it.

Figure 3.1 Handout of the task about enzymatic browning. Reproduced from ref. 119, https://doi.org/10.5565/rev/ensciencias.1469, under the terms of the CC BY 4.0 license, https://creativecommons.org/licenses/by/4.0/, Copyright 2015 Beatriz Crujeiras Pérez, María Pilar Jiménez Aleixandre.

knowledge, which means a chemical reaction involving an enzyme (in this case polyphenol oxidase) that produces brown pigments in the surface of the apple when it is exposed to oxygen, that is when the apple is cut or pealed. Hence, enzymatic activity needs to be inhibited in order to avoid browning. Considering the context of the task, framed in the alimentary industry, participants might need to consider interdisciplinary evidence when drawing their conclusions (Table 3.1).

They are supposed to write an argument including the elements represented in Figure 3.2 as a minimum.

As represented in Figure 3.2, the participants are required to consider at least two variables: (a) contact with air; and (b) different pH, since a low pH inhibits polyphenol oxidase's activity. Moreover, the solution is developed through the connection of empirical and theoretical knowledge that enables coordinating data and claims through justifications.

Table 3.1 Interdisciplinary evidence to be considered for solving the problem of browning.

Type	Evidence
Chemical	An acid preserver inhibits enzymatic activity
Biological	A mixture of water and sugar enables cell dehydration and prevents apple from browning
Domestic	Orange and lemon juices are commonly added to fruit salads to keep their initial taste
Commercial	Vacuum packing prevents the apple taste from being altered

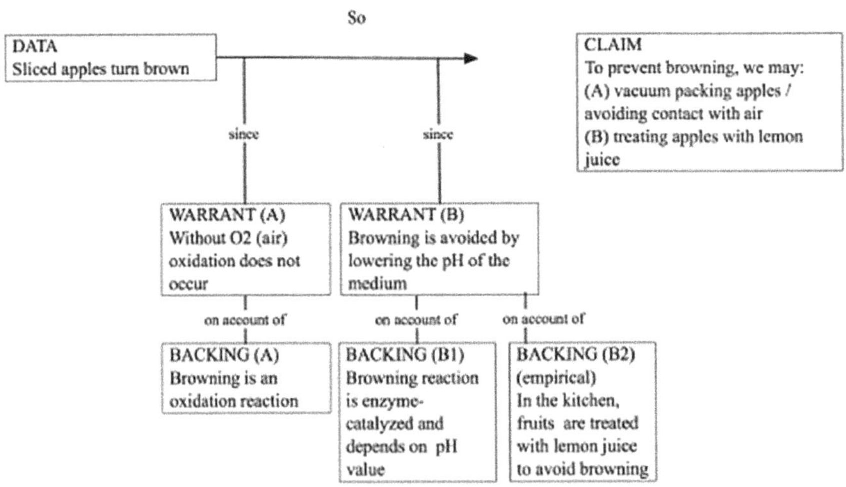

Figure 3.2 Components of one argument presented as a solution for the browning apple problem.
Reproduced from ref. 119, https://doi.org/10.5565/rev/ensciencias.1469, under the terms of the CC BY 4.0 license, https://creativecommons.org/licenses/by/4.0/, Copyright 2015 Beatriz Crujeiras Pérez, María Pilar Jiménez Aleixandre.

Table 3.2 Conclusions provided by master's students.

Conclusion	Evidence involved
Ana: "I would mix sugar with water and lemon juice. I would consider the two things pH and osmotic pressure"	Chemical–Biological
Andrés: "If it [*the apple*] is oxidized we should try to vacuum pack it as fast as possible to avoid oxidation"	Industrial
Carlos: "Considering our personal experience in preparing fruit salad, to avoid browning the apple should have an acid pH, which regulates the process"	Domestic–Chemical

Although the need for knowledge integration from biology and chemistry, in terms of interpreting browning as an enzyme-catalysed oxidation, is essential for solving this task, it is not commonly recognized by participants when the task is implemented. In Table 3.2 we summarize some examples of conclusions provided by master's students.

As summarized in Table 3.2, participants' conclusions are drawn mainly on chemical evidence, the effect of pH in the oxidation reaction, which might be due to their understanding of the browning process as a chemical reaction instead of as an interdisciplinary process involving both chemical and biological knowledge. This chemical consideration could be motivated by the fact that the browning reaction of the apple is used as an example of chemical oxidation in many student books from primary education to high school, without considering the biological knowledge involved.

3.5.2 Use of Interdisciplinary Knowledge for Evaluating Arguments in Order to Make a Decision About where to Build a Nuclear Waste Deposit

In this section, we examine the interdisciplinary knowledge used in a role play about the possibility of setting up a nuclear waste deposit in the campus facilities, in which the participants, pre-service teachers adopting different roles, have to make an informed decision about it. Participants, working in dyads, were provided with a role and a position for or against the nuclear waste deposit. There are seven roles with their corresponding positions: in favour of the deposit there are the government representatives, the local administration representatives, the managers of the company in charge of building and maintaining the deposit, and the scientists that will be working in the deposit. Against the deposit there are inhabitant representatives, environmentalists and occupational health and safety managers.

All roles were invited to a meeting for discussing and making a decision about whether to build the deposit or not. Prior to the meeting, participants have to prepare their intervention according the defined roles. To do so they were encouraged to document themselves about de assigned roles one week before the meeting.

The arguments for and against the nuclear waste storage were evaluated prior to make a final decision. Both the arguments and the knowledge involved in the decision-making process are summarized in Table 3.3.

As summarized in Table 3.3, the main arguments in favour the deposit involve economical, scientific and engineering knowledge, whereas those against the deposit appeal to health and environmental knowledge. The participants need to evaluate all the arguments and give priority to some of them in order to make an agreed decision. To do so participants need to discuss until reach a consensus. In this case the focus of discussion was about security rather than economic or scientific issues.

As summarized in Table 3.4, the main concern is about the possibility of radiation leaks (turns 21–23) raised by the inhabitants' representatives and environmental pollution (turns 26–29) posed by health and safety managers. The role of health and safety managers needs highlighting , which at the beginning was neutral, but when the discussion started changed to a position against the nuclear deposit.

At the end of the discussion the participants reach a consensus, which leads to the following decision: "*The government, both national and regional, is committed to establish some agreements with the company, in which the company is required to present reports about how the building process is taking*

Table 3.3 Pre-service teachers' arguments in favour and against the deposit.

Position	Argument	Knowledge involved
In favour of the deposit	Government representatives: "This is a unique opportunity for opening up [*the*] international market. The main reason is that this will be the first station in removing 100% of the wastes."	Economic
	Local administration representatives: "This is a good opportunity for promoting industrial growth in Galicia."	Economic
	Managers from the company in charge of building the deposit: "The deposit is completely necessary because we need a place in which store the nuclear waste from our nuclear power plant. Moreover, it would be placed in an adequately isolated area so that the radioactive residuals are completely isolated from human activities and environmental facilities. In addition, this deposit takes the residual heat emitted by the nuclear fuel. Furthermore, these residuals are stored in special ponds, being isolated by three layers of protection: the first is a concrete wall; the second is a stainless-steel container and the third one a stainless-steel capsule. Therefore, this proves that they are safely kept."	Physics
	Scientists that will be working in the deposit: "We are in favour of building the deposit because our work depends on it. Moreover, there will be plenty	Economic Scientific

Table 3.3 (*Continued*)

Position	Argument	Knowledge involved
	of trade-offs for the village inhabitants. And the deposit will be designed to support any possible risk because it will be continuously under-review. Besides it is necessary for ensuring people and environment protection because it is one way of isolating the wastes."	
Against the deposit	Inhabitant representatives: "We are against the nuclear waste deposit because it is a source of risk due to the high amount of emitted radioactivity. Besides we are afraid of developing serious illnesses and of to get our organic gardens destroyed."	Chemistry Health Environment
	Environmentalists: "We agree with the inhabitants' position because it [*the deposit*] produces respiratory diseases and cancer. In addition, it decreases life quality. Besides, waste disposals affect also to nature because sometimes they are not well controlled, and some waste is discharged to the water killing the species."	Health Environment
Neutral	Occupational health and safety managers: "Since we were not invited to participate in the design of the deposit we are not in favour or against it until we receive response to some questions."	

Table 3.4 Extract 1. Discussion about security.

Turn	Role	Transcription
21	Inhabitants representatives	Ok, but who guarantees that those [*security*] measures are effective? And what if it is a leak or something, what happens? Come on, answer!
22	Company managers	Nothing because we have the 100% of resources for recycling the waste.
23	Inhabitant representatives	If there is a leak the water will be polluted and also our gardens.
24	Health and safety managers	And how many years will take to those waste to decompose? or are they going to be always stored?
25	Company managers	No, they take hundreds or thousands of years.
26	Environmentalists	How little, right! Well 0% of risks, though! [*ironic*]
26	Health and safety managers	And in the remaining land, the crops will grow polluted.
28	Company managers	No, because the nuclear waste deposit tends to be placed in abandoned mines, not close to gardens. In the abandoned mines the land has not any activity at all.
29	Environmentalists	Yes, but the population grows, and they will get to install around it at sometime.

place. Reports about radiation emission, control of wastes and so on. In addition, we will pay continuously attention to all these reports and if the results are negative we will stop the process, always ensuring inhabitants' security."

This decision involves more security controls in terms of design, health and environment, information aligned to the interdisciplinary knowledge utilized to develop the main arguments against the deposit. To sum up, participants in this task integrated different types of knowledge in order to reach a consensus about the nuclear waste deposit construction, which included reformulated initial positions in some cases.

3.5.3 Applying an Interdisciplinary Simulation for Identifying the Best Toothpaste in Preventing Cavities

In this section we present an inquiry-based laboratory task about chemical reactions and inhibition processes, to be used in high school chemistry lessons in which students have to investigate which toothpaste would be most effective in preventing tooth decay and to draw an evidence-based conclusion.[120] The task involves applying an interdisciplinary model that simulates the process of cavities formation in the laboratory.

To investigate this issue, students have to simulate the process of cavities formation in the laboratory, which involves interpreting and applying interdisciplinary knowledge (Table 3.5).

It needs to be noted that interpreting the interdisciplinary model is crucial for planning the investigation; otherwise students will not be able to propose a method for assessing the effectiveness of toothpastes. To design and carry out the investigation, students have to understand first the correspondence between each element in the physical world and in the simulation, as well as its functioning in the laboratory context.

To do so, the role of the teacher is very important because the teaching strategies used to scaffold students for understanding and interpreting the simulation influence students' use of this model in both planning the investigation and drawing the conclusion. In this case, the teacher used questioning to scaffold students, as summarized in Table 3.6.

In this excerpt the teacher is helping students to understand the simulation prior to planning the investigation. He prompts students to identify the correspondence between each element in the simulation and in real life

Table 3.5 Description of the interdisciplinary model to be used in the investigation.

Element in the physical world	Element in the simulation
Teeth	Clamshells
Mouth	Test-tube
Acid produced after eating carbohydrates	Hydrochloric acid
Bacteria	CO_2 bubbles
Toothpaste	Toothpaste

Table 3.6 Extract 2. Interpretation of the interdisciplinary simulation for tooth decay.

Turn	Participant	Transcription
23	Teacher	Is it clear that we are going to simulate a reaction that takes place in people's mouth through another similar that will take place in a test-tube? So, instead of a tooth what are we going to use?
24	Ofelia	A clamshell
25	Teacher	And instead of the acid in the mouth?
26	Sara	Hydrochloric acid
27	Teacher	And our intention is to check what?
28	Olaia	Which one is the good toothpaste
29	Teacher	Is it the same to check which one is the best or the worst? And how can we know if a toothpaste is good or not?
30	Paula	In the number of cavities
31	Teacher	That is in the reality, but in our laboratory simulation? We are not going to produce cavities!
32	Ofelia	In the time that takes cavities to appear
33	Paula	In how many calcium layers it eliminates or reduces
34	Teacher	In how many calcium layers reduces, well it would be a possibility, but if we read the section a [in the task handout] it says: the reaction between the calcium carbonate and the hydrochloric acid produces gas releasement
35	Pilar	CO_2
36	Teacher	CO_2, very good. These are the easiest reactions to check the progress because we can see the gas released very well, right? So, we know that the reaction happens when there is gas released. If the gas is released fast the reaction will be fast. However, it the gas is released slowly, the reaction will be slow.

(turns 23–26). In addition, he guides students in framing the problem to be investigated and in selecting criteria for evaluating the effectiveness of toothpastes (turns 27–36).

The application of the interdisciplinary model is also important for interpreting the empirical results and drawing an evidence-based conclusion. The excerpt in Table 3.7 summarizes the process of interpreting results and drawing conclusions in one of the small groups of students implementing the task.

This excerpt highlights the importance of understanding the simulation, not only for planning but also for interpreting the results, in order to draw a conclusion. Students engage in argumentation about the data collected. Ofelia uses the simulation framework for comparing the obtained results for toothpaste y and the blank in terms of speed of reaction, which lead her to question the effectiveness of the toothpaste (turn 623). Another student, Olaia, supports Ofelia's claim in terms of fluoride concentration (turn 626). The application of the simulation for interpreting the results enables this group to draw an evidence-based conclusion, proposing that the toothpaste x is the best in preventing teeth from cavities because it took more time (speed of reaction) (turn 668). It needs to be noted that other groups, who did not

Table 3.7 Extract 3. Applying the interdisciplinary simulation for interpreting results in terms of speed of reaction and drawing a conclusion.

Turn	Participant	Transcription
623	Ofelia	Four minutes twenty seconds [*control sample*] and the other, y, four minutes twelve. Is it better not to brush your teeth than to use [*toothpaste*] y?
624	Olivia	I don't think so!
625	Ofelia	According to the results, which does not seem normal to me either.
626	Olaia	It would be the same, it would have as little fluoride that it is crap and it is good for nothing.
631	Ofelia	What I don't understand is how the toothpaste does protect less than not brushing your teeth, this is a bit odd.
632	Olaia	It would have too much sugar
634	Ofelia	What do you mean?
635	Olaia	Because it [*y*] has much less fluoride [*than x*].
636	Ofelia	Less than the teeth [*without toothpaste*]?
667	Óscar	The best toothpaste is x.
668	Ofelia	The best toothpaste is x, if it takes more time it will be because it has more effect [*prevents the teeth more from cavities*].

apply the simulation for interpreting the results, were not able to back their conclusions with adequate evidence or even to draw an adequate conclusion, because they were not able to recognize anomalies in the results and therefore they did not repeat the tests to obtain adequate data.

3.6 Concluding Remarks

This chapter addresses the characterization of interdisciplinary learning through argumentation and the potential of different proposals to promote knowledge integration. Interdisciplinary approaches such as STEM learning, HPS and SSI promote both knowledge integration and argumentation practices. All these proposals share curricular constraints when they are implemented that make them not to be as present as desired in chemistry lessons. These problems increase when the interdisciplinary lessons are argument-based, since students have also problems about evaluation of the evidence, NOS conceptualizations, and value-based decision-making in socio-scientific argumentation.[121] From the analysis of the interdisciplinary tasks discussed in the previous section, we can point out that engaging in interdisciplinary learning and argumentation practices is difficult, especially the knowledge integration process. However, when the context is SSI, debates and role plays are adequate means for promoting knowledge integration and argumentation.

In general, students should be familiarized with the integration process in progressing complexity, which means providing them several opportunities to engage in interdisciplinary approaches. According to Zhang and Shen,[39] students can start from basic integration and then increase the integration level so that they can have enough time to adapt to interdisciplinary thinking and develop confidence and positive attitudes in interdisciplinary learning. In the three tasks discussed in this chapter students worked organized in communities of practice.[79,122] As mentioned before, working collaboratively entails better performances in terms of knowledge transfer.[118] Moreover, the interdisciplinary tasks were implemented as part of disciplinary courses (pre-service teacher science learning or high school chemistry lessons) rather than making part of interdisciplinary courses. Therefore, the difficulties experienced by participants for integrating knowledge, such as those in the task of enzymatic browning, could be reduced by framed their learning process in a long-term interdisciplinary course. We agree with Araújo, Morais and Paiva[53] in considering that learning through unconventional teaching strategies or settings requires a longer adaptation process. Finally, another aspect to be highlighted is the need to support teachers in implementing interdisciplinary argumentation tasks, providing them resources for scaffolding students in the learning process. In-service teacher training would also be necessary for improving interdisciplinary learning.

Practical Digest

Summary of design principles for interdisciplinary and argumentation learning resources

a) *All the lesson resources provided in this chapter follow the design principles that foster students' use of evidence for sense making about the natural world:*[107]
   - Phenomenon-based: information to be used as evidence consists of empirical data (both qualitative observations and quantitative measurements) about phenomena in the natural world. In our examples we use participants' first-hand data, such as the variables that produce browning in apples or the data collected that explain the effectiveness of the toothpastes assessed. Moreover, we use also second-hand data such as the information provided to students about the nuclear waste deposit so that they could prepare the role play.
   - Transformable: students can manipulate the information provided to find patterns and evaluate the adequacy between the patterns and the competing claims. In all the three examples of teaching materials the information provided and also the data collected are transformable by students.
   - Used dialogically: information is used dialogically when students work collaboratively to make sense of it, engaging in discourse. Three tasks require students to work in groups as we present in the examples of discussions about security (Table 3.4), for interpreting the interdisciplinary simulation about tooth decay (Table 3.6) or for applying the interdisciplinary simulation to the interpretation of the data collected (Table 3.7).

b) *In an integrated approach there is no clear distinction between or within the subjects and the curriculum is organized around relevant real-world problems. There are four defining characteristics for knowledge integration:*[51] *(1) curriculum organized around problems and issues that are of personal significance in the real world; (2) use of relevant knowledge in the context of the topic; (3) knowledge that is used to study an existing problem rather than for a test or grade level outcome; (4) emphasis placed on projects and activities with real application of knowledge and problem solving.*

The lesson examples discussed in this chapter are framed in real-world problems that are familiar to the participants. The first example involves the use of interdisciplinary evidence for drawing a conclusion about how to prevent browning in apples and it is set in the context of alimentary industry. To solve the problem, the participants need to understand the browning reaction as an interdisciplinary issue involving chemical and biological knowledge. The second example involves the use of interdisciplinary knowledge for evaluating arguments in order to make a decision about where to build a nuclear waste deposit. The arguments presented by participants involved integrated knowledge from Economy, Physics, Chemistry, Environmental Sciences and Health Sciences. The third example requires the understanding and application of an interdisciplinary model that simulates the process of cavities formation in the laboratory. This model involves both chemical and biological knowledge.

c) *There are five processes of knowledge integration:*[52] *(1) eliciting students' prior ideas to create relevant connections between new and already existing ideas in a learning context; (2) adding new ideas to explore the relationships among all of their existing and new ideas, in order to eventually form connections between them; (3) distinguishing productive ideas from unproductive ones to connect scientifically relevant and normative ideas; (4) sorting out the various connections among these ideas to develop a coherent understanding of the subject; and (5) developing criteria for the relationships between ideas.*

The teaching strategies for promoting interdisciplinary argumentation in the three lesson plans included some of these processes (1, 3 and 4) that were carried out through open questions.

d) *An interdisciplinary approach improves problem-solving abilities*

For solving the three lesson plans students need to manage some problem-solving skills such as to be able to identify the issue to be investigated, to identify the type of information needed to solve the problem, to apply the concepts and principles of science and other domains, to plan the steps for reaching the solution, to propose a solution for the problem and to analyse whether their argument and evidence are adequate.

e) *An interdisciplinary learning approach helps students create deep and holistic understanding about complex phenomena*[31]

In the tasks presented, the holistic understanding has been only developed in the second example, the role play about the nuclear waste deposit. Here the participants adequately integrated different types of knowledge in their arguments in order to reach a consensus about the deposit construction, which included reformulated initial positions in some cases.

f) *Interdisciplinary learning involves many challenges. The main one is knowledge integration, due to students' tendency to isolate concepts as separate factual knowledge and difficulties connecting concepts across disciplines. This integration difficulty is one of the main concerns in STEM and STEAM approaches.*

Students in the examples experienced great difficulties in understanding interdisciplinary phenomena and in using interdisciplinary evidence and integrating it into conclusions to solve real-world problems. From the three experiences, students were able to carry out interdisciplinary argumentation in the role play. This could be explained due to the nature of the task itself, in which students were provided with some information about their roles that contained interdisciplinary evidence.

g) *Other challenges are related to the lack of teachers' preparation to implement interdisciplinary curricula,*[37] *the absence of a robust cognitive model,*[38] *the lack of effective assessments*[39] *and the fact that it is more time consuming that standard approaches.*[40]

- In the three tasks, one key challenge for teachers is selecting the appropriate scaffolding strategies to guide students in solving the tasks.
- In the browning task, although the need of knowledge integration from biology and chemistry—in terms of interpreting browning as an enzyme-catalysed oxidation—is essential for solving this task, it is not commonly recognized by participants. The interpretation of the interdisciplinary phenomena requires a great deal of scaffolding from the teacher and it takes up a great part of the time dedicated to planning the experiment. Another challenge is related to the use of interdisciplinary evidence in the final conclusion, since the participants tend to use disciplinary evidence (chemical or biological) rather than interdisciplinary evidence.
- The main challenge in the nuclear waste disposal task is understanding complex evidence and integrating them in their arguments to persuade their peers.
- In the tooth decay task, one of the most important challenges is related to understanding and interpreting the simulation. In fact, in our study the groups who did not apply the simulation for interpreting the results were not able to back their conclusions with adequate evidence or even to draw an adequate conclusion.

h) An interdisciplinary approach promotes teacher collaboration.

Although the tasks presented in this chapter involved an interdisciplinary approach, they were designed by the first author as part of disciplinary chemical content knowledge. However, after their implementation we consider that an integrated design including all the domains involved in each task would help students to solve them from an interdisciplinary perspective rather than proposing isolated disciplinary claims. This is especially relevant in the browning task, in which participants' conclusions are drawn mainly on chemical evidence.

i) In STEM education, students may develop cognitive skills while engaged in the study of specific STEM-related social or global situations.

Although the task examples discussed in this chapter are not properly STEM focused, they require students to solve everyday life issues, which could be also addressed from technology, engineering and mathematics. However, they enable students to develop similar cognitive skills than those in STEM approaches, such as self-management or systems thinking.

j) In HPS and NOS approaches the technique of role play is recommended, especially for enacting and reflecting scientific process in the classroom.

The role play technique has been used in the lesson about the nuclear waste deposit. Although the focus was on discussing viewpoints for making an informed decision that integrates different types of evidence and knowledge, this is a good opportunity for reflecting about how decisions are made in science.

k) The SSI approach involves not only scientific content but also ethical, economic or cultural aspects. It offers appropriate contexts to engage in argumentation practices.

In this chapter, the task about the nuclear waste deposit is a good resource for carrying out interdisciplinary argumentation in the context of SSI. In it, participants are required to prepare their interventions according to the assigned roles, which requires backing up their knowledge with appropriate evidence, organizing their discourse around convincing arguments and persuading their peers in order to reach a consensus for making an agreed decision.

Although the task was designed considering all the aspects highlighted above, participants' arguments focused only on the scientific and economic ones.

Acknowledgements

Work supported by the Spanish MINECO (Ministry for Economy and Competitiveness), Grant number EDU2017-82915-R.

References

1. M. B. Nakhleh, *J. Chem. Educ.*, 1992, **69**(3), 191.
2. H. Tumai, *Chem. Educ. Res. Pract.*, 2016, **17**(2), 229.
3. J. K. Gilbert, *Int. J. Sci. Educ.*, 2006, **28**(9), 957.
4. J. Osborne and S. Collins, *Pupils' and Parents' Views of the School Science Curriculum*, King's College, London, 2000.
5. D. A. Roberts and L. Ostman, *Problems of Meaning in Science Curriculum*, Teachers College Press, New York, 1998.
6. S. Celik, *Austr. J. Teach. Educ.*, 2014, **39**(1), 1.
7. G. Tsaparlis, *Chem. Educ. Res. Pract.*, 2000, **1**(1), 161.
8. J. K. Gilbert and D. F. Treagust, *Multiple Representations in Chemical Education*, Springer, The Netherlands, 2009.
9. Organisation for Economic Cooperation and Development, OECD, *PISA 2015 Assessment and Analytical Framework: Science, Reading, Mathematic and Financial Literacy*, OECD Publishing, Paris, 2016.
10. National Research Council, NRC, *Education for Life and Work: Developing Transferable Knowledge and Skills in the 21st Century, Committee on Defining Deeper Learning and 21st Century Skills*, ed. J. W. Pellegrino and M. L. Hilton, The National Academies Press, Washington, DC, 2012.
11. EACEA/Eurydice, *Science Education in Europe: National Policies Practices and Research*, Education, Audiovisual and Culture Executive Agency, 2011.
12. J. Broggy, J. O'Reilly and S. Erduran, in *Science Education: An International Course Companion*, ed. B. Akpan and K. S. Taber, Interdisciplinarity and Science Education, Sense Publishers, The Netherlands, 2017, pp. 81–90.
13. J. Osborne, *Science*, 2010, **328**(5977), 463.
14. A. Tiberghien, in *Argumentation in Science Education: Perspectives from Classroom-based Research*, ed. S. Erduran and M. P. Jimenez-Aleixandre, Springer, Dordrecht, The Netherlands, 2008, p. ix.
15. V. Boix Mansilla and E. D. Duraising, *J. High. Educ.*, 2007, **78**(2), 215.
16. C. Golding, Integrating the disciplines: Successful interdisciplinary subjects, *Centre for the Study of Higher Education*, University of Melbourne, 2009.
17. B. St Clair and D. L. Hough, ERIC Document Reproduction Service No. 373 056, 1992.
18. N. G. Lederman and M. L. Niess, *Sch. Sci. Math.*, 1997, **97**(2), 57.
19. J. Shen, S. Sung and D. Zhang, *Int. J. Sci. Educ.*, 2015, **37**(17), 2809.
20. D. Rhoten and A. Parker, *Science*, 2004, **306**(5704), 2046.
21. Organisation for Economic Cooperation and Development, OECD, *PISA 2015 Results (Volume I): Excellence and Equity in Education*, OECD Publishing, Paris, 2016.
22. J. L. Polman, A. Newman, E. W. Saul and C. Farrar, *Sci. Educ.*, 2014, **98**, 766.
23. United Nations Environment Programme, UNEP, *UNEP Frontiers 2016 Report: Emerging Issues of Environmental Concern*, United Nations Environment Programme, Nairobi, 2016.

24. R. W. Bybee, *Achieving Scientific Literacy: From Purposes to Practice*, Heinemann, Portsmouth, 1997.
25. G. E. DeBoer, *J. Res. Sci. Teach.*, 2000, **37**, 582–601.
26. D. A. Roberts, in *Exploring the Landscape of Scientific Literacy*, ed. C. Linder, L. Östman, D. A. Roberts, P.-O. Wickman, G. Erickson and A. MacKinnon, Competing visions of scientific literacy: Influence of a science curriculum policy image, Routledge, New York, 2011, pp. 11–27.
27. J. Holman, *Educ. Chem.*, 2002, **39**, 12.
28. Y. Schwartz, R. Ben-Zvi and A. Hofstein, *Chem. Educ. Res. Pract.*, 2006, 7(4), 203.
29. L. Marcu, *Fasc. Biol.*, 2007, **14**, 53.
30. W. F. McComas and H. A. Wang, *Sch. Sci. Math*, 1998, **98**(6), 340.
31. C. S. Wagner, J. D. Roessner, K. Bobb, J. T. Klein, K. W. Boyack, J. Keyton, I. Rafols and K. Börner, *J. Informetr.*, 2011, **5**(1), 14.
32. W. H. Newell, in *Interdisciplinary Education in K-12 and College*, ed. J. T. Klein, Integrating the college curriculum, College Board Publications, New York, 2002, pp. 119–137.
33. P. Horton, *Integrated Science, An Introduction and Bibliography*, Florida Institute of Technology, Melbourne, FL, 1981.
34. D. Bragaw, K. A. Bragaw and E. Smith, *Mid. Sch. J.*, 1995, **27**(2), 39.
35. S. Schaal, F. Bogner and R. Girwidz, *Res. Sci. Educ.*, 2010, **40**(3), 339.
36. H. S. You, *J. Educ. Learn*, 2017, **6**(4), 66–77.
37. C. M. Czerniak, in *Handbook of Research on Science Education*, ed. S. K. Abell and N. G. Lederman, Interdisciplinary science teaching, Lawrence Erlbaum, Mahwah, NJ, 2007, pp. 537–559.
38. J. Shen, O. L. Liu and S. Sun, *Int. J. Sci. Educ.*, 2014, **36**(11), 1773.
39. D. Zhang and J. Shen, *Int. J. Sci. Educ.*, 2015, **37**(1), 2555.
40. C. F. Quigley and D. Herro, *J. Sci. Educ. Technol.*, 2016, **25**, 410.
41. M. Honey, G. Pearson and H. Schweingruber, *STEM integration in K-12 education: Status, Prospects, and an Agenda for Research*, National Academies Press, Washington, DC, 2014.
42. V. Boix Mansilla, H. Gardner and W. Miller, in *Interdisciplinary Curriculum: Challenges to Implementation*, ed. S. Wineburg and P. Grossman, On Disciplinary Lenses and Interdisciplinary Work, Teacher College Press, New York, 2000, pp. 17–38.
43. H. S. You, C. Delgado and J. A. Marshall, *J. Res. Sci. Teach.*, 2017, DOI: 10.1002/tea.21423.
44. M. C. Linn and B. S. Eylon, Science Education: Integrating Views of Learning and Instruction, in *Handbook of educational Psychology*, ed. P. A. Alexander and P. H. Winne, Erlbaum, New York, 2006, pp. 511–544.
45. O. L. Liu, H.-S. Lee, C. Hofstetter and M. C. Linn, *Educ. Ass.*, 2008, **13**(1), 33.
46. L. Rennie, G. Venville and J. Wallace, *Knowledge that Counts in a Global Community: Exploring the Contribution of Integrated Curriculum*, Routledge, New York, 2012.

47. M. C. Linn, H.-S. Lee, R. Tinker, F. Husic and J. L. Chiu, *Science*, 2006, **313**, 1049.
48. M. C. Linn, in *The Cambridge Handbook of the Learning Sciences*, ed. R. K. Sawyer, The Knowledge Integration Perspective on Learning and Instruction, Cambridge University Press, New York, 2006, pp. 243–264.
49. J. Shen and M. C. Linn, *Int. J. Sci. Educ.*, 2011, **33**(12), 1597.
50. T. C. Mason, *J. Teach. Educ.*, 1996, **47**(4), 263.
51. J. Beane, *Mid. Sch. J.*, 1996, **28**, 6.
52. M. C. Linn, J. D. Slotta, H. Terashima, E. Stone and J. Madhok, *APFSLT*, 2010, **11**(2), 1.
53. J. L. Araújo, C. Morais and J. C. Paiva, *Chem. Educ. Res. Pract.*, 2015, **16**, 893.
54. R. Kelly, E. McLoughlin and O. E. Finlayson, *Int. J. Sci. Educ.*, 2016, **38**(11), 1766.
55. National Research Council, *STEM Learning Is Everywhere: Summary of a Convocation on Building Learning Systems*, The National Academies Press, Washington, DC, 2014.
56. V. Wong, J. Dillon and H. King, *Int. J. Sci. Educ.*, 2016, **38**(15), 2346.
57. R. Lamb, T. Akmal and K. Petrie, *J. Res. Sci. Teach.*, 2015, **52**(3), 410.
58. National Academy of Engineering (NAE) and National Research Council (NRC), *STEM Integration in K-12 Education: Status, Prospects, and an Agenda for Research. Committee on Integrated STEM education; National Academy of Engineering and National Research Council of the National Academies*, The National Academies Press, Washington, DC, 2014.
59. R. W. Bybee, The case for STEM Education: Challenges and Opportunities, *National Science Teachers Association (NSTA)*, NSTA press, Virginia, 2013.
60. National Research Council, *Exploring the Intersection of Science Education and 21st Century Skills: A Workshop Summary*, The National Academies Press, Washington, DC, 2010.
61. B. Winterman and G. M. Malacinski, *Int. J. Art. Sci.*, 2015, **8**(2), 423.
62. I. Ghanbari, *Int. J. Art. Sci.*, 2015, **16**(7), 1.
63. L. D. English, *Int. J. Sci. Math. Educ.*, 2017, **15**(suppl), S5.
64. D. Höttecke and C. C. Silva, *Sci. Educ.*, 2011, **20**, 293.
65. S. D. Kolstø, *Sci. Educ.*, 2008, **17**, 977.
66. K. Kampourakis, *Sci. Educ.*, 2017, **26**, 1.
67. W.-Z. Shi, *Eur. J. Math. Sci. Tech. Educ.*, 2015, **11**(3), 551.
68. M. Matthews, *Science Teaching—the Role of History and Philosophy of Science*, Routledge, New York, 1994.
69. K. G. Witz and L. Hyunju, *J. Curriculum. Stud.*, 2009, **41**(3), 409.
70. H. A. Wang and D. D. Marsh, *Sci. Educ.*, 2002, **11**(2), 169.
71. L. Dibattista and F. Morgese, *Sci. Educ.*, 2013, **22**, 543.
72. F. Kubli, *Sci. Educ.*, 2005, **14**(6), 501.
73. D. Metz, S. Klassen, B. McMillan, M. Clough and J. Olson, *Sci. Educ.*, 2007, **16**, 313.
74. A. Henke and D. Höttecke, *Sci. Educ.*, 2015, **24**, 349.

75. J. Osborne, S. Collins, M. Ratcliffe, R. Millar and R. Duschl, *J. Res. Sci. Teach.*, 2003, **40**, 692.
76. F. Abd-El-Khalick and N. G. Lederman, *Int. J. Sci. Educ.*, 2000, **22**, 665.
77. M. Niaz and A. Maza, *Nature of Science in General Chemistry Textbooks*, Springer, Dordrecht, 2011.
78. F. Abd-El-Khalick, *Int. J. Sci. Educ.*, 2012, **34**(3), 353.
79. R. Duschl, *Rev. Res. Educ.*, 2008, **32**, 268.
80. R. A. Duschl and R. Grandy, *Sci. Educ.*, 2013, **22**, 2109.
81. D. Allchin, *Sci. Educ.*, 2011, **95**(3), 518.
82. S. Erduran and Z. R. Dagher, *Reconceptualizing the Nature of Science for Science Education: Scientific Knowledge, Practices and Other Family Categories*, Springer, Dordrecht, 2014.
83. D. Allchin, *Teaching the Nature of Science: Perspectives and Resources*, SHiPS Education Press, Saint Paul, MN, 2013.
84. M. Kousathana, M. Demerouti and G. Tsaparlis, *Sci. Educ.*, 2005, **14**, 173.
85. M. F. Paixao and A. Cachapuz, *Chem. Educ. Res. Pract.*, 2000, **1**(2), 201.
86. S. Erduran, *Sci. Educ.*, 2005, **14**, 161.
87. M. Niaz, *Sci. Educ.*, 2009, **18**, 43.
88. A. Guerra, M. Braga and J. C. Reiss, *Sci. Educ.*, 2013, **22**, 1458.
89. T. D. Sadler, *Cult. Stud. Sci. Educ.*, 2009, **4**, 697.
90. D. L. Zeidler, S. M. Applebaum and T. F. Sadler, in *Socio-scientific Issues in Science Classrooms: Teaching, Learning and Research*, ed. T. D. Sadler, Enacting a socioscientific issues classroom: Transformative transformations, Springer, Dordrecht, 2011, pp. 277–306.
91. M. P. Jiménez-Aleixandre and P. Brocos, in *Interpersonal Argumentation in Educational and Professional Contexts*, ed. F. Arcidiacono and A. Bova, Processes of negotiation in socio-scientific argumentation about vegetarianism in teacher education, Springer, Dordrecht, 2017, pp. 117–139.
92. C. Oulton, J. Dillon and M. Grace, *Int. J. Sci. Educ.*, 2004, **26**(4), 411.
93. S. D. Kolstø and M. Ratcliffe, in *Argumentation in Science Education: Perspectives from Classroom-based Research*, ed. S. Erduran and M. P. Jiménez-Aleixandre, Social aspects of argumentation, Springer, Dordrecht, 2008, pp. 117–136.
94. S. D. Kolstø, *Sci. Educ.*, 2001, **85**(3), 291.
95. M. Stolz, T. Witteck, R. Marks and I. Eilks, *Eur. J. Math. Sci. Tech. Educ.*, 2013, **9**(4), 361.
96. I. Eilks, *Chem. Educ. Res. Pract. Eur.*, 2002, **3**, 77.
97. J. L. Eastwood, W. M. Schlegel and K. L. Cook, in *Socio-scientific Issues in the Classroom: Teaching, Learning and Research*, ed. T. D. Sadler, Effects of an interdisciplinary program on students' reasoning with socio-scientific issues and perceptions of their learning experiences, Springer, Dordrecht, 2011, pp. 89–126.
98. S. E. Barrett and M. Nieswandt, *J. Res. Sci. Teach.*, 2010, **47**(4), 380.
99. G. L. Cole, PhD thesis, University of Georgia, 2011.

100. S. D. Kolstø, B. Bungum, E. Arnesen, A. Isnes, T. Kristensen, K. Mathiassen, I. Mestad, A. Quale, A. S. V. Tonning and M. Ulvik, *Sci. Educ.*, 2006, **90**(4), 632.
101. S. Markic and P. Childs, *Chem. Educ. Res. Pract.*, 2016, **17**, 434.
102. M. P. Jiménez Aleixandre and S. Erduran, in *Argumentation in Science Education: Perspectives from Classroom-based Research*, Argumentation in science education: An overview, Springer, Dordrecht, 2008, pp. 3–27.
103. J. A. Greene, W. A. Sandoval and I. Bråten, in *Handbook of Epistemic Cognition*, ed. J. A. Greene, W. A. Sandoval and I. Bråten, An introduction to epistemic cognition, Routledge, New York, 2016, pp. 1–13.
104. K. Iordanou, P. Kendeou and K. Beker, in *Handbook of Epistemic Cognition*, ed. J. A. Greene, W. A. Sandoval and I. Bråten, Argumentative reasoning, Routledge, New York, 2016, pp. 39–53.
105. M. P. Jiménez-Aleixandre and B. Crujeiras, in *Science Education: An International Course Companion*, ed. B. Akpan and K. S. Taber, Epistemic practices and scientific practices in science education, Sense Publishers, The Netherlands, 2017, pp. 69–80.
106. V. Sampson and M. R. Blanchard, *J. Res. Sci. Teach.*, 2012, **49**(9), 1122.
107. K. L. McNeill and L. Berland, *J. Res. Sci. Teach.*, 2017, **54**(5), 672.
108. A. R. Cavagnetto, *Rev. Educ. Res.*, 2010, **80**(3), 336.
109. A. Pabuccu and S. Erduran, *Int. J. Sci. Educ.*, 2017, **39**(9), 1154.
110. E. Kaya, *Int. J. Sci. Educ.*, 2013, **35**(7), 1139.
111. M. Aydeniz and A. Dogan, *Chem. Educ. Res. Pract.*, 2016, **17**, 111.
112. C. Cigdemoglu, H. O. Arslan and A. Cam, *Chem. Educ. Res. Pract.*, 2017, **18**, 288.
113. N. Becker, C. Rasmussen, G. Sweeney, M. Wawro, M. Towns and R. Cole, *Chem. Educ. Res. Pract.*, 2013, **14**, 81.
114. R. S. Bowen, D. R. Picard, S. Verberne-Sutton and C. J. Brame, *J. Chem. Educ.*, 2018, **95**(1), 108.
115. U. Kulatunga, R. S. Moog and J. E. Lewis, *J. Res. Sci. Teach.*, 2013, **50**(10), 1207.
116. H. Jin, C. E. Mehl and D. H. Lan, *J. Res. Sci. Teach.*, 2015, **52**(8), 1132.
117. M. Karpudewan, W.-M. Roth and D. Sinniah, *Chem. Educ. Res. Pract.*, 2016, **17**, 893–901.
118. V. Sampson and D. Clark, *Sci. Educ.*, 2009, **93**, 448.
119. B. Crujeiras and M. P. Jiménez, *Ens. Cienc.*, 2015, **33**(1), 63.
120. B. Crujeiras-Pérez and M. P. Jiménez-Aleixandre, *Chem. Educ. Res. Pract.*, 2017, **18**, 99.
121. T. D. Sadler, *J. Res. Sci. Teach.*, 2004, **41**(5), 513.
122. M. P. Jimenez-Aleixandre and C. Pereiro-Muñoz, *Int. J. Sci. Educ.*, 2002, **24**(11), 1171.

CHAPTER 4

Lesson Resources and Teaching Strategies on Argumentation for Secondary Chemistry Education

ALISON CULLINANE*[a] AND ANNE O'DWYER*[b]

[a] University of Oxford, Oxford, UK; [b] Mary Immaculate College, Limerick, Ireland
*Email: alison.cullinane@education.ox.ac.uk; anne.odwyer@mic.ul.ie

4.1 Introduction

Numerous studies have shown that young people are departing from secondary school with a limited sense of the nature of scientific knowledge, a low-level understanding of science and how ideas are produced, evaluated and revised in science.[1–3] Furthermore, much of learners' difficulty with chemistry at secondary level is due to the multi-dimensional nature of the subject.[4] The abstract concepts of submicroscopic and symbolic representations used in chemistry increase the difficulty for many novice learners. There is a close relationship between argumentation and learning science. Instilling argumentation in the classroom from an early stage will allow students to develop the skills and aptitude for argumentation. And like developing any skill, students need practice in order to gain the knowledge, understanding and the confidence to use these skills, which literature indicates will lead to an understanding of the practices and ways of thinking of scientists.[3]

Advances in Chemistry Education Series No. 2
Argumentation in Chemistry Education: Research, Policy and Practice
Edited by Sibel Erduran

Published by the Royal Society of Chemistry, www.rsc.org

Learning to reason in science involves developing the skill of learning to construct arguments by linking evidence with and to ideas and theories.[5] Goodman[6] claims that there is no concept more central to logic and critical thinking than the concept of an argument. Nevertheless studies have shown that current classroom practices give little opportunity for young people to develop their ability to construct arguments.[2,3] Literature historically illustrates that science teachers' use of discussion and argument as tools for teaching and learning are limited, due to a lack of training in managing the process and uncertainty of its value in learning science.[7] Due to its increased recognition, more contemporary research and development efforts have moved to produce resources and strategies to support teachers in teaching argumentation. Many of these argumentation resources support educational goals such as understanding how we know and why we have belief in certain scientific claims and the very essence of the nature of science knowledge itself.[8–10] Such understanding is fundamental to knowing how the claims of the scientific community fit into our society. As science educators, our ultimate goal is to produce scientific citizens that are capable of evaluating evidence and making informed decisions. Argumentation provides a useful life skill to examine these claims in the chemistry classroom.

This chapter will suggest some domain-specific strategies. The first strategy will illustrate concept cartoons, the second will discuss the use of evidence-based reasoning and the third will discuss the use of current debates found on social media platforms as a stimulus for an argumentation lesson. As the internet and social media play a more prominent role in our lives, it is important to understand what is truthful and what is "fake news." Erduran[11] outlines some generic framework strategies from research that facilitates classroom discussion (p. 259). These strategies include arranging statements, concept maps, constructing arguments to designing experiments. Some of these generic approaches are used in this publication. These may be useful to support teachers who want to incorporate argumentation into other chemistry topics.

4.2 Concept Cartoons

Concept cartoons are very familiar to many in science education. They present colourful images of characters discussing science concepts and the images often present correct science claims, alongside well-established misconceptions that students are shown to possess. These are often promoted for use in the primary classroom. However, both authors have used and promoted their use in the lower secondary chemistry classroom by facilitating higher-order discussion and argument about the claims made in the cartoon images. Several researchers have identified difficulties in applying Toulmin's[12] schema that underpins many argumentation activities.[13,14] Although it draws on the work of the Toulmin model, the *Downing Model* of argumentation[13] is more often used with concept cartoons as it focuses on the nature of interactions between students and places a greater focus on the socio-cultural context. This model has seven levels[13] (p. 23). They are included here as they may be useful to teachers to help identify the extent of student interaction and argumentation in the classroom:

Level 1: Pupils are unable or unwilling to enter into discussion.
Level 2: Pupils make a claim to knowledge.
Level 3: Pupils begin to offer grounds to support their claims.
Level 4: Pupils offer further evidence to support their claims.
Level 5: Pupils respond to ideas from others in the group.
Level 6: Pupils are able to sustain an argument in a variety of ways.
Level 7: Pupils evaluate the evidence and make judgements.

The authors advocate the use of concept cartoons to promote argumentation, given that many researchers have shown the usefulness of the strategy in helping students to generate dialogue, clarify and explain their thinking, justify their answers and consider alternative explanations.[5,15] Concept cartoons have been identified as a useful classroom strategy to increase learners' levels of motivation and involvement in learning, by eliciting challenges and developing learners' ideas.[15] The visual presentation of a context-based chemistry concept make concept cartoons effective in capturing learners' attention and motivation and have the potential to legitimise classroom argumentation by presenting a number of plausible arguments about an engaging concept in visual form, which facilitates learners in extending the argument.

Keogh & Naylor[15] developed concept cartoons as a classroom strategy to clarify the relationship between constructivist models of learning, scientific epistemology and classroom practice. The core elements (minimal written language, visual images, alternative concepts and questions) of a concept cartoon are common to many teaching strategies exploring conceptual understanding.[16,17] Their concept cartoons are designed in the format of multiple choice questions, where the cartoon and image integrates the written text (most often as speech bubbles) and a visual stimulus to support the learners' perception of the very often abstract chemical concept. *'The use of dialogue creates the opportunity to present alternative ideas, including one or more views which are scientifically acceptable'*[15] (p. 432). The lack of agreement among the characters in the cartoon promotes argumentation.[9]

4.2.1 Lesson Ideas Using a Concept Cartoon

The concept cartoon shown in Figure 4.1 investigates the particulate nature of matter and how the changing states of matter could be explored. Changes in the states of matter is a common introductory topic in many chemistry curricula worldwide.

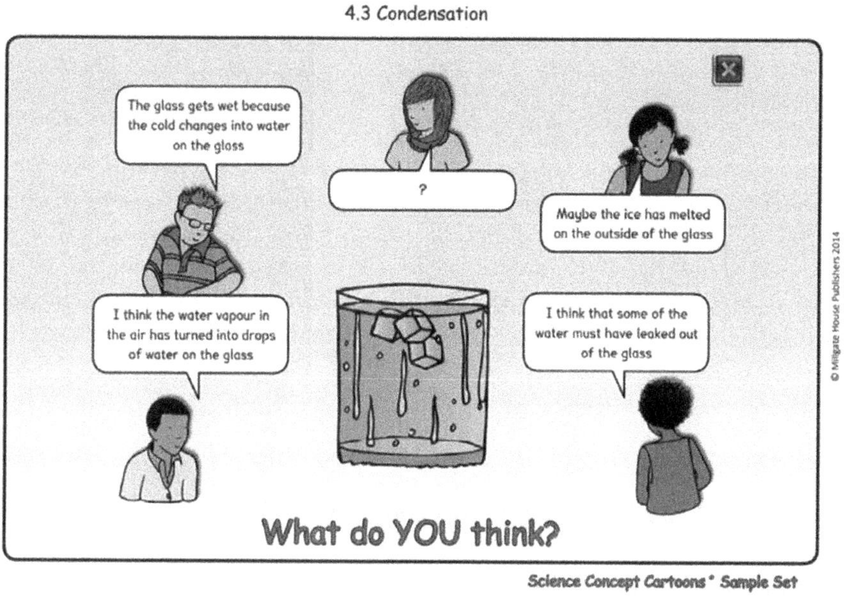

Figure 4.1 Condensation concept cartoon.[18]
Reproduced from ref. 18 with permission from Millgate House Publishers.

Keogh & Naylor[15] outline the structure of a lesson based on the use of a concept cartoon:

- *a brief introduction to the activity;*
- *an invitation to the learners to reflect on the concept cartoon(s);*
- *an invitation to the learners to discuss in groups what they think and why;*
- *interaction and intervention by the teacher (as appropriate) during the teaching session;*
- *practical investigation or research-based activity to follow up the learners' ideas as appropriate, encouraged and supported by the teacher as necessary;*
- *a whole class plenary to share and challenge ideas.*

To follow this recommended structure, the lesson begins by presenting the concept cartoon (Figure 4.1) to promote a group discussion. This initial discussion can be supported by the teacher, by asking the students to consider the ideas expressed by each character in the cartoon. This specific example also includes a blank speech bubble. This is useful in many concept cartoons as it allows students to express their own ideas, if they differ from the ideas of the other characters. It is important that the teacher is non-judgemental at this stage and to welcome students' suggestions. Naylor *et al.*[13] have found that following the opening statements (where the different points of view are outlined), students tend to build on each other's contributions in order to reach a shared understanding, through dialogical and interactive conversations. It is notable that the '*thinking out loud*' conversations often tend to be overlapping, rather than following a monological chain of reasoning. Naylor *et al.*[13] identified that students were capable of engaging in and sustaining an argument without the teacher's direction or intervention. This has implications for the teacher's role when using this classroom strategy. It was observed that when the teacher became involved in the student group discussions, this often interrupted the students' chain of reasoning and their confidence in sharing their ideas, perhaps due to the power dynamic. The students typically explored the alternative viewpoints in a collaborative process to build a consensus and understanding. The uncertainty developed by this discussion is productive.

4.2.2 Ideas for Argumentation in Lower Secondary

Useful resources to explore ideas at lower secondary level could simply be the use of small mirrors. By allowing the students to explore what happens when they breathe on a mirror, they may observe droplets (condensation). Relatable connections may be made to students' spectacles. Any student who wears spectacles will be able to share experiences of when they 'fog up,' even if not able at this point to explain why. Through the hands-on exploration using tangible resources, their ideas should be challenged or consolidated. In engaging with the concept cartoon, learners have the opportunity to experience cognitive conflict, as they are presented with conflicting but apparently plausible ideas.[15] This cognitive conflict can be used as the stimulus to nurture classroom discussion and lead to hypothesis development and exploratory investigations. During and after this exploration phase, the students may change their minds and ideas to inform their later investigations to develop a more accurate understanding.

This concept cartoon has the capability to promote argumentation style classroom investigations. For those unfamiliar with the concept depicted in Figure 4.1, in cool temperatures the water vapour from the air will condense when it hits the cold surface of the glass. It is possible to replicate the image represented in the cartoon safely and using a series of questions students could observe a cold empty glass out of a freezer which is filled with ice. Does the outside of the glass get wet? If it does, where could the water have come from? The teacher could probe further thinking by asking does it happen wherever you put the glass, for example indoors and outdoors? This could perhaps lead to the students planning and implementing a fair test with the glass and ice cubes by asking questions such as does it make a difference if we change where the glass of ice is left? Students could compare the amount of condensation present when in different places, and varying conditions such as still air, moving air, cold air, warm air, dry air and moist air. As well as scaffolding students' understandings about changes in states of matter at different temperatures, this lesson also evidences the presence of water vapour in the air. To extend and consolidate the learning, the students should be encouraged to use the observations from the investigations with the glass of ice in the classroom to explain the presence of condensation in other contexts. The students could be shown images of other everyday contexts related to this concept and asked to explain such concepts as to why are there droplets of water inside the classroom windows or inside the car window when people are wearing damp or wet clothing.

4.2.3 Ideas for Argumentation in Upper Secondary

The same image (Figure 4.1) could be used to initiate argumentation in the upper levels of secondary school by changing the ideas expressed in the speech bubbles. Extensive research into learners' alternative understandings in chemistry education has provided a diversity of concepts and alternative concepts for potential development of concept cartoons. Concept cartoons can help to accommodate for differentiation in the classroom. Possible alternative conceptions may relate to the particulate nature of matter. Examples include: *The water molecules break up when the water evaporates* (misconception about intramolecular bonds); *The water disappears after it evaporates* (misconception about states of matter); *The water can change from a liquid to gas and back to a liquid again, like it can freeze, thaw and refreeze* (changing states of matter-changing intermolecular bonds); *The water molecule, H_2O, is the same in ice, water and steam* (intramolecular bonds are not broken in changing states of matter). Naylor & Keogh[18] have published a number of concept cartoons appropriate for classroom use, and while many of these may be more appropriate for lower secondary level there is potential to adapt appropriately for upper secondary level. Concepts include chemical reactions, temperature change in chemical reactions, and properties of materials (investigating conductors, insulators, *etc.*). These concepts encourage discussion about the conditions that cause chemical reaction and change. The context of a rusty chain on a bicycle that is always parked outdoors could be used to frame the cartoon.

4.3 Evidence-based Reasoning Approach

Having argumentation tasks that are domain-specific to chemistry rather than domain-general provides a better platform to improve chemistry knowledge and understanding and improve ways of thinking.[2] Recent research has investigated learners' argumentation and understanding of chemical reactions,[19] behaviour of gases,[20] particulate nature of matter[21] and in acids and bases[22] to name but a few. Toulmin's[12] model of argumentation can be used to support the analysis of students' reasoning. The Evidence-based Reasoning (EBR) Framework[23] combines Toulmin's[12] argument pattern as well as assessing the quality of the learners' reasoning. The EBR framework has diverse potential in:

(a) *supporting students' and teachers' understanding of the process of scientific reasoning;*

(b) *modelling exemplary scientific reasoning;*
(c) *diagnosing problems and identifying pitfalls affecting student reasoning as it develops; and*
(d) *assessing scientific reasoning in the classroom both formatively and summatively.*
(Ref. 23, p. 124).

The EBR framework provides a description using theoretical statements, backed by scientific evidence to evaluate the quality of a claim. The authors of this chapter agree with Brown and colleagues'[23] intentions for the model. Its purpose is not to show how scientific knowledge is or should be generated; instead, the EBR framework describes the use of evidence in scientific reasoning to make a claim. The authors have chosen to include this framework as it has potential to help teachers to identify the type of scientific argumentation in their students' work and in classroom discourse.

4.3.1 Lesson Ideas using Evidence-based Reasoning

Erduran and Pabuccu[24] have created a resource booklet '*Bonding Chemistry and Argument: Teaching and Learning Argumentation through Chemistry Stories.*' This resource contains six exemplar lesson activities aimed at secondary chemistry education. Their approach was informed by the need to motivate students by linking chemistry knowledge to everyday contexts through the use of imaginative and creative stories.[24] Each lesson is developed around a context of a story. The reason for this is to ensure the context of the lesson is meaningful for the learners, and thus supports their pursuits in wanting to investigate and to better understand the chemistry knowledge in question. The authors recognise the importance of the co-ordination of theory and evidence through justification and reasoning in framing the argumentation activities.[24] The publication presents six activities: Activity 1: Halloween Crush!, Activity 2: Holiday in Dubai, Activity 3: Mysterious Diaries, Activity 4: Alkanes in Stress, Activity 5: Organic Potions, and Activity 6: Cheating on the exam. The activities aim to integrate the promotion of argumentation skills and provide teachers with some examples and guidelines for structuring the lessons in ways that would support EBR. Erduran & Pabaccu[24] have outlined the benefits of developing the ability to engage in EBR in relation to both scientific and socio-scientific issues.

Figure 4.2 shows an example of a scientific argument, adapted from the *Halloween Crush* lesson which explores concepts relating to air pressure, particulate nature of matter and behaviour of gases. Given the data relating to gas behaviour, these are examples of the students' warrants, backing and rebuttal to the claim about the crushed can that they find.[20] Figure 4.2 shows how the data was explicitly linked with the claim through warrants. The statement '*The shape of the can would not change if the number of molecules entering the can became equal to the number of molecules leaving the can*' is a

Figure 4.2 A scientific argument supported by Toulmin's framework[12] and an example concept about why an oil can crushes.[20]

rebuttal, and creates an opposition to the justification used in the initial argument.

Other resources for secondary chemistry to support argumentation in the laboratory include *Argument-Driven Inquiry in Chemistry*.[25] This publication from the *National Science Teachers Association* (NSTA) has authentic laboratory investigations to support learners' argumentation. Another example is a simple laboratory investigation using Gummy Bears. This lesson has been exemplified to help students recognise the importance of collaboration, communication, skepticism and peer-review as part of scientific practices.[26] In the Gummy Bear lesson, students participate in a scenario-based laboratory activity (investigating the diffusion of water into Gummy Bears) designed to help them define qualities that result in reliable and meaningful scientific research. By having students conduct an investigation that gives highly variable results within and between laboratory teams, students learn the importance of making strong arguments in science as they use evidence and reasoning to support their claims. They also develop their communication and collaboration skills by evaluating each other's claims.

4.4 Writing Frames

Using a writing frame is similar in some ways to evidence-based reasoning, as students are presented with the cultural tools and conventions of the science community as evidence used to support the learning process.[23] Adding to these tools and conventions, students use the writing frame to develop their arguments. Osborne *et al.*[2] developed a writing frame which proposed the enhancement of the quality of argumentation of science topics in schools (see Figure 4.3). If educators are new to argumentation activities, this framework provides a simplistic approach to initiate students' involvement.

Introducing the writing frame helps to structure tasks and allows the student to proceed in developing their argumentation and self-learning skills. Erduran and Villamañán[27] operationalised the writing frame shown in Figure 4.3 to examine written arguments from 130 university engineering students in the context of the thermodynamics principles around refrigeration. While their work focused on third-level engineering education, the argumentation strategies are transferable to the secondary chemistry classroom. The study found that students needed increased explicit support to help improve their argumentation writing. When this was noted in their study, they expanded the writing frame with a set of reflective questions designed to aid students to consider data collection and interpretation to target certain aspects of their arguments. Their adaptations also included a meta-level reflection where the students can self- and peer-assess the arguments suggested. Incorporation of these kinds of questions with upper secondary level chemistry is a useful strategy, as they are important for students to cement their learning.[28] Figure 4.4 is an adaptation of the writing frame that incorporates both these studies and include the original

My Argument
My Idea is that....
My reasons are that....
Arguments against my idea might be that...
I would convince somebody that does not believe me by...
The evidence I would use to convince them that ...

Figure 4.3 Writing frame to help scaffold the quality of arguments in science lessons.
Adapted from ref. 2 with permission from the authors.

writing frame with also the self and peer assessment strategies from Erduran and Villamañán.[27]

The directional flow of this writing frame pictured in Figure 4.4 begins with the box "my idea is that..." and progresses through to the final box. The students are then directed to self-assess by reflecting on their argument. This also provides an opportunity for peer assessment where their peers can examine and reflect on the argument and provide feedback or a rebuttal to their arguments. This writing frame will be used in the following section which provides strategies that can be used to promote argumentation in the secondary chemistry classroom. The lesson ideas include interrogating social media to develop argumentation and using the writing frame to help support students in developing their arguments in these areas. Due to issues that may emerge relating to use of social media in the secondary classroom,

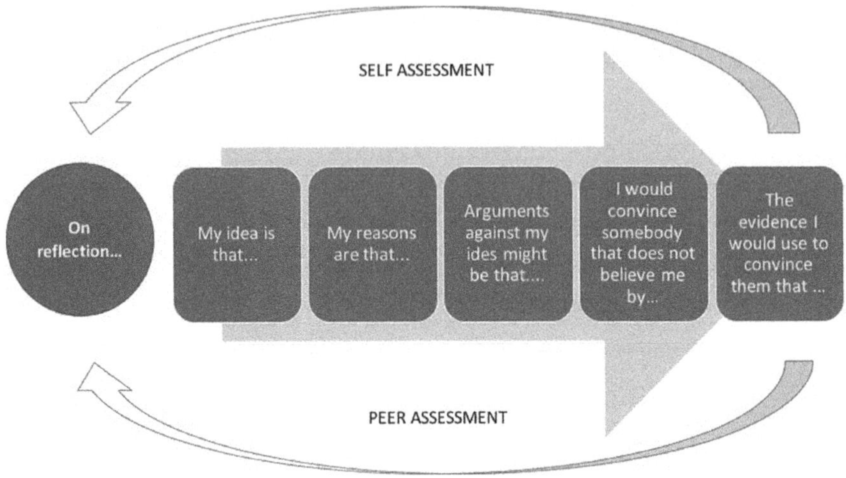

Figure 4.4 Writing frame.[27]
Adapted from ref. 2 with permission from the authors.

screenshots of tweets can be used to capture the images or develop some images that replicate some Twitter feeds. The use of these platforms therefore should not be a deterrent for teachers, who can easily present the images from social media rather than using the platforms themselves. Alternatively, it is possible to design our own 'Twitter feeds' to fit the arguments you want your students to interrogate. Due to copyright issues around the use of Twitter posts, the tweets are described in the text. The scenarios presented are scrutinized and interrogated using the writing frame.

4.5 Social Media

This section will examine some contemporary socio-scientific issues that chemistry students may encounter on social media. Presenting ideas like this is especially pertinent for educating students to be scientifically literate citizens, who will have to face hard decisions regarding socio-scientific problems in today's democracies.[29] This approach of using social media to capture the students' interest using everyday contexts acts as a stimulus to engage.[30] Many will already utilise these social media platforms. These look at the current "Coal *versus* Renewable Energy" debate, and the "Great Plastic Race." As social media plays such a large part in students' lives, using this platform as a tool for argumentation is a novel way to show young people that they need to question the data and opinions presented to them on these platforms. It also acts as evidence to showcase how chemistry permeates so many aspects of their everyday life. This practice is useful to highlight how science is a human enterprise, is socially constructed exercise through discourse, and how this practice plays a central role in establishing knowledge claims.[3] It is also a way of showing how science and argumentation can have real-life consequences in society, and how the public needs satisfactory science content knowledge and understanding about scientific knowledge itself; such as theories, laws and models. Actively studying science may not gain these outcomes alone. Explicit attention needs to be drawn to these big ideas while teaching them.[31]

4.5.1 Lesson Idea: Coal Versus Renewable Energy

The first lesson idea will look at the current debate around coal *versus* renewable energy. The main driver of this current debate is legislation being drafted in the US to increase coal production and their recent departure from The Paris Agreement. The main instigator in all of this is US President Donald Trump. This activity examines his tweets that exclaims joy over his legislation for increased coal production.[32] It states that "it is finally happening for our great clean coal miners." This is presented over an image that gives some statistics about US coal production: "Up 7.8% past year. Down 31% last 10 years." This is followed by the hashtag *#EndingWarOnCoal.* Immediately underneath the image, someone comments "Still with the coal? Dude, Move on." And in response to that tweet, another respondent states "How rude to belittle coal miners who need jobs." This one tweet presents several aspects of chemistry education that can be examined as part of organic chemistry in secondary chemistry education. Attention can be drawn to the comment of one tweeter in response to this tweet, which states "how rude to belittle coal miners who need jobs." Without turning the lesson into a social science lesson, exploring if this is something that should be thought of in this debate would allow students to see a real instance of how science and politics are impacting on the daily lives of others. In contrast to this, tweets from the International Renewable Energy Agency (IRENA) are presented.[33] This is an intergovernmental organisation that supports countries in their transition to a sustainable energy future. It states in the tweet that "*Meeting the goals of the #ParisAgreement means reducing carbon emissions from the energy sector.*" It presents an image of a chart graphing the projected energy-related CO_2 emissions until 2050. This tweet illustrates how renewable energy and energy efficiency can provide over 90% of the reduction in energy-related CO_2 emissions.[33]

Prior to undertaking this task, many aspects of chemistry can be taught, such as organic chemistry, sources of hydrocarbons, and the modification and combustion of alkanes. Although it may be beyond the specification of many secondary chemistry curricula this exercise could examine some aspects of the chemistry of coal; for example its chemical structure (contains predominantly condensed polycyclic aromatic rings) and how this relates to the conversion of coal to conventional liquid fuels, condensation reactions and how it relates to liquefying coal, the extreme pressures and temperatures required for coal conversion, how organic sulphur is much more difficult to remove than organic oxygen, among other things.[34]

As well as this, examining the ozone layer and the chlorination of alkanes, which are often part of secondary school chemistry curricula, would be useful for the students' understanding of this debate. There are different chemical processes when looking at renewable energy that can be examined prior to undertaking this task, such as the carbon cycle, chemistry of biofuels and the chemistry in battery development. The writing frame is used to support students in using the above tweets to present their arguments on these two differing sides. The teacher can present the above tweets to the students and ask them to use the knowledge they have developed over the course to support their arguments.

4.5.2 Lesson Idea: The Great Plastic Race

This lesson idea draws students' attention to the environmental effects of plastic. A recent announcement was made on the creation of a mutant enzyme that was captioned as a breakthrough that could solve the global plastic pollution crisis.[35] This is a widespread epidemic that is very topical at the moment due to the ecological devastation it is having on the earth.[37] The first tweet which is used to present the first claim is from a popular science page on Twitter called *Science Friday*. This tweet states "In 2016, scientists discovered that a bacterium had evolved to produce an enzyme that enabled it to eat plastic bottles. Now, they have discovered a way to tweak that enzyme to do the work 20 percent faster. Could it be the next recycling solution?"[35] This tweet presents the article with information about the scientific discovery of the enzyme found to digest plastic. This could be used in a lesson alongside news articles reporting the unknown consequences and possible risks this could cause to the climate such as greenhouse gas emissions, like the one presented in Figure 4.5. This figure was designed to capture legitimate current counter-arguments found in the media, which mainly centre around the unknown consequences of the enzyme.

Again, the writing frame (Figure 4.4) can be used to support students in developing arguments in the debate. This real-life socio-scientific issue has many learning possibilities, which include looking at the chemistry

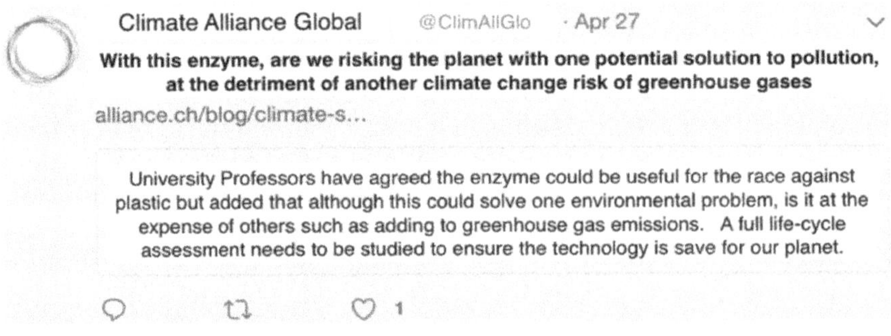

Figure 4.5 Tweet for the argumentation activity on the global plastic debate.

of plastics, plastic symbols and recycling codes and the biochemistry of enzymes, and presenting this contemporary issue allows for potential augmentation activities. One such debate facing the scientific community is weighing up the benefits of the enzymes digesting plastic, compared to the potential consequences this process could have on the emissions of greenhouse gases or the degradation of other materials that are useful. Showcasing the socio-scientific issue through Twitter or other social media platforms captures the students' attention. It gives a real-life context to what they are learning and how the development of this knowledge can have an impact on their lives and living conditions on a global scale. This can be presented to them at a local level, where they see plastic litter around their school as an initial frame of reference for discussion.

4.6 Conclusions

This chapter highlights some useful strategies that can be used to support argumentation in secondary chemistry education. Offering students the opportunity to engage in argument, where they can oppose and/or support each other's claims, promotes the development of their critical thought processes, knowledge, beliefs and reasoning.[36] Having resources that can target and develop student argumentation skills are key for teachers to understand what argumentation is and how it can be developed with students in their classrooms.[3] Argumentation gives students the skills to weigh up various claims to see if they have value scientifically in their world and society. In our evolving education systems and developments in chemistry, no programme of study can provide all the scientific content needed for our future citizens and chemistry graduates. However, being able to recognise the strengths and limitations of chemical arguments provides students with skills far reaching their classroom practice. Although we have presented selected classroom strategies and lesson resources for secondary chemistry education, the authors advocate teachers' innovation and modification of these where applicable to be used beyond the context of the specific exemplars presented in this chapter.

References

1. A. Cullinane and M. Liston, Review of the Leaving Certificate biology examination papers (1999–2008) using Bloom's Taxonomy – an investigation of the cognitive demands of the examination, *Irish Educ. Stud.*, 2016, **35**(3), 249–267.
2. J. Osborne, S. Erduran, S. Simon and M. Monk, Enhancing the quality of argument in school science, *School Sci. Rev.*, 2001, **82**(301), 63–70.
3. R. Driver, P. Newton and J. Osborne, Establishing the norms of scientific argumentation in classrooms, *Sci. Educ.*, 2000, **84**, 287–312.
4. A. H. Johnstone, Why is Science difficult to learn? Things are seldom what they seem, *J. Comput. Assist. Learn.*, 1991, **74**(3), 262–268.

5. J. Wellington and J. Osborne, *Language and Literacy in Science Education*, Open University, Buckingham, UK, 2001.
6. J. Goodman, On Defining Argument, *Argumentation*, 2018, DOI: 10.1007/s10503-018-9457-y.
7. J. Solomon, About argument and discussion, *School Sci. Rev.*, 1998, **80**(291), 57–62.
8. *Argumentation in Science Education: Perspectives from Classroom-based Research*, ed. S. Erduran and M. P. Jimenez-Aleixandre, Springer, Dordrecht, 2008.
9. M. P. Jimenez-Aleixandre, A. B. Rodriguez and R. A. Duschl, "Doing the lesson" or "doing science": argument in high school genetics, *Sci. Educ.*, 2000, **84**(6), 757–792.
10. A. Zohar and F. Nemet, Fostering students' knowledge and argumentation skills through dilemmas in human genetics, *J. Res. Sci. Teach.*, 2002, **39**(1), 35–62.
11. S. Erduran, Breaking the law: promoting domain-specificity in chemical education in the context of arguing about the periodic law, *Found. Chem.*, 2007, **9**(3), 247–263.
12. S. Toulmin, *The Uses of Argument*, Cambridge University Press, Cambridge, UK, 1958.
13. S. Naylor, B. Keogh and B. Downing, Argumentation and Primary Science, *Res. Sci. Educ.*, 2007, **37**, 17–39.
14. S. Simon, S. Erduran and J. Osborne, Enhancing the quality of argumentation in school science. Paper presented at the *National Association for Research in Science Teaching*, April 7–10, New Orleans, USA, 2002.
15. B. Keogh and S. Naylor, Concept cartoons, teaching and learning in science: an evaluation, *Int. J. Sci. Educ.*, 1999, **2**(4), 431–446.
16. R. Osborne and J. Gilbert, A method for investigating concept understanding in science, *Eur. J. Sci. Educ.*, 1980, **2**, 311–321.
17. T. Russell, W. Harlen and D. Watt, Children's ideas about evaporation and condensation, *Int. J. Sci. Educ.*, 1989, **11**, 566–576.
18. S. Naylor and B. Keogh, *Concept Cartoons in Science Education*, Millgate House Publishers, UK, 2000.
19. M. L. Weinrich and V. Talanquer, Mapping students' modes of reasoning when thinking about chemical reactions used to make a desired product, *Chem. Educ. Res. Pract.*, 2016, **17**(2), 394–406.
20. A. Pabuccu and S. Erduran, Investigating students' engagement in epistemic and narrative practices of chemistry in the context of a story on gas behavior, *Chem. Educ. Res. Pract.*, 2016, **17**, 523–531.
21. N. Becker, C. Rasmussen, G. Sweeney, M. Wawro, M. Towns and R. Cole, Reasoning using particulate nature of matter: An example of a socio-chemical norm in a university-level physical chemistry class, *Chem. Educ. Res. Pract.*, 2013, **14**, 81–94.
22. C. Cigdemoglu, H. O. Arslan and A. Cam, Argumentation to foster pre-service science teachers' knowledge, competency, and attitude on

the domains of chemical literacy of acids and bases, *Chem. Educ. Res. Pract.*, 2017, **18**, 288–303.
23. N. J. Brown, E. M. Furtak, M. Timms, S. O. Nagashima and M. Wilson, The Evidence-Based Reasoning Framework: Assessing Scientific Reasoning, *Educ. Assess.*, 2010, **15**(3–4), 123–141.
24. S. Erduran and A. Pabuccu, *Bonding chemistry and argument: supporting the teaching and learning of argumentation through chemistry storie.* Teacher and Student Resource, 2012, retrieved October 4th, 2015, from http://www.bristol.ac.uk/education/news/2012/63.html.
25. V. Sampson, P. Carafano, P. Enderle, S. Fannin, J. Grooms, S. A. Southerland, C. Stallworth and K. Williams, *Argument-Driven Inquiry in Chemistry: Lab Investigations for Grades 9-12, e-book*, National Science Teachers Association (NSTA); 2014.
26. Northwest Association for Biomedical Research. *The Social Nature of Scientific Research*, available online: https://nwabr.org/sites/default/files/NWABR_Social_Nature_Scientific_Research.pdf [Accessed 19th April 2018].
27. S. Erduran and R. M. Villamañán, Cool Argument: Engineering Students' Written Arguments about Thermodynamic in the Context of the Peltier Effect in Refrigeration, *Educ. Quím.*, 2009, 119–125.
28. A. Zohar and A. Ben-David, Explicit teaching of meta-strategic knowledge in authentic classroom situations', *Metacognit. Learn.*, 2008, **3**, 59–82.
29. G. Irzik and R. Nola, New Directions for Nature of Science Research, in *International Handbook of Research in History, Philosophy and Science Teaching*, ed. M. Matthews, Springer, Dordrecht, 2014, pp. 999–1021.
30. B. Graba, 2016, #OrganelleWars: A Model for Using Social Media in the Science Classroom, NSTA Blog [online] http://nstacommunities.org/blog/2016/05/06/organellewars-a-model-for-using-social-media-in-the-science-classroom/ [Accessed May 22nd 2018].
31. F. Abd-El-Khalick and N. Lederman, Improving science teachers' conceptions of the nature of science: a critical review of the literature, *Int. J. Sci. Educ.*, 2000, **22**(7), 665–702.
32. D. Trump, (@realDonalTrump) It's finally happening for our great clean coal miners [Twitter] 31st October 2017, 12.41pm, available: https://twitter.com/realdonaldtrump/status/925447637708496897?lang=en (Accessed 19th July 2018).
33. IRENA. (@IRENA) Meeting the goals of the #ParisAgreement means reducing carbon emissions from the energy sector Renewables + energy efficiency can do 94% of the job — @IRENA's Global Energy Transformation report explains how. [Twitter] 1st May 2018 8.40pm, available: https://twitter.com/IRENA/status/991522668255072257 [Accessed 25th May 2018].
34. S. Whitehurst, *A Primer on the Chemistry and Constitution of Coal Organic Chemistry of Coal: Proceedings of ACS Symposium Series*, American Chemical Society, Washington, 1978.

35. Science Friday (@scifri) In 2016, scientists discovered that a bacterium had evolved to produce an enzyme that enabled it to eat plastic bottles. Now, they have discovered a way to tweak that enzyme to do the work 20 percent faster. Could it be the next recycling solution? [Twitter], 24th April 2018 5.03pm, available: https://twitter.com/scifri/status/988931343491846145 [Accessed 12th May 2018].
36. C. Lee Gannets on Little Skellig are using plastic to build their nests. *The Journal*, available: http://www.thejournal.ie/gannets-plastic-nests-at-little-skellig-4059655-Jun2018/ [Accessed 30th June 2018].
37. V. Quinn, *Critical Thinking in Young Minds*, David Fulton, London, 1997.

CHAPTER 5

Using Computer Technology to Support the Teaching and Learning of Argumentation in Chemistry

J. BRYAN HENDERSON*[a] AND JONATHAN F. OSBORNE[b]

[a] Arizona State University, Mary Lou Fulton Teachers College, Tempe, AZ, USA; [b] Stanford University, Graduate School of Education, Stanford, CA, USA
*Email: jbryanh@asu.edu

5.1 The Intersection of Computer Technology and Scientific Argumentation

While the affordances of both technology and evidence-based argumentation are central to the work of the professional scientific community, the use of technology to promote the learning of science through argumentation is a relatively recent educational practice. This chapter provides examples of how technology can support chemistry students' engagement in argumentation, as well as how technology can aid analysis of student argumentation by both educational researchers and chemistry instructors. The earliest incarnations of argumentation technology were used primarily as an aid to educational researchers in the empirical analysis of recorded episodes of argumentation. However, given the importance that a new generation of science standards[1,2] have placed on evidence-based argumentation, the time

Advances in Chemistry Education Series No. 2
Argumentation in Chemistry Education: Research, Policy and Practice
Edited by Sibel Erduran

Published by the Royal Society of Chemistry, www.rsc.org

is right to consider how to adapt old technologies and create new ones to support educational praxis. In short, how can modern technology support the learning of chemistry through argumentation?

Efforts over the past several decades to use computer technology in support of student construction, critique, representation, and sharing of arguments has taken on various names, such as Computer-Supported Collaborative Argumentation (CSCA)[3] and Argumentation-Based Computer-Supported Collaborative Learning (ABCSCL).[4] Over this time, multiple overviews of research at the intersection of computer technology and learning through argumentation have been compiled.[4–8] However, as reviews of computer-supported argumentation tend to have a research focus, some technologies that have shown promise in the research literature are not necessarily straightforward for chemistry instructors to access (*e.g. Belvedere*[9] and *Digalo*[10]). As this chapter seeks to offer a practical perspective, we will reflect on some of the affordances of technology-enhanced argumentation emerging from the reviews of computer-supported argumentation. However, we choose to do this solely by pointing to technology that chemistry instructors can access easily through a basic web search at the time of this publication. We follow this review by highlighting several contemporary projects developing practical, next-generation technology to support science argumentation.

5.2 Online Learning Communities

Technical interfaces which allow teachers and students to connect virtually have the potential to support chemistry argumentation in ways that might be more cumbersome in traditional classrooms. These online tools provide instructors with the ability to set specific constraints on how student communication is guided and scaffolded through the processes of argumentation.[3] More specifically, online learning communities afford instructors customization of important factors such as the modalities students use for interaction, as well as how student groups are formed.[6]

5.2.1 Asynchronous Versus Synchronous Argumentation

Whereas the traditional chemistry classroom is marked by *synchronous* interactions from teachers-to-students and students-to-students, online learning environments make possible both synchronous and *asynchronous* interactions. The classic example of an asynchronous interaction is the online "message board" feature found in various Learning Management Systems (LMS). The Royal Society of Chemistry's *MyRSC*[†] provides a chemistry-themed, asynchronous online learning community. Whether creating a new conversation thread or responding to a post made by another contributor to the message board, asynchronous contributions are typically

[†]my.rsc.org

written, which allows students time to review and revise written posts prior to sharing with the learning community. In contrast, synchronous interfaces, such as "live chats" that are available from web-through conferencing applications, are more commonly speech-based and permit real-time interactions.

5.2.1.1 *A Matter of* Time?

Research has found discrepancies in the nature of arguments that take place in these different forums. For example, Hsi and Hoadley[11] found that asynchronous argumentation was more equitable than synchronous argumentation when it came to the number of students that ultimately participated. Hence, students that might be reticent to participate in real-time arguments might be more prone to chime in during asynchronous debate. Munneke, Andriessen, Kirschner, & Kanselaar[12] compared an asynchronous condition (discussion board) to a synchronous condition (chat box) and concluded that students using the asynchronous discussion board demonstrated more accurate use of evidence and scientific concepts in their arguments. However, this same study found that students using a synchronous chat box would elaborate more on their standpoints and that their argumentative exchanges with others would last longer than students using the asynchronous message board. It follows that while an asynchronous online community may afford greater accuracy in the use of evidence and conceptual reasoning as students have more time to reflect upon their developing standpoints, this does not, however, guarantee a deeper discussion with others.

5.2.1.2 *A Matter of* Modality?

If having more time to carefully review and revise your standpoint does not ensure deep interaction with others, the more immediate feedback that comes from synchronous argumentation may be a key ingredient to promote more extended argumentative exchanges. This is consistent with the work of Kuhn and Moore,[13] which compared students constructing arguments *via* written essays to students arguing about the same topic with others through dialogue. Two major findings were that: (1) essay writers dedicated an average of only 4% of their evidence-based claims to critique opposing positions, while in contrast, an average of 33% of evidence-based claims offered by dialogue participants did seek to weaken opposing standpoints, and (2) an average of 82% of the evidence cited by essay writers was based on information provided to them at the beginning of the activity, whereas the same information only comprised an average of 20% of the evidence offered by dialogue participants. In short, speaking may provide a student more immediate feedback from others, which makes students more immediately aware of differing standpoints, which results in more authentic disagreements that last longer. These relatively longer disagreements, in turn, require students to draw increasingly on more prior knowledge (as opposed to merely the information provided to them) and makes the activity more genuinely dialogic.

As asynchronous argumentation typically occurs through written correspondences while synchronous argumentation is more apt to involve oral communication, the Kuhn and Moore[13] research helps explain the Munneke *et al.*[12] findings that there may be a trade-off of accuracy *versus* more prolonged and deeper interactions for asynchronous *versus* synchronous argumentation, respectively. However, recent examination of the role of both writing and talking when learning science through argumentation suggests that the two modalities can mutually reinforce each other, *i.e.*, that writing and talking function best when used *simultaneously*, as opposed to in sequence.[14] It follows that given the different modalities typically associated with asynchronous (writing) and synchronous (talking) argumentation, a combination of opportunities for both asynchronous and synchronous interactions might be an optimal way to promote science argumentation through online learning communities.

5.2.2 Strategic Composition of Groups

Online learning communities can make it easier for chemistry instructors to determine which students interact with whom based on various strategic factors, including gender, level of background knowledge, and original opinions or positions on topics up for debate.[4] Research suggests that group composition can affect the nature of argumentation that ultimately takes place. Jeong and Davidson-Shivers,[15] for instance, reported that females tended to offer relatively fewer challenges and rebuttals than their male counterparts. As for background knowledge, results are unclear as to whether it is better to group students of similar or different abilities. Many argumentation studies presume that heterogeneous groupings are preferable, as it increases the likelihood that the group will contain a broader spectrum of viewpoints.[4] However, the "truth wins" account[16] suggests that since a group is likely to contain at least one member that has a competence level greater than the median competence level for all students, more competent members can potentially dominate group discussion. Such a scenario results in the appearance that heterogeneous group work is superior to individual work, but what the group puts forward might only reflect the thinking of the most able members. Indeed, Sampson and Clark's[17] study of high school chemistry students engaged in argumentation found that only 28% of students originally working in groups to generate an argument could generate a similar argument on their own at a later time.

Fortunately, "truth wins" scenarios, where the more able students dominate, can be mitigated by heterogeneous groupings where each student brings unique knowledge or skills to the group such that the group cannot succeed without utilizing the unique expertise offered by each group member.[18] Extending this idea to argumentation, by ensuring different group members bring different positions into a conversation, it is less likely that only a single participant will dominate the argument. Studies that use online learning community software to group students with differing standpoints is an approach known as "personally seeded discussions".[19] The *TELS: Probing Your Surroundings* project[20] developed technology to personally seed discussions.[6] TELS technology has since been incorporated into *WISE*[‡], which is a free and open-source platform to support science learning.[21] Clark and Sampson[19] provide an example of how to personally seed discussions using WISE, where after students have constructed individual explanations of data patterns, software groups students with differing explanations and provides support to critique differences and work toward consensus.

5.3 Digital Scaffolding

Arguments in chemistry often require abstract models to make sense of phenomena which are not intuitive to students. In addition to the complexity of working with abstract models, making arguments to support or critique models carries its own intrinsic cognitive load. Indeed, the cognitive load necessary to keep track of claims, evidence, and reasoning can become complex in an argument rather quickly.[22] Fortunately, modern computer technology allows students to organize and track the various moving parts of models and arguments, which can provide valuable metacognitive feedback.

5.3.1 Accessing and Organizing Evidence

The *WISE* platform[21] contains free chemistry lessons hosted in a guided, digital environment. At the click of a button, students have autonomy to choose which aspects of the digital lesson count as evidence that supports a culminating summary they are to write for a report to an agency such as the EPA, NOAA, *etc.* More specifically, Figure 5.1 depicts the *Idea Basket* that is common in WISE chemistry lessons. The Idea Basket not only lets students flag what they consider to be important pieces of evidence in their investigation, but the system allows students to label and tag the evidence they are logging to ease organization.

5.3.2 Visualization Tools

The *WISE* platform[21] also includes animated visualizations in its chemistry lessons. Some of these visualizations are interactive, where students can digitally manipulate models. Figure 5.2 depicts a chemistry lesson visualization through WISE that models the behavior of three different molecular structures. Students are able to visualize the behavior of each structure, and then make an argument—armed with annotated evidence in their Idea Basket—for which structure behaves as a detergent molecule.

[‡]https://wise.berkeley.edu/

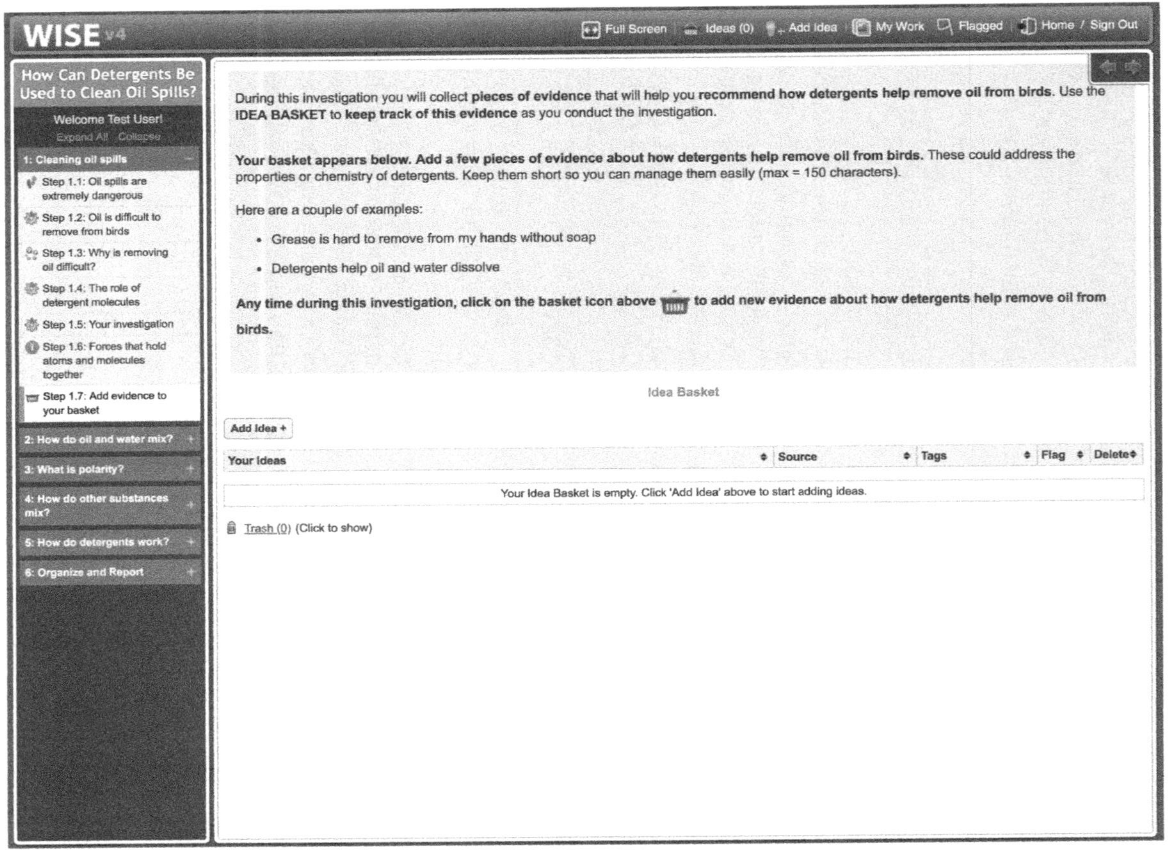

Figure 5.1 Screenshot of the *WISE Idea Basket*. Clicking on the "Add Idea" button allows students to label and tag the evidence they are logging to ease organization.

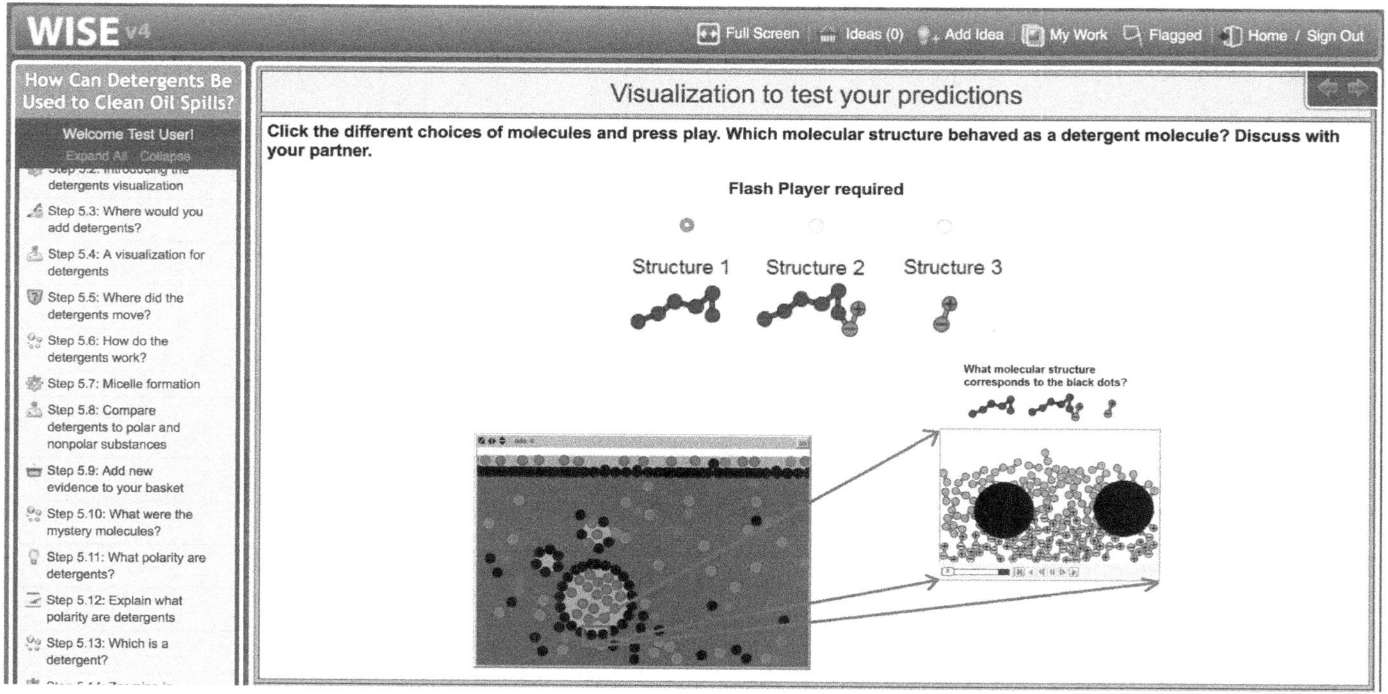

Figure 5.2 Chemistry lesson visualization through *WISE*. This visualization models the behavior of three different molecular structures. Students are able to visualize the behavior of each structure, and then make an argument for which structure behaves as a detergent molecule.

5.3.3 Argument Mapping

Perhaps the most readily available form of technology to support argumentation is argument mapping software. Argument mapping software mitigates for the increased cognitive load required for developing arguments by providing graphical representations of how evidence and reasoning support and critique claims. Not only can students use argument mapping software to visualize how the various components of an argument are connected, but they can also move various argument components around on the computer screen. This can allow students to sort and rank different types of evidence, for example. *Argunet*[§], *Araucaria*[¶], and *DebateGraph*[‖] are three examples of free and easy to access argument mapping software. Figure 5.3 provides a screenshot from Argunet depicting three different kinds of argument structure.

Having provided an overview of the intersection of technology and argumentation over the past several decades, we now provide several more detailed examples of contemporary projects developing next-generation technology to provide teachers and students with practical supports for learning science through argumentation.

5.4 Supporting Argumentation through Classroom Response Technology

5.4.1 Braincandy

Most educational curricula and assessments focus primarily on the correct answers students are expected to provide. However, decades of educational research show that students hold many preconceptions about the world that can be difficult to change.[23–25] *Braincandy*** are preconception-based questions focused on STEM content.[26] The incorrect answers to Braincandy questions (Figure 5.4) are based on popular preconceptions, and they are phrased in the form of competing claims by fictitious characters. The use of fictitious characters is based on research suggesting that students are less prone to confirmation bias when externally adjudicating standpoints that are not originally their own.[27] More specifically, when a student selects from a multiple-choice question, that selection becomes their claim, and as that claim becomes their personal standpoint, subsequent argumentation with others may focus more on subjective defense of their standpoint than objective pursuit of the best answer. As Mercier and Sperber point out, however, students are actually quite adept adjudicators when posed situations in which they are to decide upon the merits of standpoints from others. Hence, the subtle difference in how Braincandy questions are posed – students

§www.argunet.org
¶araucaria.computing.dundee.ac.uk
‖debategraph.org
**www.braincandy.org

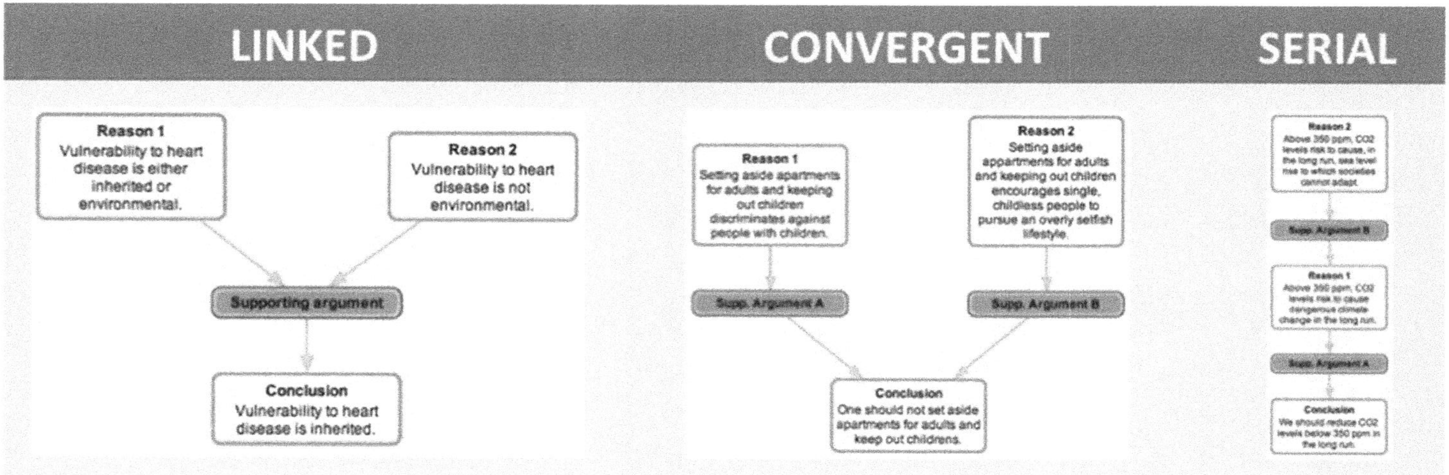

Figure 5.3 Screenshot from *Argunet* argument mapping software depicting three different types of argument structure. Reproduced from http://www.argunet.org/ with permission from Gregor Betz.

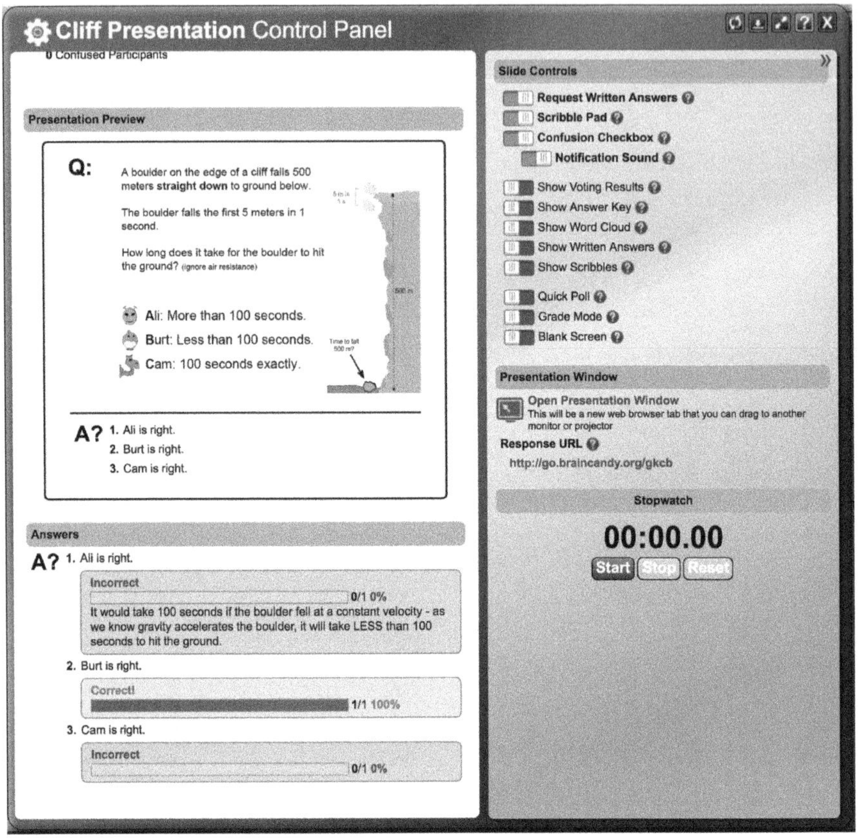

Figure 5.4 The *Braincandy* teacher dashboard. Preview of the current class question is in the upper left. Annotations for correct and incorrect answers are found on the bottom left. The upper right shows the different tools a teacher can toggle on or off to solicit various forms of feedback from students. Real-time tabulations of student responses are displayed on the bottom right. Also on the bottom right, a click on Open Presentation Window allows student feedback to be projected to the front of the classroom. Meanwhile, the Response URL is where students are directed to submit their responses – free of charge and with no login account necessary.

adjudicate the claims of others (*i.e.*, the fictitious characters) instead of defending their own multiple-choice selection.

Braincandy questions are complimented with multiple technological tools for sharing student thinking, and as all contributions are anonymous, sharing is more frequent and more honest. Similar to handheld student response systems, *i.e. clickers*,[28,29] students can answer questions provided by the instructor with any web-enabled device. Beyond a clicker system, students can indicate confusion, answer open-ended questions, and provide anonymous feedback directly to the instructor (see Figures 5.4 and 5.5).

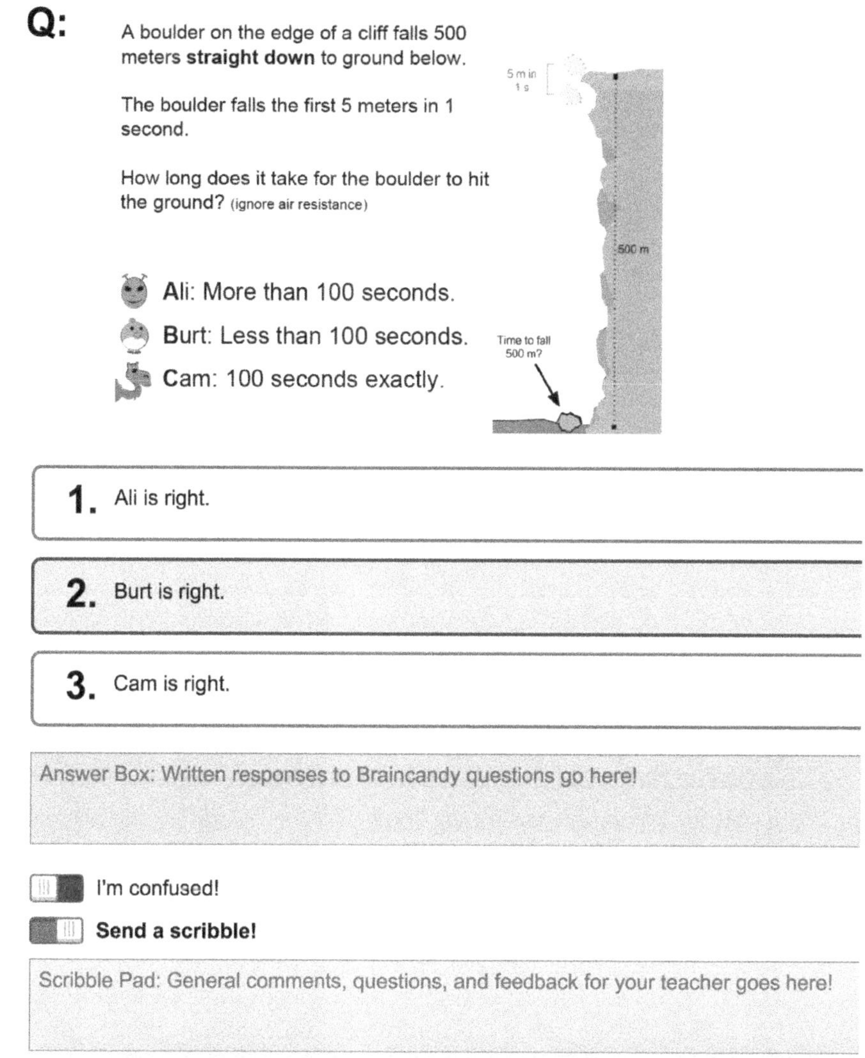

Figure 5.5 The *Braincandy* student interface. For multiple-choice questions, students simply click on one of the answer boxes, and are free to change their vote as often as they wish (the highlighted answer indicates the current response being sent to the instructor). The Answer Box allows students to submit written responses to open-ended questions. Scribble Pad allows students to scribble down a message to the instructor at any time, and students are also able to notify the instructor whenever they might be confused (a notification sound is played for the instructor, as well as a count of how many students report being confused). In all cases, the feedback students provide their instructor is completely anonymous, thereby providing a safe space for students to articulate their current thinking, regardless of what stage that thinking might be at and/or how confident the student might feel in their current thinking.

Additionally, instructors can read results, confusion level, and written feedback on a live dashboard and display the results to students through a multitude of different visual representations.

The Braincandy system poses questions that students answer under the veil of anonymity, which provides a safe space for students to volunteer prior thinking they might normally be reticent to share out loud. With more frequent and honest student contributions in this safe space, differing viewpoints are more common, and Braincandy visualization tools allow students to see this uncertainty. Classroom uncertainty sets the stage for students to talk critically to each other with the goal of reaching consensus. More specifically, instead of merely telling students the correct answer, an instructor can frame the uncertainty as an unresolved problem best addressed through the process of argumentation. Hence, the Braincandy system creates authentic opportunities for classroom talk by making prior thinking not a hindrance, but rather, an integral part of constructing more nuanced thinking. The Braincandy system is freely available to chemistry instructors, where they can share their Braincandy questions and provide feedback on how the questions are written. Indeed, writing Braincandy questions that promote authentic argumentation can be much easier said than done, and hence the benefit of accessing questions by other chemistry instructors using the system. Furthermore, the system tracks student responses to each Braincandy question, providing data on how questions perform.

5.5 Supporting Argumentation through Computerized Assessment

How assessments are designed has a pervasive influence on how teachers teach.[30,31] Hence, classroom opportunities for students to construct and critique scientific arguments will likely increase in frequency if argument proficiency can be assessed by instructors in a valid, reliable, and practical fashion. We now provide several examples of how computer technology can support the assessment of scientific argumentation across multiple modalities. These examples include explanations as to how these modern assessments of scientific argumentation can support the teaching of scientific argumentation in a *practical* fashion.

5.5.1 *DiALoG:* Supporting the Assessment of Oral Arguments

Talking is thinking. There is a large corpus of literature surrounding how talking plays a key role in our brain's acquisition, storage, and retrieval of information.[32] It follows that if we want our students to think deeply and critically in the classroom, we need to support them to talk critically in the classroom.[33] Unfortunately, while talk abounds in many classrooms, students are often not provided regular, deliberate opportunities to talk critically with others.[34] While we have seen the potential affordances of speaking and listening during synchronous argumentation *via* online learning communities, science classrooms typically do not have the structured activities for speaking and listening that they do for reading and writing. This is because, in part, valid and reliable assessments of reading and writing are far more common than assessments of speaking and listening.

The *DiALoG* project[††] is focused on developing and testing a digital assessment of oral classroom argumentation that will be of practical use for instructors of any STEM discipline, including chemistry.[35,36] In order to make the DiALoG (Diagnosing Argumentation Levels of Groups) assessment practically useful for teachers, psychometric development converged on two bundles of assessment items that were not too long for practical use, and yet robust enough to reasonably account for the range of competencies that can be observed when students are engaged in oral argumentation.

While the specifics of the psychometric development of the DiALoG assessment[35,36] is beyond the scope of this chapter, it is important to note that the DiALoG assessment accounts for multiple aspects of classroom talk. In short, a large number of items based on potentially important aspects of oral argumentation were used to score video episodes of classroom argumentation. This included the 19-item *Assessment of Scientific Argumentation inside the Classroom (ASAC)*[37] as well as literature on *pragma-dialectics*.[38] The number of items was then winnowed down to eight using factor analysis.

Iterative refinement of wording and bundling for remaining DiALoG items was guided by the Michaels, O'Connor, and Resnick[33] work on *Accountable Talk*. Figure 5.6 depicts a schematic of the theoretical framework guiding the

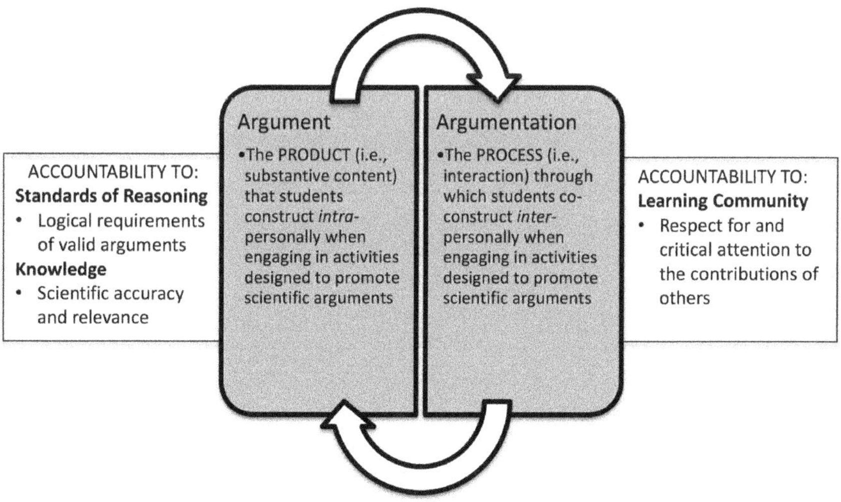

Figure 5.6 *DiALoG* theoretical framework linking argument *product* and the argumentation *process* to Accountable Talk.

[††]National Science Foundation, DRK-12 #1621441 and #1621496

bundling of DiALoG items. Arrows denote the reflexive relationship between the different intrapersonal and interpersonal aspects of classroom arguments. The Michaels *et al.*[33] notions of *Accountability to the Standards of Reasoning* and *Accountability to Knowledge* guided bundling of four items relevant to the *product* of an argument (left side of Figure 5.6). That is, in order to evaluate the substantive content of classroom talk, DiALoG contains items probing the degree to which students were accountable for both the logical requirements of a valid argument in addition to the scientific accuracy and relevance of their utterances. As for assessment of the dialectical *process* that yields an argument product, the bundling of four additional items was guided by the Michaels *et al.* notion of *Accountability to the Learning Community*, which emphasizes respect for and critical attention to the contributions of others so that ideas can be built upon one another (right side of Figure 5.6).

Each box in Figure 5.7 represents one of the eight DiALoG items that teachers use to score oral classroom argumentation. Items are criterion-referenced. For example, the item purporting to measure the degree to which students co-construct knowledge begins with the following statement: *Students seemed to construct or refine their thinking in ways that build on what others had to say*. Users then rate on a 3-point scale how descriptive various statements were of the classroom argumentation they just observed (0 = "Blank stares"; 1 = "There, but rare"; 3 = "There, everywhere"). The arrow in Figure 5.7 suggests a progression, *i.e.* that proficiency of lower items is necessary to ultimately achieve proficiency of higher items. As it is possible that students generate arguments that are logically valid, but lack soundness as the arguments digress from the science topic at hand, the Relevance item score is multiplied by the sum of the Claims, Evidence, and Reasoning item scores. Similarly, while some students may be exhibiting

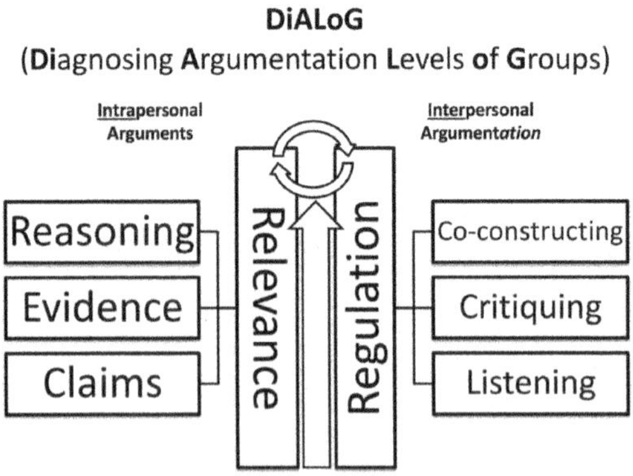

Figure 5.7 *DiALoG* item structure. The arrow suggests a progression, *i.e.* that proficiency of lower items is necessary to ultimately achieve proficiency of higher items. The Relevance score is multiplied by the sum of the Claims, Evidence, and Reasoning scores, while the Regulation score is multiplied by the sum of the Listening, Critiquing, and Co-Constructing scores.

signs of better-quality interpersonal argumentation on a specific issue, it is another skill to do so cognizant of the goals and participation of the group as a whole. Hence, the Regulation item score is multiplied by the sum of the Listening, Critiquing, and Co-Constructing item scores. The four items on the left of Figure 5.7 represent a bundle for *Intra*personal Arguments (*i.e.*, the *product*), while the four items on the right of Figure 5.7 represent a bundle for *Inter*personal Argumentation (*i.e.*, the *process*).

To make scoring practical and user-friendly, the DiALoG project is developing a tablet-based application for teachers to enter their scores flexibly with the touch of a finger, as opposed to entering, deleting, and re-entering scores when they change their minds. The application instantly calculates total scores and has the potential to be a powerful feedback device for teachers to quickly identify the strengths and weaknesses of the classroom talk they are observing. This parsimony positions DiALoG as a viable candidate to overcome some of the longstanding barriers for teachers to assess verbal arguments in their classrooms. Figure 5.8 provides a visualization of the current version of the DiALoG user interface, which is based on insights from teachers piloting the instrument in their classrooms.

5.5.1.1 Development of Responsive Mini-Lessons (RMLs)

Recognizing that formative assessment is most powerful when teachers take action on the information they gather, the DiALoG project is developing a portfolio of instructional suggestions aligned with the DiALoG assessment. These *Responsive Mini-Lessons* (RMLs) correspond to a range of possible scores for the different items measured by DiALoG. Customized RMLs are being designed for unique combinations of high and low scores on the eight DiALoG items. RMLs are 30–60 minutes in length and designed to be used as a follow-up to argumentation episodes for which the DiALoG instrument's use has revealed a need for further support.

5.5.1.2 Pilot Testing with Teachers

As DiALoG was first developed in a controlled setting observing video episodes of classroom argumentation, the project goal of creating a tool that is practically useful for instructors has naturally led to pilot testing with teachers in different classrooms. Professional development, classroom observations, interviews, and surveys of pilot teachers has allowed triangulation on multiple important considerations for future development of the DiALoG assessment instrument and the accompanying RMLs. Some of the most prominent findings are summarized below.

The DiALoG instrument helps teachers shift to a student-centered mindset. The DiALoG assessment and its accompanying RML lessons are helping teachers shift from teacher-centered to a more student-centered form of instruction. The instrument prompts restrain teachers from interjecting into student

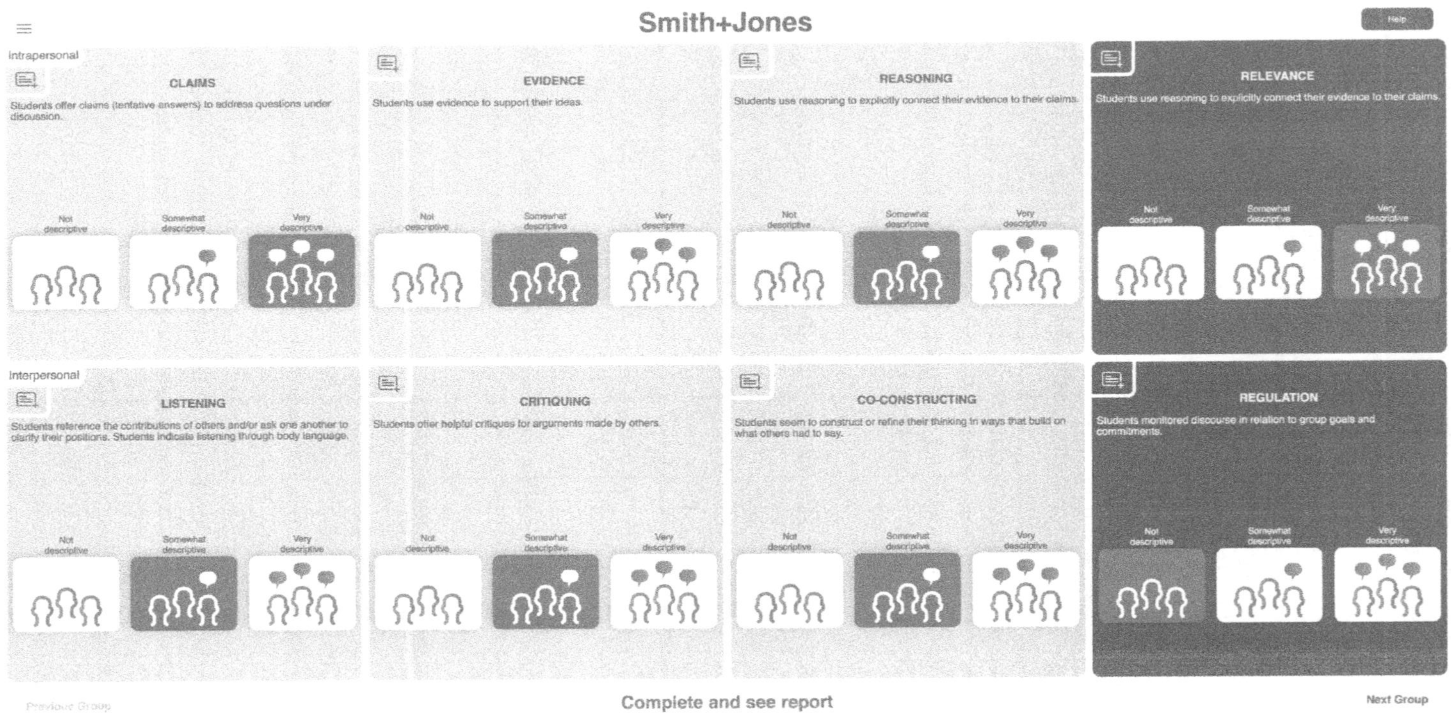

Figure 5.8 The most recent version of the *DiALoG* user interface, based on feedback from pilot teachers. To reduce clutter, items are minimized in the background, with the interface zooming in on the item most recently selected by the touch of a finger. Users can type in their own annotations for each item, and a breakdown of scores by both the *Intra*personal (left) and *Inter*personal (right) dimensions is updated in real time at the center of the screen.

conversations as readily as they are accustomed to, and when interjections are made, they are based more on feedback from the assessment.

Improvements to the DiALoG user interface are not just an issue of ease of use, but also depth of use. Teachers need additional support to integrate the eight DiALoG items during assessment. Many of the items assessed by DiALoG happen simultaneously, and hence, the most recent version of the DiALoG user interface (Figure 5.8) includes the ability to score all eight items on a single screen. Furthermore, some teachers prefer to move about the classroom and attempt to drop in on multiple student discussion groups. Thus, future versions of DiALoG user interface will likely permit teachers to toggle between multiple simultaneous scoring sessions corresponding to different groups of students.

There is, however, *variation in how teachers interpret DiALoG scores*. While DiALoG was developed as a formative assessment to guide selection of follow-up RMLs, multiple teachers during pilot testing used the assessment in a summative fashion. When instructors find themselves preoccupied with tabulating how many students are participating and for how long, working from a perceived need to assign individual grades in a higher stakes summative context can divert instructor attention away from more important macro-level trends in classroom talk. Future versions of the DiALoG assessment need to emphasize that the primary purpose of the instrument is for formative assessment.

5.5.2 Supporting the Assessment of Written Arguments

Argumentation requires complex reasoning. The construction of an argument requires a student to assemble a claim, evidence, and warrant in a coherent manner. While it is possible to ask students to choose from an array of each element, a more naturalistic approach requires students to display the ability to compose extended constructed responses. This is even more true of the ability to engage in argumentation which, at its highest level,[22] requires students to construct counter arguments to another's argument and to evaluate the merits of competing claims. To date, the only method of evaluating such competence with argumentation has been to use human graders. Four problems exist with such an approach. First, even with elaborated scoring rubrics, it is impossible to remove the element of human subjective judgment. The consequence is that commonly the best reliability that can be achieved with two human graders is around 70% agreement. Second, the use of human graders for such purposes is time consuming when conducted by teachers and expensive when conducted for the purposes of external assessment. The combination of time and/or expense is a prohibition on its use. Third, human graders are also often influenced by irrelevant factors such as spelling, punctuation, and the coherence of the prose which means that essentially good answers that might, for instance, be expressed in poor English by a second language learner are downgraded. Yet, if one of the goals of using argumentation in the classroom is to develop student's competency with scientific argumentation then it is essential to have the means to assess any student's competency in a reliable and valid manner. Thus, an inevitable corollary is the assessment of constructed responses requiring the use of scientific argumentation.

The natural field to turn to achieve this goal is automated lexical analysis. Today, advances in both technology and measurement research now make it feasible to apply these techniques in instructional settings with the potential to have substantial educational impact.[39–47] Overall, this work has shown that: (1) it is possible to create computerized scoring models that predict human scoring with inter-rater reliability (IRR) measures approaching that of well-trained expert raters; (2) that these types of questions reveal the heterogeneity of student thinking that cannot be revealed by traditional multiple-choice items; and (3) we can capture, represent, and analyze this multidimensional information in a variety of ways that provide instructors rich insights into student thinking.

The basic approach in the use of machine rating is to use an initial sample of responses which have been human-graded to teach the machine the 'features' of correct responses. Such a sample is then used to build automated models that can score complex constructed-response items. For instance, in our work to date we have examined how successfully the computer might score a series of prompts about what happens when sugar is placed in water. At this phase in the questioning, this item reveals whether or not students can provide a warrant for the evidence supporting Mary's claim shown in Figure 5.9.

For the analysis, we selected the subset of the data from students who identified Mary as having the correct response ($n = 257$). These were scored by experts using a two-level rubric. Level 1: Student selects Mary but does not explain how the evidence supports the claim ($n = 37$). Level 2: Student selects Mary and provides an explicit warrant explaining how the evidence supports the claim ($n = 220$). We then built a model that could be used to score student responses that used seven categories: Mary's idea, conservation of matter, dissolve, evidence, sugar, taste, and taste sweet or sugar. The model

You said that Evidence 1 supports: Mary

Complete the last sentence below to explain how the evidence supports Mary's idea.

Mary says: *I think the sugar is still there.*

Evidence 1: After stirring, each student tastes the water. They both agree the water tastes sweet.

This evidence supports Mary's idea because. . .

Figure 5.9 Example of an item requiring a student to provide a warrant in support of a hypothetical claim.

Table 5.1 Table matrix of results from computer-scored analysis.

		Predicted score		
		1	2	Total
Expert score	1	**20**	17	37
	2	6	**215**	221
Total		26	232	258

correctly classified 90.7% of the cases (chi-square = 92.183, df = 1, $p < 0.001$; Kappa = 0.59). The classification accuracy is shown in Table 5.1. The numbers of correctly predicted scores are shown in bold.

Another example is the use of c-rater-ML, an automated scoring engine developed by Educational Testing Service, for scoring eight science inquiry items that require students to use evidence to explain complex phenomena. In this study, automated scoring showed satisfactory agreement with human scoring for all test takers as well as specific subgroups. In their work[43,48] each argumentation response was scored on a seven-point scale ranging from zero to six, representing:

Score 0: Blank or off-task responses
Score 1: Incorrect claim, data, or reasoning is mentioned
Score 2: Restatement of the claim they chose in the prior multiple-choice claim prompt
Score 3: Scientifically relevant but not fully elaborated statements related to data or reasoning
Score 4: Scientifically valid, relevant, and fully elaborated data citation without the mention of reasoning
Score 5: Scientifically valid, relevant, and fully elaborated reasoning statement without the mention of data
Score 6: Scientifically valid, relevant, and fully elaborated data and reasoning

These examples suggest that this technology offers a promising approach to scoring constructed-response science items with the potential to increase the use of such items both in instruction and assessment.

Another approach we are exploring is whether it is possible to use the affordances of computer technology to design questions which require the complex skill of argumentation and yet can be scored without the use of human graders. Computerized technology now permits students to highlight elements of text – for example, so as to identify a warrant, a claim or a piece of evidence; the use of drop-down menus to select choices in the construction of, for instance, an *if* [choice 1] *then* [choice 2] *therefore* [choice 3]; or the selection and sequencing of sentences so as to construct an argument or counter argument (see Figure 5.10).

More importantly, the use of machine scoring affords the possibility of providing instant feedback to students on the nature and quality of their

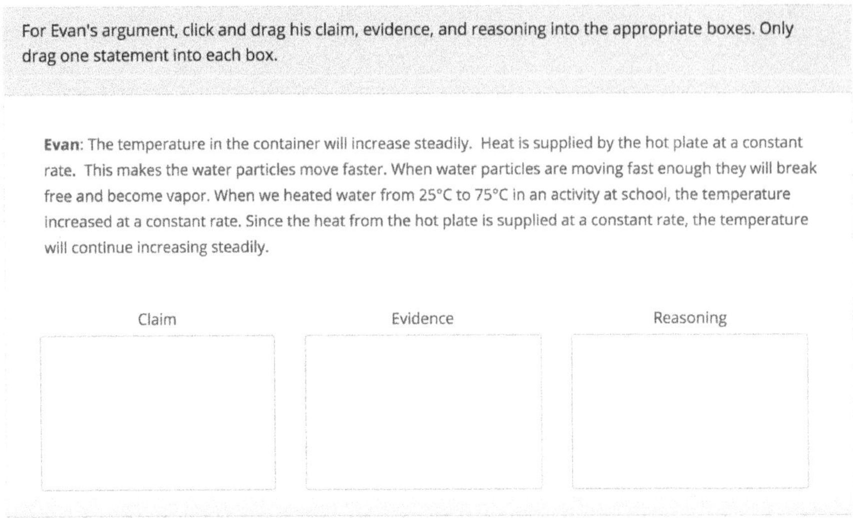

Figure 5.10 Drop-down menus permit students to construct an argument that can be scored by computer technology.

response. As research on formative assessment has shown, such feedback is critical to improving students' ability with any competency – particularly a complex one required for scientific argumentation. Machine scoring affords the possibility of providing such feedback then and there. In Zhu *et al.*'s work,[47] they were able to provide feedback of the following form immediately:

0. You have not explained your claim yet. Can you include scientific evidence and reasoning that explain your claim?
1. Your claim, evidence, or reasoning was either inconsistent with scientific views or was unclear. Can you modify or elaborate your explanation?
2. You made a claim without an explanation. Can you include scientific evidence and reasoning that support your claim?
3. You identified some climate-related factors associated with temperature. Can you include evidence and reasoning that explain the associations?
4. You included evidence for your claim. Can you elaborate how or why the evidence supports your claim?
5. You included scientific reasoning that explained your claim. Can you add evidence to support your reasoning?
6. You included evidence and reasoning to support your claim. Great job!

While such feedback may not be as rich as the quality that a good teacher might provide, it is considerably better than no feedback or even, one might argue, post-hoc feedback that comes several days later. And meta-analyses of research have shown that feedback is one of the top ten influences on improving the quality of student learning.[49]

5.6 Implications for Chemistry Education

As a STEM discipline, chemistry education stands to benefit from reform efforts emphasizing the learning of science through greater use of evidence-based argumentation. In essence, chemistry is an argument for a view of a world which consists only of a limited number of elemental substances which then combine and interact to form all other substances. The behavior of these substances can be determined and predicted by measuring and modeling their interactions. These interactions often happen at a scale that is far too small to be intuitive to students, and hence, our chemistry students essentially take on faith many of the chemical explanations made by their teachers. Computer simulations are one way of helping not only to visualize what is being discussed but also a means of investigating its behavior and at least seeing some of the evidence, if not first hand, then modeled in a form which can be investigated. The Concord Consortium[‡‡], for instance, offers a variety of tools that support modeling and simulation. In the example shown in Figure 5.11, the effect of phase changes at the molecular level can be explored by running the simulation and investigating what causes the phase changes with the barrier in place and with it removed. In this example, the simulation can be run with the barrier present and with it removed helping to build a visual picture of what happens at the molecular level.

Computer technology then affords the opportunity for students to visualize chemical phenomena and the models used to understand them, and does so with a degree of efficiency and sophistication that has not been

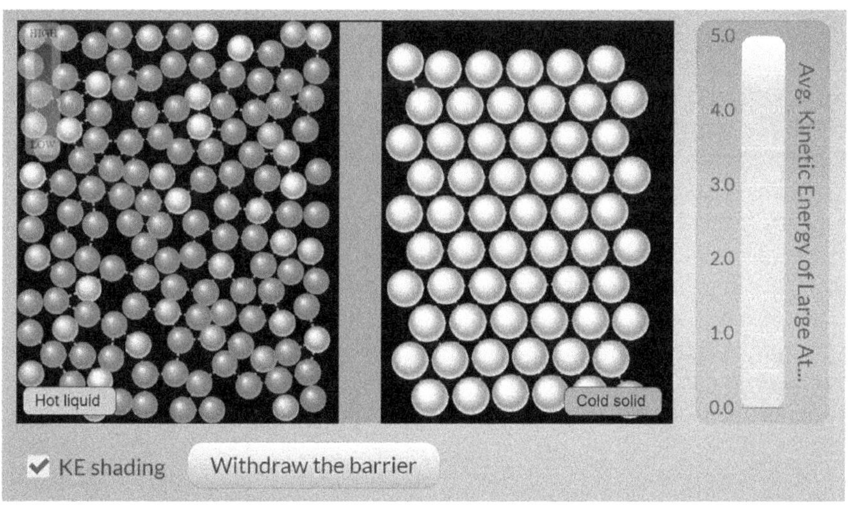

Figure 5.11 Example of a molecular simulation offered by the Concord Consortium. Reproduced from www.concord.org under the terms of the CC BY 4.0 license, https://creativecommons.org/licenses/by/4.0/.

[‡‡]https://concord.org

previously possible. Similarly, as abstract chemical models can quickly become complex, so too can the structure of arguments supporting or critiquing these models. Modern technology helps to provide evidence for the ideas that we wish students to understand, as well as map the complexity of developing arguments by providing simulations where they can explore, as in the example in Figure 5.11, of what the effect might be of raising the temperature or removing the barrier. Such simulations then provide evidence that supports any argument they might wish to make about, in this case physical changes. Thus, computer technology can provide an efficient online platform for students to compare and critique their understandings of models, both asynchronously and synchronously and test out their hypotheses about chemical and physical interactions.

In addition, teacher explanations of chemical phenomena are often presented as both uncontested and certain, creating a perception of chemistry as being merely a collection of unequivocal facts and rules to be learnt. One way to overcome this misperception of chemistry is for students to grapple with the data from which our understanding of the world is built – investigating data for themselves. In short, students need to explore data sets and argue about their interpretation and meaning. Here modern technology can help. For instance, it is now possible to take data sets and investigate them in ways that were much more difficult before. Tuva Labs[§§] offer a powerful tool for analyzing data that circumvents the complexity of using tools such as Excel. In the very basic data set in Figure 5.12 (more complex ones are available), it is possible for students to simply plot any one variable against another by dragging the variable named on the left to the x or y axis. The data is then presented visually in a form where it can be interrogated. In this case, the chart shows a two-dimensional matrix of what the data show when "types of element" (metal, non-metal, semi-metal) are plotted against "the natural state". Figure 5.12 shows that the second data point of the box which is the union of "semi-metals" and "solids" has been selected and identified in the "table view" automatically as silicon. Looking at each of the data points, students can then start to build a picture of what it is about these elements which makes them difficult to classify as either a metal or a non-metal rather than learning the characteristics of the elements as received wisdom.

In summary, computer-supported platforms offer a variety of tools which can support argumentation. Not only do they provide better evidence for what we want students to believe, they also provide tools to support students engaging in more elaborated reasoning building connections to what they already know. Providing digital platforms for both inquiry and argumentation in chemistry is just the first step, however, as chemistry teachers and students need to be supported in the use of developmentally appropriate progressions in both argumentation skills and the structures to support argumentation. A key area is assessment of student performance in

[§§]https://tuvalabs.com/

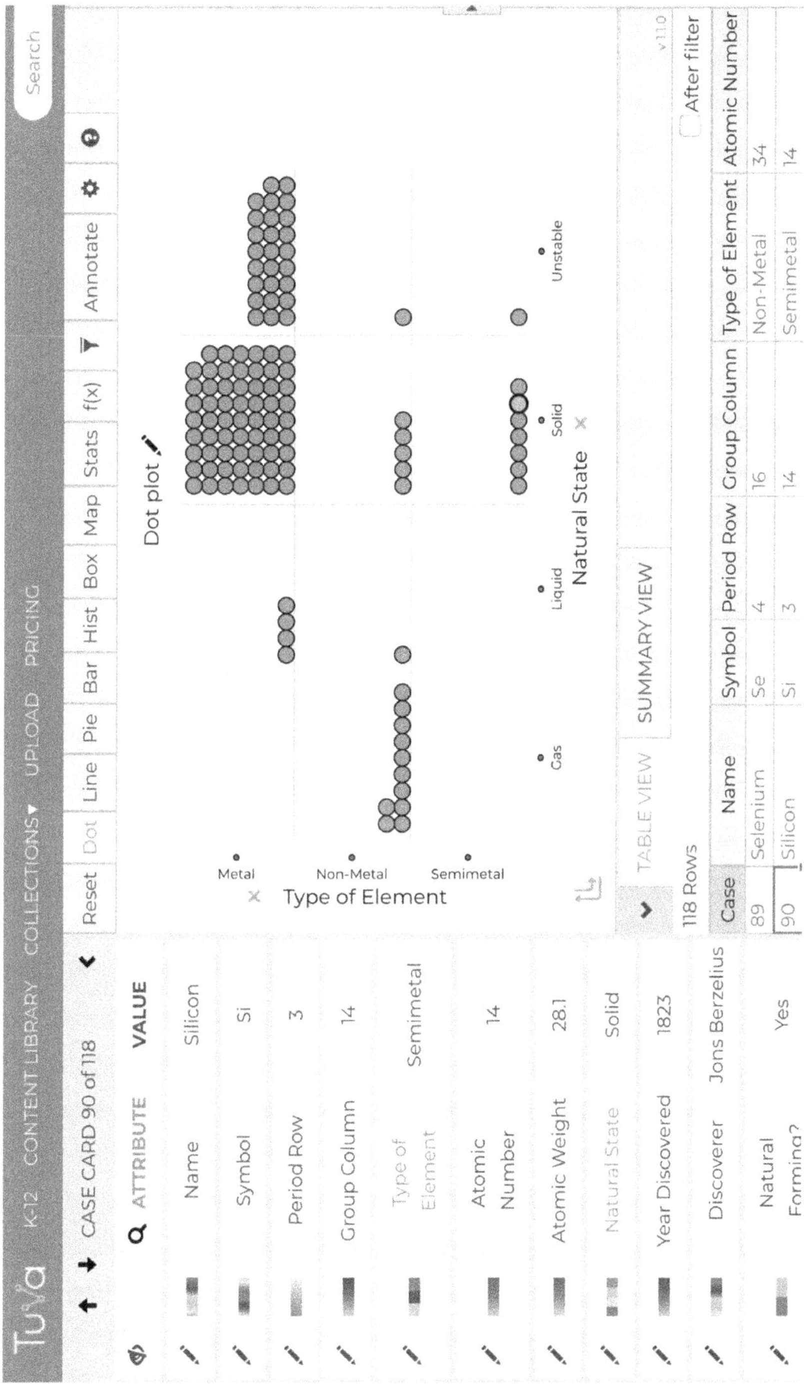

Figure 5.12 Tuva Labs offer powerful data analysis features that are user-friendly for students. This plot was created on Tuva. Go to www.tuvalabs.com/k12/ to learn more.

students' ability to argue from evidence. Advances in the computerized assessment of scientific argumentation discussed in this chapter offer chemistry educators an efficient means to assess two notoriously cumbersome things to measure in a valid and reliable fashion – classroom talk and constructed responses to open-ended items. These advances have the potential to scaffold and improve the quality of the argumentation process in the chemistry classroom, while expanding the possibilities for what we can ask students to produce at the end of class.

Practical Digest

As this chapter referenced a multitude of freely-available digital technologies that have potential to support the teaching and learning of argumentation in chemistry, this final section provides a digest of many of the technologies we highlighted. This digest is provided through Table 5.2,

Table 5.2 Practical digest of freely-available digital technologies highlighted in this chapter.

Technology	Description	Chapter reference	Practical reference
MyRSC	A chemistry-themed, asynchronous online learning community. Students communicate through writing *via* a threaded "message board."	5.2.1	my.rsc.org
WISE	Contains free chemistry lessons hosted in a guided, digital environment. This includes supports for students to access and organize evidence, as well as interactive visualizations of non-intuitive chemical models.	5.3.1; 5.3.2	wise.berkeley.edu
Argunet *Araucaria* *DebateGraph*	Argument mapping software that allows students to visualize and manipulate the structure of the arguments they construct and critique.	5.3.3	argunet.org araucaria.computing.dundee.ac.uk debategraph.org
Braincandy	Cloud-based classroom response technology where students anonymously answer questions based on common student preconceptions. Visualization tools show students there is classroom uncertainty – uncertainty that can set the stage for peer-to-peer argumentation with the goal of reaching classroom consensus.	5.4.1	braincandy.org

Table 5.2 (*Continued*)

Technology	Description	Chapter reference	Practical reference
DiALoG	A tool being developed for teachers to assess the quality of oral argumentation as it happens in their classrooms. Responsive Mini-Lessons (RMLs) accompany different combinations of scores allowing teachers to take action on the assessment information.	5.5.1	Will be free for instructors at conclusion of research and development phase.
Concord Consortium	Offers a variety of tools that support modeling and simulation.	5.6	concord.org
Tuva Labs	A powerful tool for analyzing data that is much more user-friendly for students than more conventional data analysis software.	5.6	tuvalabs.com

which makes it easy to reference the name and primary purpose of the technology, as well as where the technology can be referenced in this chapter and where it can be obtained for practical use.

Acknowledgements

Research and development of the DiALoG tool is supported by the National Science Foundation under Grant Nos. #1621441 and #1621496: Supporting Teacher Practice to Facilitate and Assess Oral Scientific Argumentation: Embedding a Real-Time Assessment of Speaking and Listening into an Argumentation-Rich Curriculum. Work on automated assessment of argumentation items is supported by NSF grant #1561149 and by the Institute of Education Sciences, U.S. Department of Education, through Grant R305A160320 to University of California, Berkeley. The opinions expressed are those of the authors and do not represent views of the Institute or the U.S. Department of Education.

References

1. Common Core State Standards Initiative, *Common Core State Standards for English Language Arts & Literacy in History/Social Studies, Science, and Technical Subjects*, 2011.
2. NGSS Lead States, *Next Generation Science Standards: For States, by States*, The National Academies Press, Washington, DC, 2013.
3. A. Jeong and J. Lee, *Br. J. Educ. Technol.*, 2008, **39**(4), 651.

4. O. Noroozi, A. Weinberger, H. J. Biemans, M. Mulder and M. Chizari, *Educ. Res. Rev.*, 2012, 7(2), 79.
5. D. Clark, V. Sampson, K. Stegmann, M. Marttunen, I. Kollar and J. Janssen, *et al.*, *E-Collab. Knowl. Constr. Learn. Comput.-Support Virtual Environ.*, 2010, 1.
6. D. Clark, K. Stegmann, A. Weinberger, M. Menekse and G. Erkens, *Argumentation in Science Education*, ed. S. Erduran and M. Jiménez-Aleixandre, Springer, 2007, pp. 217–243.
7. N. Pinkwart and B. M. McLaren, *Educational Technologies for Teaching Argumentation Skills*, Bentham Science Publishers, 2012.
8. O. Scheuer, F. Loll, N. Pinkwart and B. M. McLaren, *Int. J. Comput.-Support Collab. Learn.*, 2010, **5**(1), 43.
9. A. L. Veerman, *Arguing to Learn: Confronting Cognitions in Computer-supported Collaborative Learning Environments*, ed. J. Andriessen, M. Baker and D. Suthers, Kluwer, Dordrecht, 2003, pp. 117–143.
10. N. M. Mirza, V. Tartas, A.-N. Perret-Clermont and J.-F. de Pietro, *Int. J. Comput.-Support Collab. Learn.*, 2007, **2**(2–3), 247.
11. S. Hsi and C. M. Hoadley, *J. Sci. Educ. Technol.*, 1997, **6**(1), 23.
12. L. Munneke, J. Andriessen, P. Kirschner and G. Kanselaar, *Proceedings of the 8th International Conference on Computer Supported Collaborative Learning*, 2007, pp. 532–541.
13. D. Kuhn and W. Moore, *Learn. Res. Pract.*, 2015, **1**(1), 66.
14. Y.-C. Chen, S. Park and B. Hand, *Cogn. Instr.*, 2016, **34**(2), 100.
15. A. Jeong and G. V. Davidson-Shivers, *Educ. Technol. Res. Dev.*, 2006, **54**(6), 543.
16. D. L. Schwartz, *J. Learn. Sci.*, 1995, **4**(3), 321.
17. V. Sampson and D. Clark, *Sci. Educ.*, 2009, **93**(3), 448.
18. N. Rummel, H. Spada and S. Hauser, *Int. J. Comput.-Support Collab. Learn.*, 2009, **4**(1), 69.
19. D. Clark and V. Sampson, *Int. J. Sci. Educ.*, 2007, **29**(3), 253.
20. D. Clark, *Internet Environ. Sci. Educ.*, 2004, 175.
21. M. C. Linn, D. Clark and J. D. Slotta, *Sci. Educ.*, 2003, **87**(4), 517.
22. J. F. Osborne, J. B. Henderson, A. MacPherson, E. Szu, A. Wild and S.-Y. Yao, *J. Res. Sci. Teach.*, 2016, **53**(6), 821.
23. M. T. H. Chi, *J. Learn. Sci.*, 2005, **14**(2), 161.
24. J. Confrey, *Rev. Res. Educ.*, 1990, **16**, 3.
25. R. Duit, *International Encyclopedia of Education*, ed. T. N. Postlethwaite and T. Husen, Pergamon Press, New York, 2nd edn, 1994, pp. 4648–4652.
26. K. Close and J. B. Henderson, *Braincandy: A Cloud-Based Platform Providing Students Authentic, Engaging, and SAFE Spaces to Articulate and Refine Oral Argumentation*, 12th International Conference on Computer Supported Collaborative Learning, Philadelphia, 2017.
27. H. Mercier and D. Sperber, *Behav. Brain Sci.*, 2011, **34**(2), 57.
28. N. J. Hunsu, O. Adesope and D. J. Bayly, *Comput. Educ.*, 2016, **94**, 102.

29. A. R. Trees and M. H. Jackson, *Learn. Media Technol.*, 2007, **32**(1), 21.
30. H. Burkhardt, *"What You Test Is What You Get." The Dynamics of Curriculum Change in Developments in School Mathematics Worldwide*, University of Chicago School Mathematics Project, Chicago, 1987.
31. A. Desoete, *Metacognit. Learn.*, 2008, **3**(3), 189.
32. C. B. Cazden, *Classroom Discourse: The Language of Teaching and Learning*, Heinemann, Portsmouth, 2nd edn, 2001.
33. S. Michaels, C. O'Connor and L. B. Resnick, *Stud. Philos. Educ.*, 2008, **27**(4), 283.
34. N. Mercer, R. Wegerif and L. Dawes, *Br. Educ. Res. J.*, 1999, **25**(1), 95.
35. J. B. Henderson, C. Fenton, K. L. McNeill, P. D. Pearson and J. Barber, *DiALoG: A Practical Instrument Designed for the Assessment of Verbal Classroom Argumentation in Real Time*, 2015 Annual Meeting of the National Association for Research in Science Teaching, Chicago, 2015.
36. P. D. Pearson, A. M. Knight, M. A. Cannady, J. B. Henderson and K. L. McNeill, *Theory Pract.*, 2015, **54**(3), 228.
37. V. Sampson, P. J. Enderle and J. P. Walker, *Perspectives on Scientific Argumentation*, ed. M. S. Khine, Springer, Dordrecht, 2012, pp. 235–264.
38. F. H. Van Eemeren and R. Grootendorst, *A Systematic Theory of Argumentation: The Pragma-Dialectical Approach*, Cambridge University Press, Cambridge, 2003, vol. 14.
39. R. E. Bennett, *Autom. Scoring Complex Tasks Comput.-Based Test.*, 2006, 403.
40. M. Ha, R. H. Nehm, M. Urban-Lurain and J. R. Merrill, *CBE-Life Sci. Educ.*, 2011, **10**(4), 379.
41. K. C. Haudek, J. J. Kaplan, J. Knight, T. Long, J. Merrill, A. Munn, R. H. Nehm, M. K. Smith and M. Urban-Lurain, *CBE-Life Sci. Educ.*, 2011, **10**(2), 149–155.
42. K. C. Haudek, L. B. Prevost, R. A. Moscarella, J. Merrill and M. Urban-Lurain, *CBE-Life Sci. Educ.*, 2012, **11**(3), 283.
43. O. L. Liu, J. A. Rios, M. Heilman, L. Gerard and M. C. Linn, *J. Res. Sci. Teach.*, 2016, **53**(2), 215.
44. R. H. Nehm, M. Ha and R. Mayfield, *J. Sci. Educ. Technol.*, 2012, **21**(1), 183.
45. J. E. Opfer, R. H. Nehm and M. Ha, *J. Res. Sci. Teach.*, 2012, **49**(6), 744.
46. M. Urban-Lurain, R. A. Moscarella, K. C. Haudek, E. Giese, D. F. Sibley and J. E. Merrill, *Proceedings of the Frontiers in Education Conference*, 2009, pp. 1–6.
47. M. Zhu, H.-S. Lee, T. Wang, O. L. Liu, V. Belur and A. Pallant, *Int. J. Sci. Educ.*, 2017, **39**(12), 1648.
48. O. L. Liu, C. Brew, J. Blackmore, L. Gerard, J. Madhok and M. C. Linn, *Educ. Meas. Issues Pract.*, 2014, **33**(2), 19.
49. J. Hattie, *Visible Learning: A Synthesis of Over 800 Meta-Analyses Relating to Achievement*, Routledge, London, 2008.

CHAPTER 6

Assessment of Argumentation in Chemistry: A Model for Designing Items

NG YEE PING DIANA

Oxford University Centre for Educational Assessment, University of Oxford, 15 Norham Gardens, Oxford, OX2 6PY, UK
Email: Diana.Ng@education.ox.ac.uk

6.1 Introduction

Over the past decade, argumentation has become a widely espoused and researched practice in school science and science education studies.[1,2] Educational initiatives and curricular documents of many countries reflect this interest and emphasis.[3] In some recently revised curricular or newly implemented initiatives, there is a progressive introduction of argumentation notions to build proficiency in pupils who start formal schooling and advance through primary and secondary grades over time. However, as evidenced from these same documents and programmes, there is a notable lack of prescribed guidance about the assessment of argumentation abilities demonstrated by typical pupils in these grades.

The paucity of guidelines reflects the state and constraints of research on defining how these abilities should be measured as learners become increasingly more skilled. Recent research has focused on assessing the process of the argumentation discourse and its outcomes such as written or oral student-generated arguments.[4,5] In contrast, few have investigated the

Advances in Chemistry Education Series No. 2
Argumentation in Chemistry Education: Research, Policy and Practice
Edited by Sibel Erduran

Published by the Royal Society of Chemistry, www.rsc.org

assessment of argumentation abilities which gradually develop as a result of instructional exposure.[6] Also, recent assessments have tended to highlight the disciplinary norms of argumentation (*i.e.*, identifying features of educationally sound scientific arguments), and have largely originated from research settings, which could prove challenging for standard classroom adaption and implementation.[4,7] Specifically, the diagnostic and analytical frameworks underlying these assessments investigated the structural patterns, justification content, or conceptual and epistemological presentation of argumentation outcomes.[5,8] The theoretical complexities and inherent limitations of these various frameworks can pose issues for general school application by teachers.[4] Other impediments to successful implementation include pupils' ethnic and linguistic backgrounds, as well as teacher readiness and aptitude in mediating intricate scientific practices such as argumentation and reasoning.[9–12] It is well-documented, for instance, that pupils' cultural backgrounds, everyday experiences and linguistic competencies impact the growth of their skills with scientific and academic discourses, which includes scientific argumentation.[13] Unfortunately, there is currently limited research about the assessment of scientific argumentation abilities in diverse pupil populations typical of general school settings.[14]

In response to this gap, more recent research centred on developing and validating learning progressions of scientific argumentation have emerged. Learning progressions are models of continua of deepening expertise, knowledge and skills, with the potential to serve as frameworks for the design of valid assessments.[15] However, the distinctive differences among the various learning progressions proposed for scientific argumentation present difficulties when findings are transferred for application to the actual classroom.[16] These differences include the conceptual basis, nature of instructional contexts investigated (*e.g.*, depth and type of scientific content), time span covered, and proof of concept approaches (*e.g.*, through empirical validation). Notably, some of these differences also account for the limited impact of more established content-based learning progressions on curricular content design.[17] Although research on learning progressions in scientific argumentation is in its nascent phase, there are promising findings to inform the design of assessments that measure proficiency in argumentation with scientific content and skills (see, *e.g.*, Osborne *et al.*).[6]

This chapter reviews current understandings about suitable test features for assessing argumentation skills of various proficiencies and proposes item conceptualisation and design guidelines. These guidelines draw on analyses of curricular documents from England and Australia as well as learning progression research. To illustrate the application of the guidelines, the remaining chapter focuses on item exemplars assessing substantive scientific concepts from a common chemistry topic. Limitations and potential from current research to practice are discussed in the conclusion.

6.2 Argumentation in Science Education

Contemporary argumentation theory arose from seminal publications in logic and philosophy, such as Toulmin's 1958 treatise *The Uses of Arguments*, in the last century.[19] However, its adoption in science education only took place during the last twenty years as fresh insights about thinking and learning surfaced from multiple research disciplines.[20] These multi-disciplinary perspectives regard science as an entity socially constructed from the products of declarative, procedural, and epistemic inquiry.[21,22] Regulated by normative activities and shared values of the science community, this entity functions collectively to advance the goals of science.[23,24] Working together to promote its cause, scientists use reasoning, discourse and rhetorical tools to uphold epistemological truths about what scientific knowledge should be.[25] For instance, the acts of defending claims, assessing evidence and legitimising knowledge occur when scientists actively engage in various combinations of dialogic, dialectic and discursive processes.

As acknowledgement of the inadequacy and pitfalls of the "positivistic" paradigm took root within the science community, science educators were encouraged to expose pupils to critical analysis and debate with competing ideas, theories and methodologies.[26] Pupils' scientific epistemology, or understanding about the nature of science, is thought to deepen as they participate in advancing claims and counterclaims backed by evidence.[27] Besides increasing epistemic understanding, honing thinking and reasoning with discourse-based tasks leads to social construction of knowledge and deepens conceptual understanding.[21,22,28] The co-construction, negotiation and active engagement with the viewpoints of others mirror authentic practices of scientists and build pupils' capacity for scientific debate.[29] In short, argumentation is an ideal pedagogical tool.[30,31]

Within science education, argumentation has not been precisely defined but carries various interpretations.[7,24] One widely-cited account of argumentation discusses its sense-making function within and between individuals.[32] At the individual level, the emergence of beliefs and scientific understanding involves internalising a chain of reasoned arguments. Between individuals, argumentation is one common approach to communicate views, which are oftentimes opposing. A second perspective considers the role of argumentation in the process of influencing another *vis-à-vis* the production of a reasoned analysis – the product.[33] Many studies in science education adopt this perspective to understand and help learners construct the products of argumentation. Interestingly, Toulmin's (1958) classic exposition on the structural elements of an informal argument, consisting of *data*, *claim*, *warrant*, *rebuttal*, and *qualifier*, is the most common analytical framework used among these studies.[34] A third way of viewing argumentation relates to its supposed benefits to science learners.[19] As noted earlier, argumentation can serve to help learners construct disciplinary scientific knowledge and increase their epistemic fluency.

Argumentation can also improve written and spoken scientific literacy, leading to richer enculturation of scientific practices and values.[35] Other advantages of argumentation include changes to pupils' understanding of scientific concepts and more appropriate or sharpened reasoning faculties much akin to expert cognition.[36,37]

6.3 Science Curriculum and Argumentation

Given the centrality of argumentation as an important pedagogical and cognitive tool, there is a growing emphasis on its promotion in the science curricula and standards of many countries.[3,19] For instance, the curricula or standards documents of some countries who participated in the most recent Trends in Mathematics and Science Study (TIMSS), made explicit or indirect references to argumentation.[38] TIMSS is a large-scale testing programme whose assessment framework heavily promotes argumentation skills such as critical analysis of evidence and justification of claims.[19]

This chapter analyses published-in-English curricular documents of countries which introduced differentiated argumentation notions across year groups in programmes and initiatives implemented within the past five years. To facilitate the analysis, there must be availability of test items from previously administered assessments or item exemplars accompanying the curricular and standards. Therefore, though the documents of four countries provide detailed information in English about the range of argumentation abilities expected of primary and secondary pupils, this chapter focuses only on Australia and England. Excluded from the analysis are New Zealand's curricular documents because item exemplars are not easily assessible. Also, the current implementation of assessments aligned to the Next Generation Science Standards (NGSS) in the United States means that exemplars are being developed and publicly inaccessible.[39,40]

The newly implemented Australian Curriculum, for five- to sixteen-year-old pupils, carries an explicit reference to the acquisition of argumentation as a curriculum aim.[41] To meet this aim, one of its three strands in the curriculum, known as the *Inquiry Skills*, involves "evaluating claims, investigating ideas, solving problems, drawing valid conclusions, and developing evidence-based arguments" (p. 9).[41] Although there is no explicit definition of argumentation in the curriculum, a closer reading of the learning outcomes suggests a developmental sequence of related concepts across the three strands. For instance, in the sub-strand of *Communicating*, the expectations of Years 7 to 8 pupils include the ability to draw on evidence to present solutions to problems using a variety of scientific discourse, representations and technology. In Years 9 to 10, pupils advance to "constructing evidence-based arguments" (p. 105).[42] This manner of ordering increasingly sophisticated content from lower to higher grades share many similarities to well-known curriculum design and pedagogical approaches such as the spiral curriculum and cognitively guided instruction.[43–45]

The aim is to enable learners to develop an appreciation and gradual facility with argumentation in science.[45]

The articulation of argumentation concepts such as *evidence*, *claim*, *argument*, and *reasoning* across school years in the Australian Curriculum is likewise noted in the recently revised National Curriculum of England.[46] Specifically, the attainment targets of age-appropriate argumentation in the science programme, *Working Scientifically*, lays down "the understanding of the nature, processes and methods of science for each year group" (p. 169).[47] Competencies introduced in earlier grades support the development of differentiated and advanced abilities for subsequent levels. For instance, a Lower Key Stage 2 specification that pupils use simple evidence to address questions is a precursor for the identification of scientific evidence to support or rebut arguments in a subsequent stage.[48]

At every Key Stage, there is parallel development of age-appropriate competencies such as raising questions, gathering data by observation, analysing and explaining the data or phenomena, alongside the abilities to argue and evaluate evidence.[49] As these targets specify process skills, procedural and epistemic understanding, they are concurrently taught to reinforce understanding.[50,51] Clustering skills in teaching and learning, which is carried on from the previous version of the National Curriculum, is important for promoting critical discourse and examination of data and evidence.[49,52] Significantly, the skill sets supporting argumentation promoted in both current and previous versions of the National Curriculum are remarkably similar, which suggests their continued relevance.[53] Evidence for this similarity surfaced from Smith's classification of skills promoted in the *Working Scientifically* programme using a model of inquiry-based learning proposed by Harlen and Qualter.[48,53] See Table 6.1 for Smith's classification and correspondence to targeted argumentation abilities. Clustering

Table 6.1 Skills in the National Curriculum and correspondence to potential argumentation abilities.[49,53,82]

Skills in the 2014 National Curriculum	Potential argumentation abilities targeted
Raising questions, predicting and planning (setting up investigations)	Understanding the roles of data, evidence, theories and explanations in scientific knowledge and their relationships with each other; justifying procedures and choices of experimental design
Gathering evidence by observing and using information sources (collecting data from investigations)	Generating and applying criteria for evaluation of evidence
Analysing, interpreting and explaining evidence (making conclusions)	Generating and applying criteria for evaluation of evidence; constructing evidence-based arguments
Communicating, arguing, reflecting and evaluating evidence (reporting, reflecting and applying)	Constructing, presenting and defending arguments

attainment targets to support the competencies of using evidence and evaluating claims is also apparent in the Australian Curriculum. The sub-strands forming the *Inquiry Skills* strand outline supporting targets such as the skills of questioning, data processing and communication of ideas.

The science curricula of England and Australia converge on two other note-worthy aspects. First, there is a similar range of argumentation abilities prescribed for attainment. Initial abilities pertain to the understanding and uses of scientific evidence. These fundamental abilities gain complexity at higher grade levels and include the evaluation of evidence and claims, as well as the construction and evaluation of arguments. Second, the grade specifications for the simplest and most complex argumentation skills are nearly identical. Therefore, the entry point to build argumentation proficiency in English and Australian pupils occurs when they are seven to nine and eight to ten respectively. Correspondingly, exposure to complex argumentation occurs in the culminating phases of secondary schooling between ages 14 and 16. The Australian science curriculum additionally expects its graduating pupils to be able to evaluate and distinguish between "ethical, social and economic arguments" (p. 119).[41]

Differences in the curricula can be seen in the articulation, organisation and sequence of argumentation-related learning outcomes. These differences affect the content and format of assessment at specific grades. For instance, *claim* appears for the first and only time in the Key Stage 4 preface of England's National Curriculum, where pupils are to "evaluate claims based on science through critical analysis of the methodology, evidence and conclusions, both qualitatively and quantitatively" (p. 213).[47] In comparison, the term's first mention in the Australian Curriculum occurs in Years 7–8, which targets the appraisal of claims with theoretical and investigative evidence. Another notable difference is the articulation of argumentation notions in the culminating stages of secondary schooling (*i.e.*, Years 9–10 *vis-à-vis* Key Stage 4). Unlike the Australian Curriculum, which provides detailed expectations for Years 9–10 pupils, there is less explicit articulation for Key Stage 4. The distinction arose from the dominant form of assessment associated with each curriculum; the Australian Curriculum is characterised by school-based assessments while Key Stage 4 pupils pursue subject qualifications from external awarding organisations. These organisations design their specifications using interpretations of the National Curriculum.[54] See Tables 6.2 and 6.3 for a content analysis of the National Curriculum and the Australian Curriculum in relation to *evidence*, *claim* and *argument*.

6.4 Assessment of Science Curricula and Argumentation

Given the primacy of argumentation evidenced in these two curricular documents, there is a conspicuous lack of curricular guidance about the assessment of argumentation abilities with scientific content. In the case of the Australian Curriculum, support and guidelines are critical as

Table 6.2 Statutory requirements in the National Curriculum related to *evidence, claim and argument.*[a]

National Curriculum in England (statutory requirements) – Key Stages 1 to 4				
Programme of Study	Lower Key Stage 2 (Years 3 and 4; Ages 7–9)	Upper Key Stage 2 (Years 5 and 6; Ages 9–11)	Key Stage 3 (Years 7 to 9; Ages 11–14)	Key Stage 4 (Years 10 and 11; Ages 14–16)
Preface		Pupils should draw conclusions based on their data and observations, use **evidence** to justify their ideas, and use their scientific knowledge and understanding to explain their findings	Pupils should understand that science is about working objectively, modifying explanations to take account of new **evidence** and ideas and subjecting results to peer review	Students develop their ability to **evaluate claims** based on science through critical analysis of the methodology, **evidence** and conclusions, both qualitatively and quantitatively
Working scientifically	Using straightforward scientific **evidence** to answer questions or to support their findings	Identifying scientific **evidence** that has been used to support or refute ideas or **arguments**	Understand that scientific methods and theories develop as earlier explanations are modified to take account of new **evidence** and ideas, together with the importance of publishing results and peer review	Students should be taught so that they develop understanding and first-hand experience of explaining everyday and technological applications of science; evaluating associated personal, social, economic and environmental implications; and making decisions based on the evaluation of **evidence** and **arguments**
Subject content		Give reasons, based on **evidence** from comparative and fair tests, for the particular uses of everyday materials, including metals, wood and plastic		Students should be taught: the **evidence** for evolution; the **evidence** for composition and evolution of the Earth's atmosphere since its formation; **evidence**, and uncertainties in **evidence**, for additional anthropogenic causes of climate change; the nuclear model and its development in the light of changing **evidence**

[a]*Note:* Adapted from DfE.[46]

Table 6.3 Content descriptions in the Australian Curriculum related to *evidence, claim and argument.*[a]

Australian Curriculum – Foundation to Year 10				
Strand	Primary Year 3–4 (Ages 8–10)	Primary Year 5–6 (Ages 10–12)	Secondary Year 7–8 (Ages 12–14)	Secondary Year 9–10 (Ages 14–16)
Science understanding Sub-strands: 1. Biological sciences 2. Chemical sciences 3. Earth and space sciences 4. Physical sciences			*Year 8 only* Chemical change involves substances reacting to form new substances Elaboration: Identifying **evidence** that a chemical change has taken place [Sub-strand: 2]	*Year 10 only* The theory of evolution by natural selection explains the diversity of living things and is supported by a range of scientific **evidence** [Sub-strand: 1] *Year 10 only* The universe contains features including galaxies, stars and solar systems, and the Big Bang theory can be used to explain the origin of the universe Elaboration: Identifying the **evidence** supporting the Big Bang theory, such as Edwin Hubble's observations and the detection of microwave radiation [Sub-strand: 3]
Science *as a human endeavour* Sub-strands: 5. Nature and development of science	*Year 4 only* Science involves making predictions and describing patterns and relationships	Science involves testing predictions by gathering data and using **evidence** to develop explanations of events and	Scientific knowledge has changed peoples' understanding of the world and is refined as new **evidence** becomes available	Scientific understanding, including models and theories, is contestable and is refined over time through a process of review by the scientific community

Table 6.3 (*Continued*)

Australian Curriculum – Foundation to Year 10				
Strand	Primary Year 3–4 (Ages 8–10)	Primary Year 5–6 (Ages 10–12)	Secondary Year 7–8 (Ages 12–14)	Secondary Year 9–10 (Ages 14–16)
6. Use and influence of science	Elaboration: Exploring ways in which scientists gather **evidence** for their ideas and develop explanations [Sub-strand 5]	phenomena and reflects historical and cultural contributions [Sub-strand 5]	[Sub-strand 5]	Elaboration: *Year 9 only* Investigating how the theory of plate tectonics developed, based on **evidence** from sea-floor spreading and occurrence of earthquakes and volcanic activity *Year 10 only* Considering the role of different sources of **evidence** including biochemical, anatomical and fossil **evidence** for evolution by natural selection; Investigating the development of the periodic table and how this was dependent on experimental **evidence** at the time [Sub-strand 5] People use scientific knowledge to evaluate whether they accept **claims**, explanations or predictions, and advances in science can affect people's lives, including generating new career opportunities [Sub-strand 6]

Science inquiry Sub-strands: 7. Questioning and predicting 8. Planning and conducting 9. Processing and analysing data and information 10. Evaluating 11. Communicating	Represent and communicate observations, ideas and findings using formal and informal representations Elaboration: Using simple explanations and **arguments**, reports or graphical representations to communicate ideas to other students [Sub-strand: 11]	Compare data with predictions and use as **evidence** in developing explanations [Sub-strand: 9] *Year 6 only* Communicatig ideas, explanations and processes using scientific representations in a variety of ways, including multi-modal texts Elaboration: Using a variety of communication modes, such as reports, explanations, **arguments**, debates and procedural accounts, to communicate science ideas [Sub-strand: 11]	Summarise data, from students' own investigations and secondary sources, and use scientific understanding to identify relationships and draw conclusions based on **evidence** [Sub-strand: 9] Use scientific knowledge and findings from investigations to evaluate **claims** based on **evidence** [Sub-strand: 10] Communicate ideas, findings and **evidence**-based solutions to problems using scientific language, and representations, using digital technologies as appropriate [Sub-strand: 11]	Use knowledge of scientific concepts to draw conclusions that are consistent with **evidence** Elaboration: *Year 10 only* Using primary or secondary scientific **evidence** to support or refute a conclusion; constructing a scientific **argument** showing how their **evidence** supports their **claim** [Sub-strand: 9] Evaluate conclusions, including identifying sources of uncertainty and possible alternative explanations, and describe specific ways to improve the quality of the data [Sub-strand 10] Critically analyse the validity of information in primary and secondary sources and evaluate the approaches used to solve problems Elaboration: *Year 9 only* Describing how scientific **arguments** are used to make decisions regarding personal and community issues

Table 6.3 (*Continued*)

Australian Curriculum – Foundation to Year 10				
Strand	Primary Year 3–4 (Ages 8–10)	Primary Year 5–6 (Ages 10–12)	Secondary Year 7–8 (Ages 12–14)	Secondary Year 9–10 (Ages 14–16)
				Year 10 only Describing how scientific **arguments**, as well as ethical, economic and social **arguments**, are used to make decisions regarding personal and community issues [Sub-strand 10] Communicate scientific ideas and information for a particular purpose, including constructing **evidence-based arguments** and using appropriate scientific language, conventions and representations Elaboration: *Year 10 only* Constructing evidence-based **arguments** and engaging in debate about scientific ideas [Sub-strand: 11]

[a]*Note:* Adapted from ACARA.[41]

school-based assessments are the predominant assessments at the primary and lower secondary levels.[55] The Australian Curriculum recommends the use of a variety of teacher-designed-and-scored assessments such as experimental investigations, projects, checklists, and tests consisting of multiple choice and constructed-response item formats. Teachers also have access to the publicly-available work portfolios composed of actual student assessment for all year levels.[42] Every piece of assessment in the portfolios has an accompanied write-up outlining the classroom contexts for deployment (*e.g.*, group work), and annotations to highlight the targeted achievement standards. However, assessments designed to assess argumentation notions are few and localised to the Year 10 work portfolio.

A closer analysis of some assessments in the Year 10 portfolio suggests some instances of disparity between the intended assessment objective and the actual trait measured. For instance, the published answer to a question in an assessment intended to gauge the ability to construct an evidence-based argument[†] suggested that what was actually assessed is the ability to account for an observational trend (*i.e.*, as concentration increases the rate of reaction increases) using evidence from knowledge (*i.e.*, collision theory). In effect, the answer was an account of why the phenomena happens and not an argumentative attempt to state or support a claim.[56]

In contrast, the design of a test question in another assessment from the same portfolio ensured that the defence and rejection of the validity of explanatory hypotheses in two claims occurred through argumentation (see Figure 6.1 for the item). To answer the question, pupils needed to produce persuasive warrants for the fit between the known data and the chosen explanatory account. The warrants are not used to explain the scientific phenomenon itself. Unfortunately, the annotation accompanying this question did not highlight these argumentation features.

In the new National Curriculum in England, teachers award grades to Key Stages 1 to 3 pupils as end-of-stage summative evaluations of their performance.[54] Key Stage 4 pupils have the option of selecting separate and specialised science subject qualifications awarded by external organisations who administer the assessments. Since 2016, there has been a biennial statutory sampling at the end of Key Stage 2 across schools in England to gauge national standards.[57] Selected questions from these sampling tests, along with commentaries about notable aspects of pupils' performance, such as misconceptions, are published and available online for teachers to consult.[58] Serving as another online reference are sample test materials produced by the testing agency to demonstrate new question types and the assessment of new content.[59]

Analysis of released items from the 2016 Key Stage 2 inaugural national sampling under the new National Curriculum reveals the assessment of a

[†]The intended testing objective is evidenced from the annotations accompanying the work sample (Year 10 Above satisfactory: Work sample portfolio) of an investigation report on rates of reaction. Due to copyright restrictions, this chapter did not reproduce students' responses.

(c) Emma and Luke were asked to explain why reactions B and D have different graphs.

Luke said "Obviously there was an error in the measurement. Reaction B should have produced CO_2 at the same rate as reaction D because they both use 100 mL of solutions. Acids have water in them anyway so it makes no difference that there is 50 mL of acid and 50 mL of water."

Emma said "Well they have the same volume of solution but it's not the water that reacts with the marble chips, it's the acid. So reaction B really only has half the amount of acid as reaction D so its graph is different."

Evaluate the claims made by these two students using your knowledge of chemical reactions and factors that influence their rate. State who you agree with and why.

Figure 6.1 Excerpt from a sample assessment of a written test on chemical reactions.

simple form of argumentation competency by some items.[58] In these items, pupils determine the plausibility of the options (potential claims) using a combination of learnt content knowledge and information provided in the item stem or stimulus.[9,60] These items have a common assessment design and deploy selected response formats such as matching and true/false item structure. Without requiring pupils to articulate the underlying reasoning how the evidence relate to their choice (or exclusion) of claims, their abilities to use evidence to construct and refute arguments are not investigated. In this respect, it is unclear if there is attainment of the stipulated argumentation abilities – "identifying scientific evidence that has been used to support or refute ideas or arguments" (p. 166) expected for this Key Stage.[48] Figure 6.2 is an example of a released item demonstrating the assessment of claim-making.

6.5 Learning Progressions in Scientific Argumentation

The various definitions of learning progressions in science proposed over the past 20 years differed significantly over time.[61] However, there were two

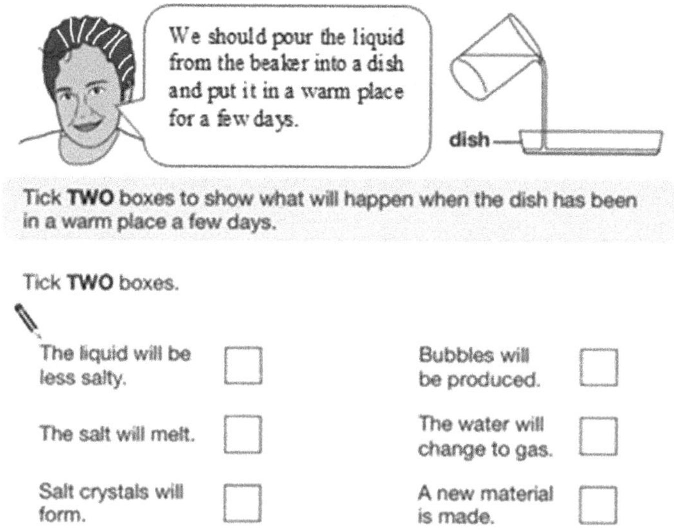

Figure 6.2 Example of a released item from the *Key Stage 2 2016 Science Sampling: Test booklet* (p. 40), by the Standards and Testing Agency, 2017, London. Contains materials developed by STA for national curriculum assessments. Contains public sector information licensed under the Open Government Licence v3.0, https://assets.publishing.service.gov.uk/government/uploads/system/uploads/attachment_data/file/648195/2016sciencesamplepublicrelease_Test_Booklet_V4.pdf.

discernible broad emphases among these learning progressions. Earlier definitions focused on the typical trajectories and learning that correspond to the learner's natural maturation and development (see *e.g.*, Forster and Masters[62]). As such, the progress of learners was monitored using descriptors of increasing proficiency characteristically displayed at distinctive developmental junctures.[63] Later definitions of learning progressions highlighted the impact of instruction on achievement of curricular goals, the transfer and applicability of the acquired abilities or content to other contexts and the tentativeness of learning routes.[45,64] For instance, the report *Taking Science to School* defined learning progressions as continua of coherent and increasingly complex learning sequences informed by the disciplinary domain and impacted by curriculum and instructional practices.[15] In another definition, Deane and Song called them "tentative maps of skill development" (p. 1), thus underscoring the provisional and evolving nature of the content pending available evidence (*e.g.*, from future research).[65]

On the whole, the majority of the learning progressions developed in science focused on scientific content such as concepts or practices (*e.g.*, see Alonzo and Steedle's work on forces and motion), with fewer studies directed solely on scientific argumentation.[66] One of the earliest learning progressions proposed for argumentation came from Berland and McNeill.[67]

Their model defined argumentation as "a reasoned piece of discourse in which a claim has been justified" (p. 772).[67] They postulated that the overall complexity of argumentation tasks depended on three interdependent dimensions: (1) supportive factors in the instructional context; (2) the requirements imposed by the argumentative product; and (3) the argumentative process. The dimension of instructional context identified four task features: the size of the expected outcome space of the task (*e.g.*, are there many plausible answers?), the size and type of data set which pupils analysed, the availability of scaffolds (*e.g.*, materials or peer support) and pupils' current state of understanding. The inclusion of the fourth feature accorded with existing research suggesting that pupils with stronger grasp of scientific concepts and principles performed better in argumentation.[68]

The second dimension of task requirements drew on Toulmin's (1958) model of informal argumentation and identified four characteristics of pupils' written or verbal argumentation discourse. Briefly, three of these characteristics examined the quality of the claims, which are answers provided by pupils in response to the posed question or problem. The criteria for evaluating claims included the manner of defending the claims (*e.g.*, are claims defended with evidence and reasoning?), reaction to counterarguments (*e.g.*, are counterclaims rebutted?), and the extent the claims resolve the problem (*e.g.*, is a causal explanation provided?). The final characteristic evaluated in pupils' argumentative product was the overall suitability and adequacy of the evidence, reasoning, and rebuttal. These three structural elements should cohere logically, use scientifically established knowledge and reasoning, and provide a convincing basis for persuasion. The final dimension in Berland and McNeill's model evaluated two qualities of argumentation discourse. The first described the initial production of claims, which subsequently advanced to appraising and revising claims. The second considered the degree of pupils' social engagement with the argumentation task and ranged from teacher engineering to pupils' intrinsic engagement.

Qualitative evidence for Berland and McNeill's model came from observational and transcript analyses of four actual science classes with fifth, seventh and twelfth graders. These classes had various forms of scaffolds (*e.g.*, writing and discussion prompts) to support the learning of curriculum units. Although the study did not investigate experimentally pupils' competence according to the dimensions of their learning progression, there was encouraging evidence that young children could construct and defend arguments with adequate support. Moreover, the study identified key features of tasks (*e.g.*, outcome space, type and size of data embedded in task) that could impact upon the nature of elicited arguments and pupils' successes at argumentation.

In a separate formulation, H.-S. Lee *et al.* devised a learning progression based on the notion of "uncertainty" or the degree of trust in or persuasive power of an argument.[69] The researchers argued that the idea of uncertainty or tentativeness is poorly represented in current conceptualisations of scientific argumentation. They highlighted that the onus of arguments and counterarguments is to provide a strong justification for scientific conclusions and claims made, keeping in mind the many forms of uncertainties present. These uncertainties range from untested theories, experimental errors, instrumental imprecision, and biases (*e.g.*, statistical, perspectives). H.-S. Lee *et al.* chose to emphasise the two structural features of "modal qualifiers" (p. 101) and "conditions of exception or rebuttal" (p. 101) found in Toulmin's (1958) framework.[18] The qualifier referred to the degree of plausibility of conclusions or claims and the conditions, or what the researchers termed as "uncertainty rationale," were the special circumstances that dictated the plausibility.

H.-S. Lee *et al.*'s learning progression had six levels. At the two lowest levels (Levels 0 and 1), products of argumentation were non-scientific and consisted of only claims, respectively. At Level 2, there was support for claims with the relevant data or concepts. The next level described pupils' abilities to coordinate claims with evidence (theory and data) through the necessary warrants, which were the reasoning to legitimise claims with evidence. The first juncture in the progression which incorporated the uncertainty element was at Level 4. To achieve this level, pupils asserted their confidence in the strength of their claims by recognising the likelihood of the warrants. At Level 5, which was the highest level of the progression, students accounted for the strength of their claims by identifying task-specific constraints imposed by the instrumentation, measurement or theory. Depending on the assessment objective, items designed to measure the levels in the progression take on different item formats. Figure 6.3 shows a question designed using H.-S. Lee *et al.*'s model.

Psychometric results from three runs of a construct validation study using Rasch modelling with 473 high school pupils led H.-S. Lee *et al.* to reject the notion of qualifiers from their argumentation construct. They concluded that pupils misinterpreted the assessment intent of the rating item used to elicit their uncertainty. Rather than qualifying the fit between content knowledge and the evidence, many pupils instead signalled how confident they were in their grasp of the required scientific content or tested material. More than two thirds of the pupils identified personal reasons (*e.g.*, inability to understand question) as sources of uncertainty instead of evidence insufficiency or experimental bias. A closer read of the questions, such as the example shown in Figure 6.3, suggested that some of the crafted stems could have contributed to the misinterpretation. Overall, the average item difficulties in the test instrument performed as hypothesised, providing evidence for the progression levels. The study covered a range of topics related to climate change, the search for life, and availability of fresh water.

Osborne *et al.* proposed the third and latest progression. They conceptualised argumentation as a "*competency* which draws on a mix of

Damming a river changes the flow of the river. The graph shows the average water flow of the Colorado River at Lee's Ferry, Arizona during 1942 and 1996. The water discharge is measured in cubic feet per second. The Glenn Canyon Dam, which created Lake Powell, just upstream of Lee's Ferry, was constructed in 1964.

The map shows the location of Lee's Ferry, as well as the Hoover Dam and the Colorado River outlet in northern Mexico. Water quality changes when a river is dammed. The salinity (amount of salt) in the water can increase when there is less water flow because the water stays longer in one place, allowing it to dissolve salts in the rocks.

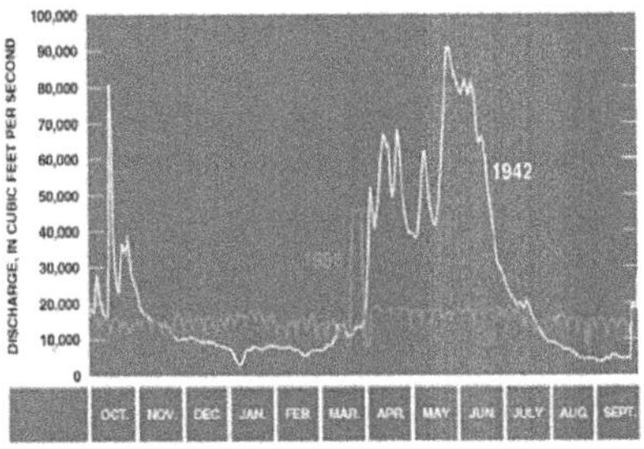

(Claim) Based on the graph, when was the salinity of the water near Lee's Ferry the highest?

- January 1942
- June 1942
- January 1996
- March 1996

(Justification) Explain your claim.

(Uncertainty rating) How certain are you about your claim based on your explanation?

(1) Not at all certain
(2)
(3)
(4)
(5) Very certain

(Uncertainty rationale) Explain what influenced your certainty rating.

Figure 6.3 A scientific argumentation task found in an online earth science module created by Pallant, Pryputniewicz, and Lee.[86] The argumentation task is developed based on the uncertainty-infused scientific argumentation framework.[69]
Image reproduced with permission from the Concord Consortium. Map reproduced from https://en.wikipedia.org/wiki/Colorado_River#/media/File:Colorado_River_basin_map.png, image credit Shannon1, under the Creative Commons Attribution-Share Alike 4.0 International license, https://creativecommons.org/licenses/by-sa/4.0/. Graph reproduced from ref. 87.

content knowledge, procedural knowledge, and epistemic knowledge" (p. 823, emphasis in original).[6] Like the previous two learning progression models, the researchers drew on Toulmin's framework. They argued that scientific argumentation involved a coordination between the construction and the critique of elements such as evidence, claims and warrants. On the basis that more coordination in tasks correlated to higher cognitive loads, they proposed three broad levels in their learning progression. Level 0, the bottom anchor of the progression, consisted mainly of singular abilities such as construction or identification of a claim. Level 1 described more sophisticated argumentation skills featuring one appropriate connection between abilities; for instance, by supporting claims with the relevant evidence through warrant identification. Tasks at the highest level (Level 2) assessed evaluative abilities and therefore involved at least two connections. For instance, the top anchor (*i.e.*, 2d) described the construction of justified counterclaims. Table 6.4 summarises the levels and sub-levels in the learning progression.

Osborne *et al.*'s learning progression underwent a systematic and iterative development and validation process over four years with about 2000 pupils, who were predominantly sixth to eighth graders. Besides using unidimensional and multidimensional Rasch modelling to ascertain the correspondence of items to the hypothesised progression, Osborne *et al.* also examined the item quality using verbal report protocols with selected pupils. However, Osborne *et al.* only reported data from a selected cohort of 490 eighth grade pupils. These pupils were administered two groups of items – one requiring scientific knowledge (scientific argumentation

Table 6.4 Learning progression of argumentation involving construction and critique.[a]

Level	Constructing	Critiquing
0		
0a	Constructing a claim	
0b		Identifying a claim made by another person
0c	Providing evidence	
0d		Identifying evidence used by another person
1a	Constructing a warrant	
1b		Identifying a warrant provided by another person
1c	Constructing a complete argument	
1d	Providing an alternative counter-argument to rebut another's claim	
2a	Providing a counter-critique of another's argument	
2b	Constructing a one-sided comparative argument	
2c	Providing a two-sided comparative argument about two competing arguments	
2d	Constructing a counter-claim with justification	

[a]*Note*: Adapted from Osborne *et al.*[6]

Something in the Air?

Imara and Alexis are wondering where the water on the outside of the glass of water with ice comes from.

They each have a different idea:

Imara: The water came over the top of the glass.

Alexis: The water came from the air.

A. How could you argue that Imara's idea is wrong?

B. How could you argue that Alexis' idea is wrong?

C. Imara thinks of a way to test the two ideas. She wipes the water off the glass and measures the height of the water in the glass. After an hour, she notices that the drops of water on the glass have returned and the height of the water in the glass has not changed.

Whose idea does this support, Imara or Alexis?
Explain why you think it supports Imara or Alexis.

Figure 6.4 A scientific argumentation item "bundle" developed by Osborne *et al.* consisting of three items sharing a common context.[6] Reproduced from ref. 6 with permission from Stanford University.

items), and the other without (general argumentation items). Contexts used in the second group were drawn from experiences familiar to eighth graders. For example, pupils used given information in an item to argue whether schools should adopt a new lunch programme advocated by the federal government. See Figure 6.4 for two scientific argumentation items from that study.

Analysed data showed that item difficulties from both groups agreed with the hypothesised learning progression, providing support for the three levels. Interestingly, the researchers also found general argumentation items to be easier than scientific argumentation items, even when both assessed the same argumentation skill. They reported that Level 2 tasks proved difficult for their sample of eighth graders, regardless of which group the item came from. Unsurprisingly, follow-up analyses using similar Level 2 tasks with a different sample of eighth and tenth graders found the older group to be more proficient.

6.6 Implications for the Conceptualisation and Design of Argumentation Items

The three learning progression models presented in the preceding section have distinctive conceptual frameworks and investigative research approaches (*e.g.*, grain-size and content tested). Each model and its corresponding validation efforts contributed unique and notable insights toward the item design process proposed later in this chapter. This section discusses three key insights – context-setting, the scientific content and the argumentation skill to be assessed.

Manipulating the context of an item is a common method employed by science item writers to assess if test-takers can apply the knowledge in different situations.[70] Item writers also hope to motivate test-takers by making abstract concepts concrete and familiar (*e.g.*, by setting items in real-world scenarios). The work of Berland and McNeill highlighted three important features about the context of argumentation items to be considered: the range of expected plausible answers; the nature and scope of data presented in the items which test-takers must analyse; and the presence of appropriate instructional scaffolds (*e.g.*, design of tasks will differ if peer-supported or facilitated in an online environment).[67] Drawing on authentic data from real-world contexts was a common feature found among the three studies. However, some of the items in H.-S. Lee's study contained novel contexts which were not within pupils' typical experiences. Arguably, the unfamilar contexts could have impacted upon performance.

Regarding the science content to be tested, H.-S. Lee *et al.* recommended that "more defined science topics…may show different relationships among the four argument element item types from those we discovered in this study" (p. 602).[69] Their conclusion underscored the importance of pupils' familiarity and ease with the scientific content assessed in scientific argumentation tasks and echoed similar recommendations from the other two studies. Possessing stronger understanding of scientific concepts and principles helped pupils perform better in argumentation.[68] More useful for item-setting is the distinction Osborne *et al.* made regarding the three types of science domain knowledge – content, procedural, and epistemic. Broadly, these knowledge types seek to describe and account for phenomena in the material realm, investigate the causes of these phenomena and validate the knowledge obtained respectively.[50] Unfortunately, although Osborne *et al.* distinguished between the types of science knowledge, their study did not investigate how each impacted upon performance. Collectively, the work of H.-S. Lee *et al.* and Osborne *et al.* suggest that argumentation items could assess the plausibility of conclusions or claims from these three scientific bodies of domain knowledge.

Finally, in terms of the argumentation skills to be assessed, the three learning progressions shared many commonalities, with each also contributing unique insights. Besides similar references to *claims*, *evidence*, *rebuttals*, and *warrants*, the simplest and most complex skills identified across the three progressions generally corresponded: all began with the construction of a claim and culminated with the ability to rebut/evaluate/counter-propose claims. Differences lie with the specificity and sequencing of abilities. For example, Osborne *et al.*'s learning progression had the most levels. Overall, each progression contributed distinctive perspectives to the assessment of argumentation such as the notions of process and product assessment, inclusion of an "uncertainty rationale" in appraising pupil's abilities, and the facilities of construction and critique. The following section explains how these perspectives provide the cornerstones of an item design model.

6.6.1 Three Cornerstones of a Model for Designing Scientific Argumentation Items

The preceding discussion highlights the challenges involved in writing items to test argumentation abilities in science. Although there are well established practices for item development in traditional content assessment, there is limited knowledge about how items can be written to elicit desired higher-order reasoning such as scientific argumentation.[6,71] Theoretically-informed guidelines for item generation provide a sound basis for controlling item indices such as difficulty and discrimination. Such guidelines also facilitate an understanding and identification of key item features (*e.g.*, item format) that provide evidence of knowledge and reasoning attainment.[72]

From an item writing perspective, the first step is to understand the knowledge and skills to be assessed. Next, it is necessary to define clear behavioural descriptors of pupils' performance as evidence of attainment.[73] The three learning progressions discussed earlier provide rich specifics about the argumentation abilities appropriate for primary and secondary pupils. It is also evident from research that pupils' grasp of the scientific concepts is critical if they are to be successful in argumentation. Recent research identifies scientific knowledge as composed of content, procedural, and epistemic domains.[51,74] Clarity about the content and reasoning to be elicited by test items provide an initial basis for item writers to focus on. However, to be able to make strong inferences about pupils' abilities, the anatomical features of the items must also be considered.[75] There should be careful selection of item features to promote demonstration of complex skills such as argumentation and reasoning. For instance, cognitive science research shows that item formats that allow for extended writing or articulation of reasoning have high fidelity to problem solving, critiquing and critical thinking abilities.[76] Findings from the learning progression research also foreground the importance of choosing scenarios appropriate for the age, typical development and experiential levels of respondents. Other item features to consider include text length, linguistic elements and visual representation.[77–79]

In summary, writing an argumentation-focused science item must factor the targeted argumentation skill, the curriculum's learning objective, and the anatomical features of the item. Together, these three item attributes form the cornerstones of the item conceptualisation and design model shown in Figure 6.5.

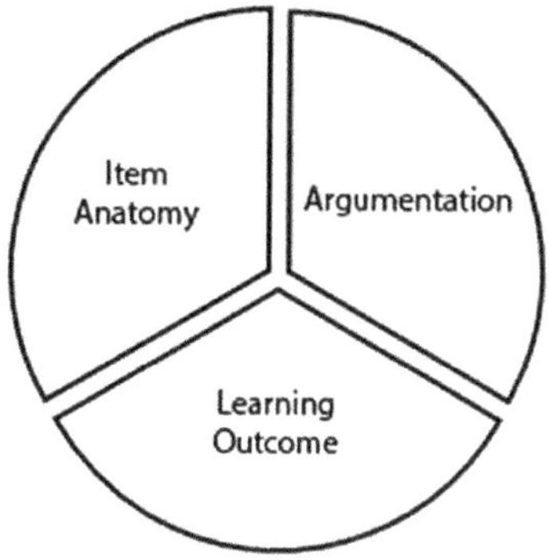

Figure 6.5 Three attributes for the conceptualisation and design of scientific argumentation items.

6.6.2 Deployment of Model for Item-setting in Chemistry

The model suggests that testing outcomes (*e.g.*, item difficulty) are a combined function of the three item attributes. Significantly, the model also indicates the approach for *controlling* some of these outcomes as more items are designed. A writer begins the construction of an item by deciding the assessment objectives (learning outcome and argumentation) and the anatomy of the item. The item's cognitive demand can be gauged using findings from current research when these decisions are made. For instance, requiring pupils to argue with anomalous data will be more challenging than making a claim with straightforward patterns in data.[80] Writers can develop an overall sense of the cognitive demand by simultaneously attending to the three item attributes. It also means that by selecting aspects of these attributes to vary, a writer can better approximate an item's performance (*e.g.*, difficulty). For a more systematic approach, the writer can select one of the three attributes to manipulate while holding the remaining constant in the design of similar items. Variations in responses thus reflect the manipulated attribute.

To illustrate the deployment of this approach, this section presents original or adapted item exemplars from two of the discussed learning progression studies (see Figures 6.3 and 6.4). All item exemplars assessed content from the molecular theory of matter, a chemistry topic commonly taught to primary and early secondary pupils. Concepts in the topic are of interest to researchers and educators as they form a foundation for understanding more advanced knowledge.[16] These fundamental concepts include properties of objects, changes of state, and evidence for the particulate structure of matter.[81] Table 6.5 provides explanations of how the exemplars (Boxes 6.1–6.5) illustrate the systematic manipulation of each of the three attributes. Two other exemplars in the Practical Digest (Boxes 6.6 and 6.7) illustrate the use of the model with item modification and scoring.

Table 6.5 Manipulation of item attribute illustrated in Boxes 6.1 to 6.5 (listed on the next few pages).

Attribute manipulated (unshaded sector)	Exemplar and explanation
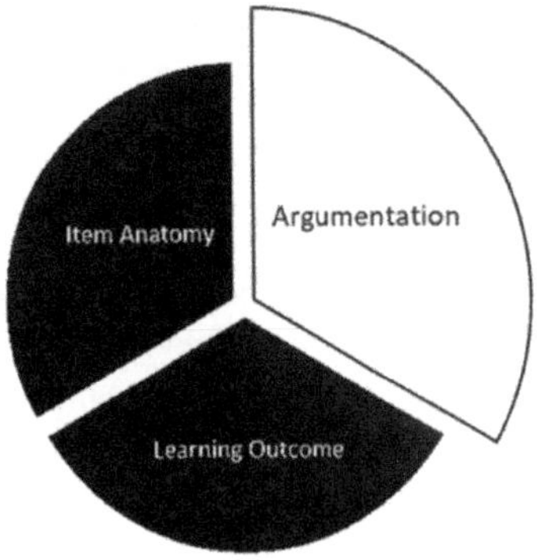	Boxes 6.1 and 6.2 illustrate how the assessment of argumentation can be executed during item design. The targeted argumentation abilities between both exemplars are deliberately different, while the tested scientific content knowledge and broad item features such as format and real-world contexts are kept similar. Findings from the learning progression research suggest that the argumentation abilities assessed in Box 6.1 are less challenging than those in Box 6.2.
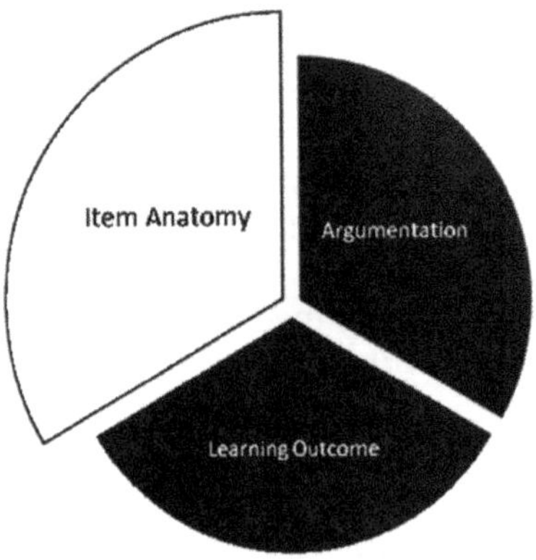	Boxes 6.3 and 6.4 assess similar content knowledge and argumentation abilities. However, adjustments to the visual representation between both exemplars resulted in differing demands. The provision of salient pictorial clues in Box 6.4 for the activation of relevant schemas and prior knowledge is likely to reduce the cognitive load imposed on respondents.[83] This postulation of an easier demand has support from similar findings in cognitive science research.
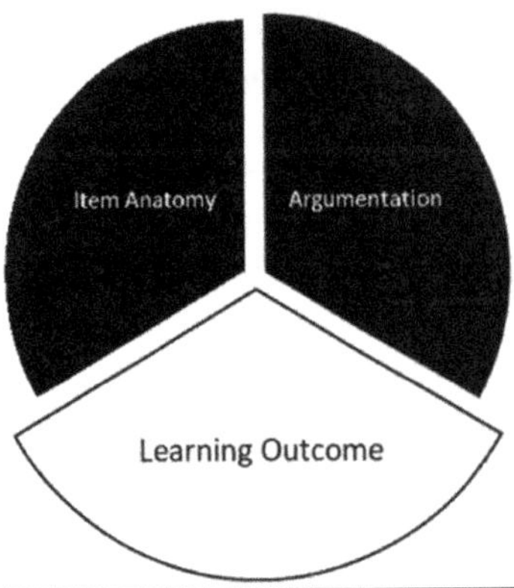	Box 6.5 contains three items, each of which assessed a different body of science knowledge. The items have constructed-response formats, share a common item context and assess the construction of claims. Findings from science education and nature of science literatures suggest that epistemic and procedural knowledge are advanced forms of understanding as compared to content knowledge.[84,85] Items (b) and (c) are thus postulated to be more difficult than (a).

Box 6.1 Construction of an evidence-based claim and argument.

Adapted from a scientific argumentation task found in an online earth science module created by Pallant, Pryputniewicz, and Lee.[86] Image adapted with permission from the Concord Consortium. Map reproduced from https://en.wikipedia.org/wiki/Colorado_River#/media/File:Colorado_River_basin_map.png, image credit Shannon1, under the Creative Commons Attribution-Share Alike 4.0 International license, https://creativecommons.org/licenses/by-sa/4.0/. Graph reproduced from ref. 87.

The map shows the location of a dam built in 1964 across a river. Study the map.

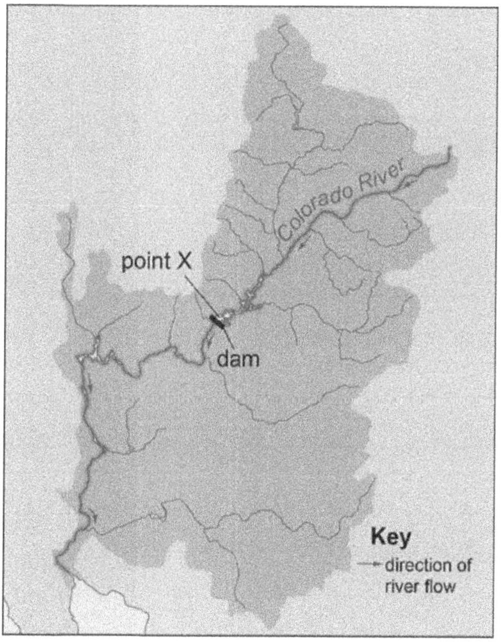

The flow rate of river water changed when it was dammed. The graph below shows the average flow rate of the river at point X in the months of

1942 and 1996. The lower the rate, the longer water stayed at a place. Study the graph.

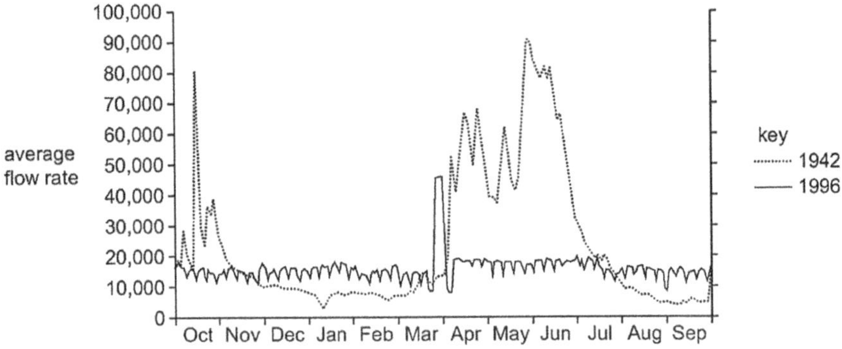

The salinity (amount of salt) in the river changed when it was dammed. The salts in the rocks next to a river could dissolve into the water.

Based on the map and graph, when was the salinity of the water at point X the highest?

Month:

Year:

Give a reason for your answer.

Box 6.2 Construction of counter-argument based on new evidence.

Adapted from a scientific argumentation task found in an online earth science module created by Pallant, Pryputniewicz, and Lee.[86] Image adapted with permission from the Concord Consortium. Map reproduced from https://en.wikipedia.org/wiki/Colorado_River#/media/File:Colorado_River_basin_map.png, image credit Shannon1, under the Creative Commons Attribution-Share Alike 4.0 International license, https://creativecommons.org/licenses/by-sa/4.0/. Graph reproduced from ref. 87.

The map shows the location of a dam built in 1964 across a river. Study the map.

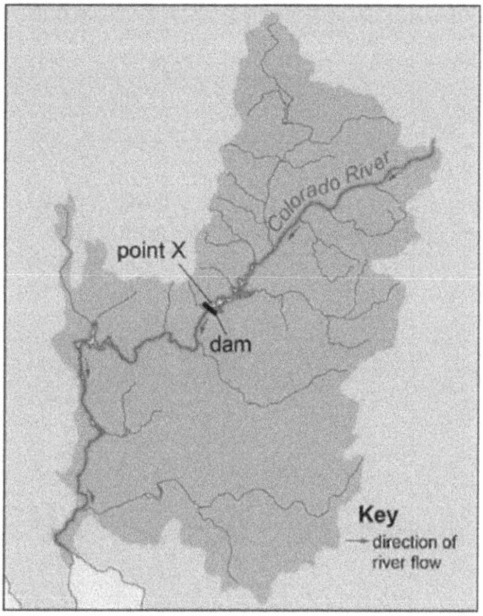

The salinity (amount of salt) in the river changed when it was dammed. The salts in the rocks of a place could dissolve into the water. John studied the following graph of the average flow rate of the river at point X in the months of 1942 and 1996. The lower the rate, the longer water stayed at a place.

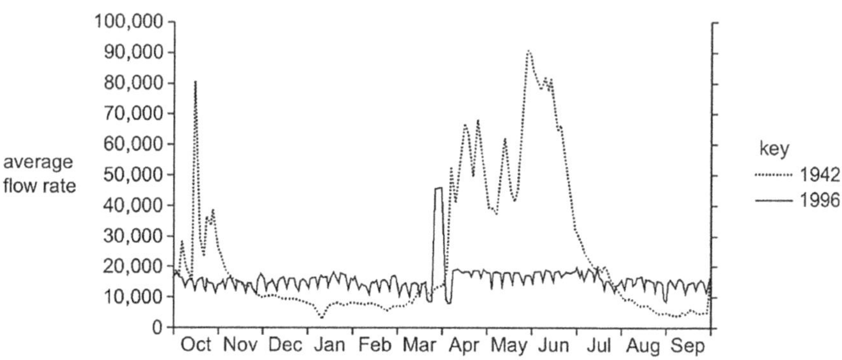

John concluded:

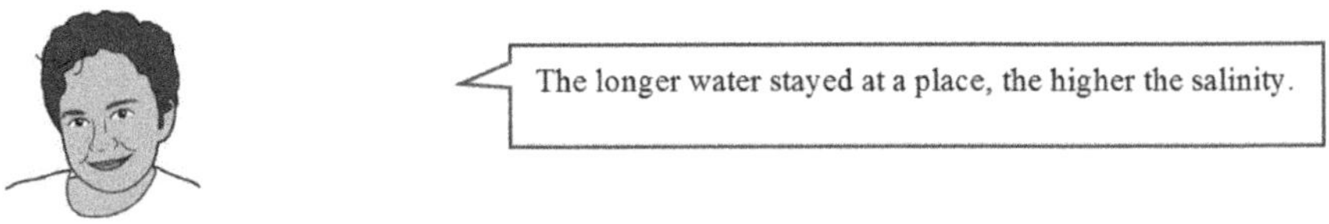

(a) Based on the graph and **John's** conclusion, when was the salinity of the water at point X the highest?

Month:

Year:

Rory disagreed with John's conclusion. The diagram shows the results of an experiment he conducted:

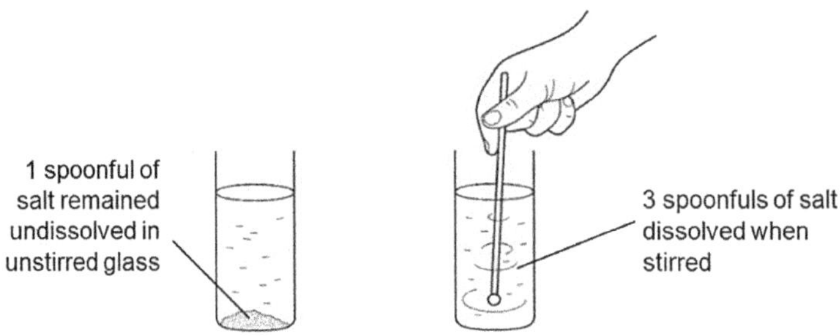

Rory concluded:

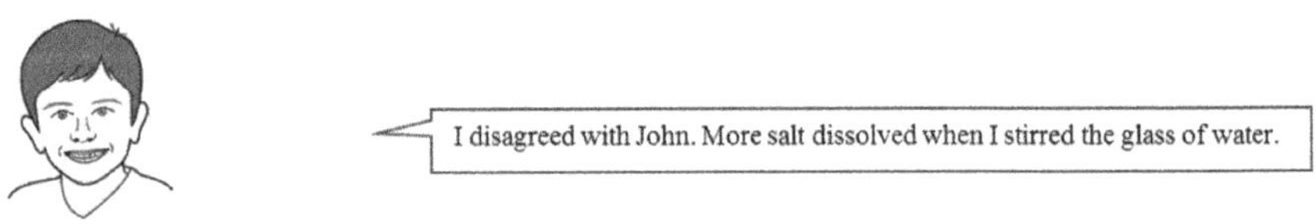

(b) Based on the graph and **Rory's** conclusion, when was the salinity of the water at point X the highest?

Month:

Year:

(c) Do you agree with John or Rory? Give your reason(s).

Box 6.3 Evaluation of evidence-based claims/explanatory hypotheses.

In the set-up of an experiment below, Rory and Alice saw water droplets formed on two covered beakers of water as shown below.

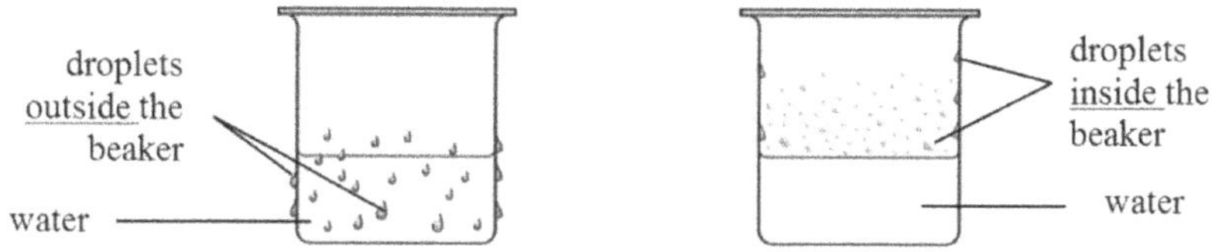

Rory said:

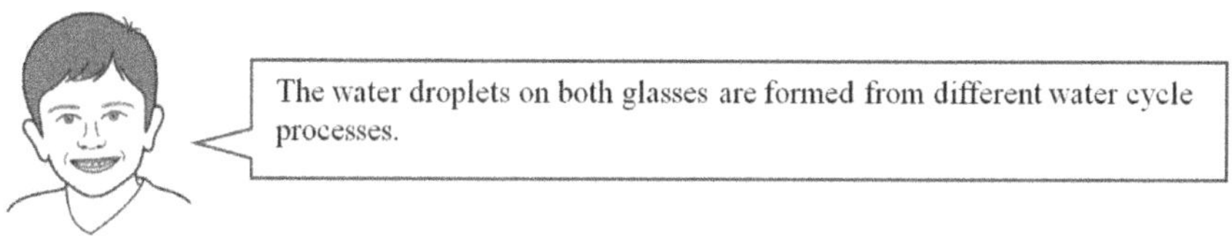

Alice said:

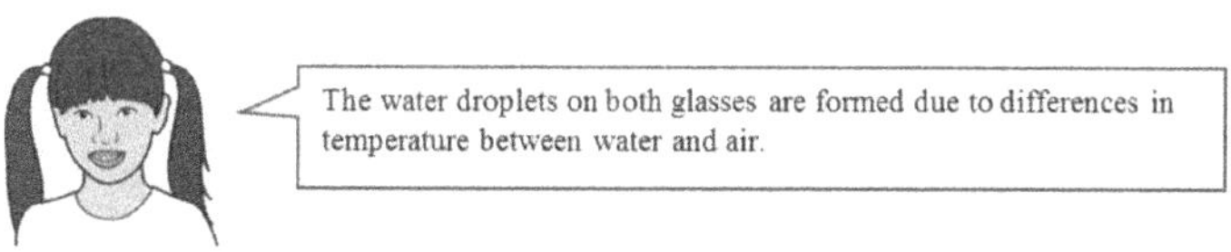

I agree/disagree with Rory. My reason(s) are

____

I agree/disagree with Alice. My reason(s) are

____

Box 6.4 Evaluation of evidence-based claims/explanatory hypotheses.

Same question as Box 6.3 but replaced with the following picture of the beakers.

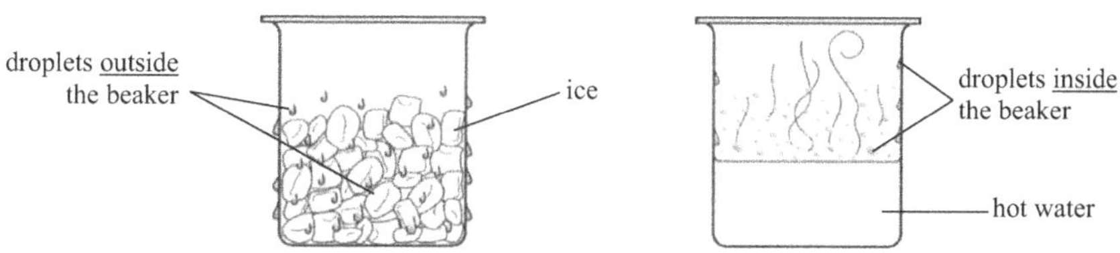

Box 6.5 Construction of evidenced-based procedural, content, and epistemic claims in (a), (b) and (c).

Betty had some marbles. She used two different methods, A and B, to find the volume of one of the marbles. She started with the equipment shown below.

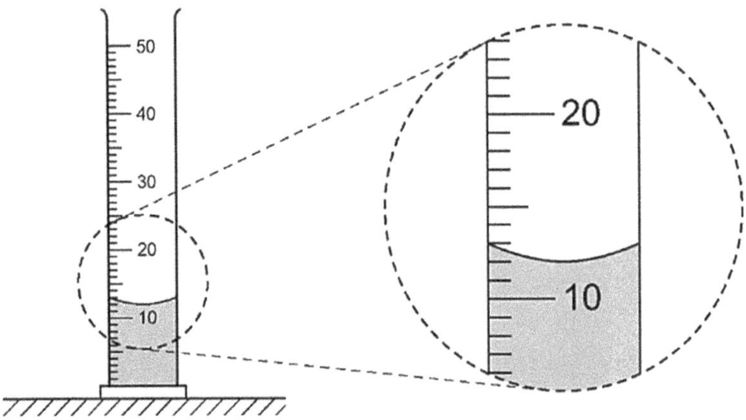

(a) What was the volume of water in the equipment?

Method A
Betty next put a marble into the equipment to find its volume. Her result was shown below.

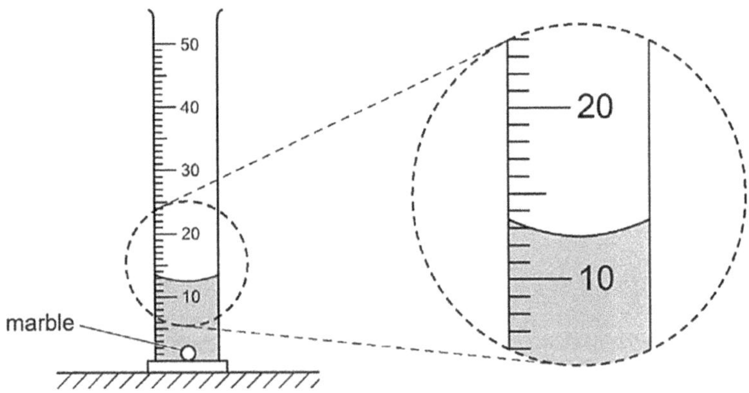

(b) State a property of the marble which allowed her to find its volume using this method.

Method B

Betty then put ten marbles in the equipment to find the volume of one of the marbles. Her result was shown below.

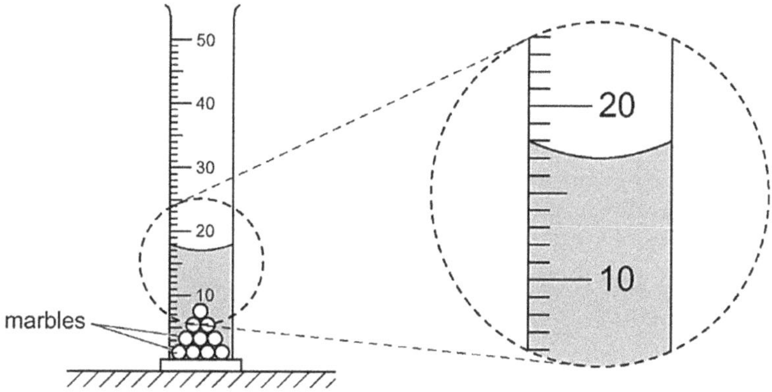

(c) Which method, A or B, produced a more accurate result? Explain your answer.

6.7 Limitations and Future Directions

This chapter analyses English and Australian primary science curricular documents to understand how these curricula addressed the assessment of argumentation skills currently. It also reports the findings of three pieces of learning progression research on scientific argumentation. A model of item design is put forth based on a synthesis of this research, and comprises three attributes of item anatomy, learning outcome, and argumentation abilities. Five item exemplars help demonstrate how manipulation of the attributes guides the writing of items.

The main limitation of this conceptual model is that it remains untested empirically. Moreover, there is little research on which anatomical features of an item are pivotal in drawing out the relevant scientific argumentation abilities. For instance, one item feature which remains ambiguous is item context. Understanding how context interacts with other factors is crucial for addressing pupils' performance in argumentation. These factors include what Osborne *et al.* rightly alerted as "cultural and socio-emotional" (p. 840) elements which alter pupils' perception of the items' contextual information.[6] A third limitation is the uncertainty over the sequence and approach of introducing argumentation notions for elementary and secondary school children as there are no clear findings from the validation attempts of two of the learning progressions. Unanswered questions relate back to crucial teaching and learning issues such as *when* (*e.g.*, which age group?), *what* (*e.g.*, what skills to teach in tandem with the curriculum?), and *how* (*e.g.*, what appropriate teaching approaches to adopt?).

Insights from research investigating the assessment of the three bodies of science domain knowledge are just emerging. How these bodies of science knowledge can be assessed argumentatively is currently unclear. This chapter presents some initial ideas using chemistry knowledge; it will be fruitful to explore possibilities in other disciplines. These suggested investigations are opportunities for exploring the ideal depth and grain-level of curricular involvement. Future work addressing gaps and potential, such as the refinement of learning progressions as well as empirical investigations of this proposed item design model, can increase understanding about the assessment of argumentation abilities within and across grade levels.

Practical Digest

Examples (Boxes 6.6 and 6.7) illustrating the modification and scoring of items according to the model. Contains public sector information licensed under the Open Government Licence v3.0, https://assets.publishing.service.gov.uk/government/uploads/system/uploads/attachment_data/file/648195/2016sciencesamplepublicrelease_Test_Booklet_V4.pdf.

Box 6.6

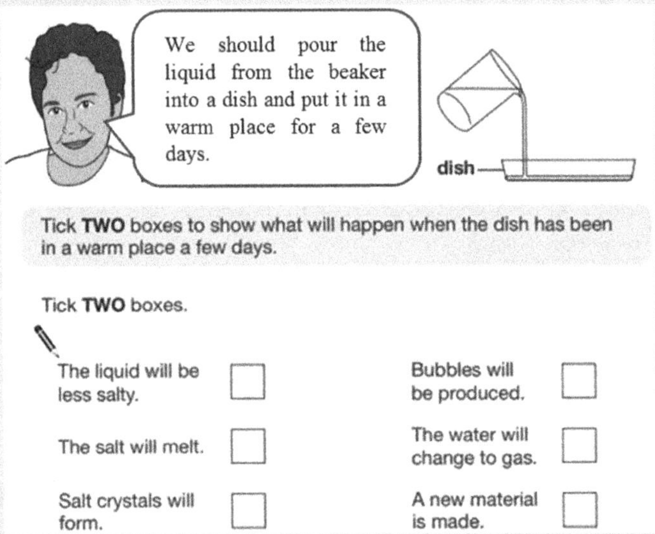

Figure 6.6 Example of a released item from the *Key Stage 2 2016 science sampling: Test booklet* (p. 40), by the Standards and Testing Agency, 2017, London: Author. Contains materials developed by STA for national curriculum assessments and licensed under Open Government Licence v. 3.0. Adapted with permission.

Annotations:
This released item is part of a four-item question and the liquid referred to contained dissolved salt. The selected-response item format facilitates claim-making (*i.e.*, what will happen to the liquid in the dish).

There are various ways to modify this item based on the item-design model to assess the argumentation ability required for this Key Stage. For example, one way to draw out the ability to support an argument with evidence is to make changes to the item anatomy by including a follow-up constructed-response item asking if the dish had been placed in a cold place instead, will the pupils select the same two boxes and give their reasons.

Box 6.7

Rory placed a metal saucepan outside his house on a cold night and made a prediction.

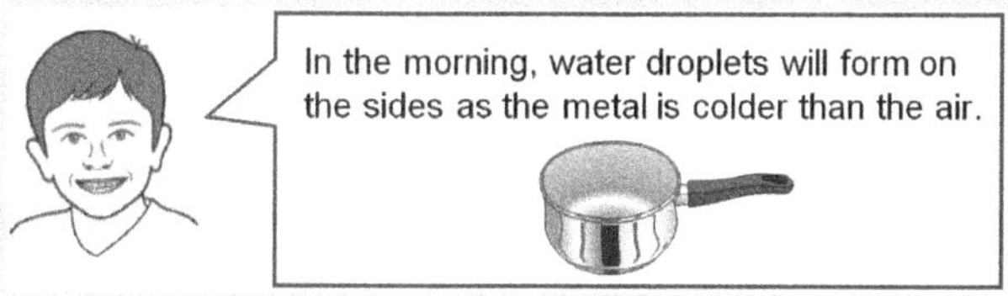

Do you agree with John? Give reason(s) for your answer.

Annotations:
This exemplar assesses pupils' ability to evaluate a claim using their prior content knowledge. From the item design model, scoring of the responses focuses on the accuracy and the depth of the content knowledge required as the claim is evaluated. Depending on the testing purpose and classroom implementation, acceptable responses range from identification of the water cycle process, and/or explanatory accounts of how differential temperatures affect the process, or even the property of materials when cooled/heated.

References

1. F. Fischer, I. Kollar, S. Ufer, B. Sodian, H. Hussmann, R. Pekrun, B. Neuhaus, B. Dorner, S. Pankofer, M. R. Fischer, J.-W. Strijbos, M. Heene and J. Eberle, *Frontline Learn. Res.*, 2014, **4**, 28–45.
2. T.-C. Lin, T.-J. Lin and C.-C. Tsai, *Int. J. Sci. Educ.*, 2014, **36**, 1346–1372.
3. S. Erduran, Y. Ozdem and J.-Y. Park, *Int. J. STEM Educ.*, 2015, **2**, 5.
4. D. B. Clark and V. Sampson, *J. Res. Sci. Teach.*, 2008, **45**, 293–321.

5. V. Sampson and D. B. Clark, *Sci. Educ.*, 2008, **92**, 447–472.
6. J. Osborne, J. B. Henderson, A. MacPherson, E. Szu, A. Wild and S.-Y. Yao, *J. Res. Sci. Teach.*, 2016, **53**, 821–846.
7. O. Lee, *Educ. Res.*, 2017, **46**, 90–102.
8. C. Rapanta, M. Garcia-Mila and S. Gilabert, *Rev. Educ. Res.*, 2013, **83**, 483–520.
9. S. Erduran, S. Simon and J. Osborne, *Sci. Educ.*, 2004, **88**, 915–933.
10. O. Lee and J. Krajcik, *J. Res. Sci. Teach.*, 2012, **49**, 271–280.
11. National Research Council, *Taking Science to School: Learning and Teaching Science in Grades K-8*, The National Academies Press, Washington, D.C., 2007.
12. C. Zembal-Saul, *Sci. Educ.*, 2009, **93**, 687–719.
13. B. A. Brown and E. Spang, *Sci. Educ.*, 2008, **92**, 708–732.
14. K. L. McNeill, *J. Res. Sci. Teach.*, 2011, **48**, 793–823.
15. National Research Council, in *Taking Science to School: Learning and Teaching Science in Grade K-8*, ed. R. A. Duschl, H. A. Schweingruber and A. W. Shouse, The National Academies Press, Washington, D.C., 2007, pp. 213–250.
16. T. Corcoran, F. A. Mosher and A. Rogat, *Learning progressions in science: An evidence-based approach to reform*, Report CPRE Research Report # RR-63, Center on Continuous Instructional Improvement, 2009.
17. L. Mohan and J. Plummer, in *Learning Progressions in Science: Current Challenges and Future Directions*, ed. A. C. Alonzo and A. W. Gotwals, Sense Publishers, Rotterdam, 2012, pp. 139–147.
18. S. E. Toulmin, *The Uses of Argument*, Cambridge University Press, 1958.
19. M. P. Jiménez-Aleixandre and S. Erduran, in *Argumentation in Science Eucation: Perspectives from Classroom-based Research*, ed. S. Erduran and M. P. Jiménez-Aleixandre, Springer, Dordrecht, 2007, pp. 3–25.
20. R. Driver, P. Newton and J. Osborne, *Sci. Educ.*, 2000, **84**, 287.
21. R. Duschl, *Rev. Res. Educ.*, 2008, **32**, 268–291.
22. J. Osborne, S. Simon, A. Christodoulou, C. Howell-Richardson and K. Richardson, *J. Res. Sci. Teach.*, 2013, **50**, 315–347.
23. R. Duschl and J. Osborne, *Stud. Sci. Educ.*, 2002, **38**, 39–72.
24. M. Garcia-Mila and C. Anderson, *Argument. Sci. Educ.: Perspect. Classroom-Based Res.*, 2008, 29–45.
25. W. A. Sandoval and B. J. Reiser, *Sci. Educ.*, 2004, **88**, 345–372.
26. P. Newton, R. Driver and J. Osborne, *Int. J. Sci. Educ.*, 1999, **21**, 553–576.
27. D. Kuhn and S. Franklin, *Handb. Child Psychol.*, *2*, 2006, DOI: 10.1002/9780470147658.chpsy0222.
28. H. Siegel, *Informal Logic*, 1995, **17**, 159–176.
29. J. A. Nielsen, *Res. Sci. Educ.*, 2013, **43**, 371–393.
30. P. C. Cardoso Mendonça and R. Justi, *Int. J. Sci. Educ.*, 2013, **35**, 2407–2434.
31. K. McCain, *Sci. Educ.*, 2015, **24**, 827–854.
32. F. H. v. Eemeren and R. Grootendorst, *A Systematic Theory of Argumentation: The Pragma-dialectical Approach*, Cambridge University Press, New York, 2004.

33. D. Kuhn and W. Udell, *Child Dev.*, 2003, **74**, 1245–1260.
34. L. A. Bricker and P. Bell, *Sci. Educ.*, 2008, **92**, 473–498.
35. G. J. Kelly, S. McDonald and P.-O. Wickman, in *Second International Handbook of Science Education*, ed. B. J. Fraser, K. G. Tobin and C. J. McRobbie, Springer, Dordrecht, 2012, pp. 281–291.
36. M. A. Finocchiaro, Arguments about arguments, *Systematic, Critical and Historical Essays in Logical Theory*, Cambridge University Press, New York, 2005.
37. C. Pontecorvo and S. Pirchio, *Hum. Dev.*, 2000, **43**, 361–363.
38. I. V. S. Mullis, M. O. Martin, S. Goh and K. Cotter, *TIMSS 2015 Encyclopedia: Education policy and curriculum in mathematics and science*, Boston College, TIMSS & PIRLS International Study Center, 2016.
39. J. Woods, Assessments 101: A policymaker's guide to K-12 assessments, Education Commission of the States, Denver, CO, 2017.
40. Ministry of Education, The New Zealand Curriculum, Report 978 0 7903 264 7, Author, Wellington, 2007.
41. Australian Curriculum and Assessment Reporting Authority [ACARA], The Australian Curriculum: Science, 2016.
42. Australian Curriculum and Assessment Reporting Authority [ACARA], Work samples, https://www.australiancurriculum.edu.au/resources/work-samples/, (accessed 23 October 2017, 2017).
43. J. S. Bruner, *The Process of Education*, Vintage, New York, 1960.
44. T. P. Carpenter, E. Fennema and M. L. Franke, *Element. School J.*, 1996, **97**, 3–20.
45. R. G. Duncan and C. E. Hmelo-Silver, *J. Res. Sci. Teach.*, 2009, **46**, 606–609.
46. Department for Education [DfE], National curriculum in England: science programmes of study, https://www.gov.uk/government/publications/national-curriculum-in-england-science-programmes-of-study/national-curriculum-in-england-science-programmes-of-study#contents, (accessed 22 Oct 2017).
47. Department for Education [DfE], *The national curriculum in England: Framework document*, 2014.
48. Department for Education [DfE], *The national curriculum in England: Key stages 1 and 2 framework document*, 2013.
49. W. Harlen and A. Qualter, *The Teaching of Science in Primary Schools*, Routledge, London, 6 edn, 2014.
50. P. Kind and J. Osborne, *Sci. Educ.*, 2016, 8–31, DOI: 10.1002/sce.21251.
51. J. Osborne, *Thinking Skills Creativity*, 2013, **10**, 265–279.
52. DfES/QCA (Department for Education and Skills/Qualifications and Curriculum Authority), *Science: the National Curriculum for England*, 2004.
53. B. L. La Velle and S. Erduran, *School Sci. Rev.*, 2007, **88**, 31–40.
54. M. Richardson and T. Isaacs, in *TIMSS 2015 Encyclopedia: Education Policy and Curriculum in Mathematics and Science*, ed. I. V. S. Mullis, M. O. Martin, S. Goh and K. Cotter, Boston College, TIMSS & PIRLS International Study Center, 2016.

55. N. Wernert and S. Thomson, in *TIMSS 2015 Encyclopedia: Education Policy and Curriculum in Mathematics and Science*, ed. I. V. S. Mullis, M. O. Martin, S. Goh and K. Cotter, Boston College, TIMSS & PIRLS International Study Center, 2016.
56. J. Osborne and A. Patterson, *Sci. Educ.*, 2011, **95**, 627–638.
57. Standards and Testing Agency, Guidance: Science sampling arrangements, https://www.gov.uk/government/publications/key-stage-2-science-sampling-tests/science-sampling-arrangements, (accessed 23 October 2017, 2017).
58. Standards and Testing Agency, 2016 key stage 2 science sampling: sample questions, mark scheme and commentary, https://www.gov.uk/government/publications/2016-key-stage-2-science-sampling-sample-questions-mark-scheme-and-commentary, (accessed 23 October 2017, 2017).
59. Standards and Testing Agency, Sample test material for 2016 national curriculum tests published, https://www.gov.uk/government/news/sample-test-material-for-2016-national-curriculum-tests-published, (accessed 23 October 2017, 2017).
60. K. L. McNeill and J. Krajcik, *J. Learn. Sci.*, 2009, **18**, 416–460.
61. A. W. Gotwals and A. C. Alonzo, in *Learning Progressions in Science: Current Challenges and Future Directions*, ed. A. C. Alonzo and A. W. Gotwals, Sense Publishers, Rotterdam, 2012.
62. M. Forster and G. Masters, in *International Encyclopedia of Education*, ed. P. Peterson, E. Baker and B. McGaw, Academic Press, Oxford, UK, 3rd edn, 2010.
63. K. Rowe, presented in part at the NSW DET K-4 Early Years of Schooling Conference, Sydney, 2006.
64. J. W. Pellegrino, L. V. DiBello and S. R. Goldman, *Educ. Psychol.*, 2016, **51**, 59–81.
65. P. Deane and Y. Song, *The key practices, discuss and debate ideas: Conceptual framework, literature review, and provisional learning progressions for argumentation*, Report ETS RR–15-33, ETS, Princeton, NJ, 2015.
66. A. C. Alonzo and J. T. Steedle, *Sci. Educ.*, 2009, **93**, 389–421.
67. L. K. Berland and K. L. McNeill, *Sci. Educ.*, 2010, **94**, 765–793.
68. K. L. McNeill, D. J. Lizotte, J. Krajcik and R. W. Marx, *J. Learn. Sci.*, 2006, **15**, 153–191.
69. H.-S. Lee, O. L. Liu, A. Pallant, K. C. Roohr, S. Pryputniewicz and Z. E. Buck, *J. Res. Sci. Teach.*, 2014, **51**, 581–605.
70. A. Ahmed and A. Pollitt, presented in part at the BERA Conference, Leeds, 2001.
71. S.-Y. Yao, *Doctor of Philosophy*, University of California, 2013.
72. T. M. Haladyna, *Automatic Item Generation: Theory and Practice*, Routledge, New York, 2013.
73. T. M. Haladyna and S. M. Downing, *Educ. Meas.: Issues and Pract.*, 2004, **23**, 17–27.
74. P. Kind, *J. Res. Sci. Teach.*, 2013, **50**, 530–560.

75. J. S. Gorin and S. E. Embretson, *Automatic Item Generation: Theory and Practice*, Routledge, New York, NY, 2013, ch. 9.
76. T. M. Haladyna and M. C. Rodriguez, *Developing and Validating Test Items*, Routledge, New York, 2013.
77. J. Stiller, S. Hartmann, S. Mathesius, P. Straube, R. Tiemann, V. Nordmeier, D. Krüger and A. U. Z. Belzen, *Assess. Evaluation Higher Educ.*, 2016, **41**, 721–732.
78. V. Crisp and E. Sweiry, *Educ. Res.*, 2006, **48**, 139–154.
79. A. Ahmed and A. Pollitt, *Assess. Educ.: Princ., Policy Pract.*, 2007, **14**, 201–232.
80. C. A. Chinn and W. F. Brewer, *Rev. Educ. Res.*, 1993, **63**, 1–49.
81. P. Black, M. Wilson and S.-Y. Yao, *Meas.: Interdiscip. Res. Perspect.*, 2011, **9**, 71–123.
82. K. Smith, *Working scientifically: A Guide for Primary Science Teachers*, Routledge, Abingdon, Oxford, 2016.
83. M. P. Cook, *Sci. Educ.*, 2006, **90**, 1073–1091.
84. W. A. Sandoval and A. Çam, *Sci. Educ.*, 2011, **95**, 383–408.
85. C. Thoermer and B. Sodian, *New Ideas Psychol.*, 2002, **20**, 263–283.
86. A. Pallant, S. Pryputniewicz and H.-S. Lee, High-Adventure Science: Will there be enough fresh water?, https://authoring.concord.org/sequences/387).
87. Daily mean discharge at Colorado River at Lee's Ferry, Arizona, 1942 and 1996 water years. U.S. Geological Survey, Department of the Interior/USGS, https://pubs.usgs.gov/fs/fs-014-00/.

CHAPTER 7

Research and Practice on Science Teachers' Continuous Professional Development in Argumentation

CARLA ZEMBAL-SAUL* AND ABHA VAISHAMPAYAN

The Pennsylvania State University, Department of Curriculum and Instruction, University Park, State College, PA 16801, USA
*Email: czem@psu.edu

7.1 Introduction

In recent decades, argumentation has been established as an important feature of school science given that it reflects the social, discursive, and dialogic practices of scientists.[1] More recently, advances in learning sciences highlight the role of argumentation as a fundamental scientific discourse that contributes to students' learning of disciplinary core ideas in deep and meaningful ways, as well as learning about how science is done and scientific knowledge is created.[2] Argumentation, however, remains elusive in school science.[3] In general, science teachers, and especially primary school teachers, have not experienced learning science in ways that move beyond an emphasis on "final form science and textbook-driven instruction" (p. 13).[3] How, then, do we support their learning of ambitious pedagogical practices for argumentation across the professional continuum – from teacher education candidates, to early career teachers, to experienced science teachers?

Advances in Chemistry Education Series No. 2
Argumentation in Chemistry Education: Research, Policy and Practice
Edited by Sibel Erduran

Published by the Royal Society of Chemistry, www.rsc.org

In the United States, the publication of the National Academies of Sciences, Engineering and Medicine (NASEM) report in 2015, *Science Teachers' Learning: Enhancing Opportunities, Creating Supportive Contexts*, recognizes the central role that teachers play in the consequential and equitable science learning of their students.[4] The consensus panel report is part of a larger reform initiative in science education that grew out of contemporary research on learning and learning progressions upon which the *Framework for K-12 Science Education*[5] and the *Next Generation Science Standards* (NGSS)[6] are based. More specifically, the *Framework* and associated standards portray a vision of student learning that foregrounds the social and epistemic practices of science and discourse for sensemaking in science. There has been an intentional effort to address teachers' continuous professional development (CPD) in ways that are aligned with this vision and that are job-embedded, coherent, and robust. Current research on science teachers' professional learning in argumentation has been informed by these contemporary science education reform efforts and *vice versa*.

The structure of this chapter reflects shifts in the landscape of science education research and practice that have had a substantial influence on teacher learning and development in argumentation. First, the manuscript addresses existing research on specialized knowledge and practices unique to science teaching. The question is raised about why argumentation is uncommon in school science when there has been significant attention to research and practice in this area for decades. Next, argumentation is considered in the context of current reform efforts in science education that foreground participation in scientific discourse and practices as a necessary condition for learning disciplinary core ideas. The demanding nature of supporting teacher development in argumentation is further exposed using conceptual, epistemic, and social lenses. Once the landscape is mapped for the reader, science teachers' learning needs are examined, followed by promising approaches to teacher learning and development – both at the program and systems levels. The chapter concludes with what the authors believe to be one of the most pressing considerations moving forward with the CPD of science teachers, that of equity in opportunity and access to high quality science learning experiences for every student.

Given the limited number of published studies on CPD in argumentation specific to Chemistry, this chapter will extend the review to science teacher professional learning in argumentation more generally. Additionally, professional development in argumentation will include attention to the broader framing of epistemic discourse and practices as they relate to sensemaking in science.

7.2 Teacher Professional Learning in Argumentation

A number of studies associated with teacher learning and development draw on pedagogical content knowledge (PCK) as a heuristic for program design, a conceptual framework for study design and data collection and analysis, and/or an explanatory framework; thus, it is important to pause and revisit the construct. In the mid-1980s, Shulman asserted that teaching for understanding is a complex endeavor that requires the transformation of knowledge from diverse domains – subject matter knowledge, general pedagogical knowledge, and knowledge of context.[7,8] Put another way, PCK is unique to teaching because it requires highly contextualized knowledge and practices necessary for making subject matter more comprehensible to others.[9] Deep subject knowledge is necessary but insufficient for effective teaching. Although teachers will never be able to develop all of the PCK they will need throughout their careers,[10] it is reasonable to support teachers in co-constructing conceptual frameworks that will scaffold their continued learning from practice.

If the focus of PCK is on the transformation of subject matter knowledge, where does argumentation as discourse genre fit? Schwab's representation of scientific knowledge as substantive and syntactic can be useful in addressing this question.[11] Substantive knowledge is what we typically think about as subject matter knowledge and includes central concepts and principles of a discipline and how they are organized. Historically, there has been a tendency in research and practice to focus on substantive knowledge at the cost of ignoring syntactic knowledge.[12] Syntactic knowledge refers to the cultural practices of science, including the "rules of the game" regarding how scientific knowledge is constructed, interrogated, and revised over time. Argumentation bridges the substantive and syntactic.[13] This notion is reflected in current reform initiatives that leverage research demonstrating how students learn core disciplinary ideas more meaningfully when simultaneously engaged in scientific practices.[3,5] However, the question of how to support teachers' development of knowledge and practices for engaging their students in scientific argumentation remains.

A decade ago, Erduran and Alexandre-Jiménez published a comprehensive volume on argumentation in school science.[14] Erduran's rationale for the essential role of argumentation as the primary discourse genre and epistemic reasoning tool for science learning and teaching is compelling and remains largely intact.[15] The educational context, however, continues to evolve, as do the pedagogical tools and resources used to promote and assess argumentation and the ways in which we study classroom discourse in science.

In the chapter on teacher professional development in argumentation, Zohar raised questions about what teachers need to know in order to implement argumentation in their classrooms.[16] She noted that the practices necessary to engage students in "high level discussions" are not scripted or embedded in curriculum. She stated, "In order to be able to respond to the unexpected events that characterize lessons that are rich in thinking and argumentation, teachers must be able to teach in an intelligent, flexible and resourceful way" (p. 246). Zohar highlighted the challenging move away from the teacher as the authority figure with the "right answers" to teaching higher order thinking and argumentation-a persistent tension between doing school and doing science.[17,18] Her review highlighted research from a series of professional development projects and teacher education programs that target teachers' PCK in argumentation.[7] Ultimately, Zohar advocated for the importance of attending to the nature of classroom discourse and the central role of scientific reasoning in learning, teaching, and teacher education.[16]

In the subsequent decade, scholars have actively contributed to the research base on teacher education and teachers' professional development in argumentation in K-16 science. Like Zohar, McNeill and Knight advocated for PCK in argumentation through their research and practice.[19] They specifically designed professional learning opportunities to support teachers' development of PCK in argumentation, which include features such as analyzing pre-existing records of classroom practice, supporting teachers in designing argumentation lessons, having teachers select and record classroom practice, and supporting teacher reflection for future practice. Findings from their research shed light on the challenges associated with integrating argumentation into school science. For example, they found that teachers struggled to analyze student talk both in terms of identifying the structural components of scientific argument and dialogic interactions. On the other hand, a majority of teachers found it easier to analyze students' writing by using the Claims–Evidence–Reasoning (CER)[20] framework, which is based on Toulmin's argument pattern.[21] In the context of school science, CER is generally accessible to learners and teachers alike.[13,22] The authors recommend that further research should focus more deeply on analyzing students' argumentation across multiple modalities, including talk and writing.[19]

Berland and McNeill developed a learning progression for scientific argumentation and studied how teachers learn to support their students' argumentation.[2] Additionally, McNeill and Krajcik investigated how practicing teachers support students' argumentation through instructional practices, such as operationalizing scientific explanation using the structure of argument, making the purpose of scientific explanation explicit, modeling argumentation and explanation construction, and connecting scientific explanations to real-life phenomena.[20]

In 2009, Zembal-Saul published the results of a longitudinal, design-based study funded by the National Science Foundation[†] known as *TESSA: Teaching Elementary School Science as Argument.*[13] In this research program, the structure of argument—evidence-based claims that are justified scientifically and collectively serve to explain phenomena—was intentionally employed as a scaffold for students' sensemaking. The resulting framework for Teaching Science as Argument (TSA) included engaging students in the language of science (*i.e.*, epistemic discourse) and public reasoning in a community of learners. TSA framing allowed for core instructional and

[†]Contract grant sponsor: National Science Foundation. Contract grant number: NSF REC 0237922. Any opinions, findings, and conclusions or recommendations expressed in this material are those of the authors and do not necessarily reflect the views of the National Science Foundation.

assessment practices to be identified, observed, rehearsed, and analyzed by preservice teachers in the program. In addition, the framework informed the design of online case-based resources to support teacher learning.

The TESSA project leveraged the use of the TSA framework across preservice elementary teachers' programs of study, which included specially designed undergraduate science content courses, a science teaching and learning course,[23,24] and field experiences with teachers from a professional development school partnership who were knowledgeable about the approach (see Practical Digest).[25] This design provided essential coherence across teacher preparation and field experience contexts. Moreover, the resulting research findings, which examined preservice teachers' developing knowledge and pedagogical practices associated with TSA, were promising. One of the key moves toward adopting aspects of the TSA framework occurred when preservice teachers shifted beyond a focus on 'superficial activities' to embrace (or even attempt) the practice of treating data as foundational for students to co-construct claims from evidence. *"What's your evidence?"* became a catchphrase for the work, and ultimately the title of a practitioner-based companion publication to TESSA research.[26]

Even with significant investments in promoting argumentation as a primary discourse genre and epistemic reasoning tool for science learning and teaching, argumentation in school science is rare.[3] Moreover, studies that attend to the nature, sources, and development of teachers' argumentation knowledge and practices are becoming less common in the literature. The following section addresses the evolving context of school science and how argumentation has been integrated into contemporary reform initiatives that emphasize scientific practices and their centrality to meaningful science learning.

7.3 Argumentation in the Context of Reform

It is impossible to address the professional learning needs of teachers without considering the role of argumentation in the context of current reform in science education. In the United States, contemporary reform initiatives are based on rigorous research syntheses of learning in science and science studies. Prominent among these is *Taking Science to School (TSS)*, in which there is strong evidence for the assertion that learners come to school with the cognitive resources necessary to engage in sophisticated forms of reasoning in science.[3,27] Subsequent studies of students' capabilities to engage in scientific reasoning, specifically in argumentation, continue to bear this out.[2,22,25,28] *TSS* boldly redefined proficiency in science as students being able to do the following (p. 36–37):

- know, use, and interpret scientific explanations of the natural world;
- generate and evaluate scientific evidence and explanations;
- understand the nature and development of scientific knowledge; and
- participate productively in scientific practices and discourse.

While students come to school ready and able to participate productively in science, educational systems (curriculum, instruction and assessment) are not prepared to meet their needs. Thus, the image of students' science learning that emerged from *TSS* remains elusive.

The *Framework for K-12 Science Education*[5] and *Next Generation Science Standards*,[6] which are described in more depth elsewhere in this volume, were informed by the aforementioned research syntheses. In contrast with the lockstep "scientific method" that is predominant in school science, scientific and engineering practices are intended to reflect the "spheres of activity of scientists and engineers"[5] – investigating real world phenomena (asking questions, making observations, and collecting data; evaluating across "spheres" (argument and critique); and developing explanations and solutions (models, theories, hypotheses) (p. 45). Argumentation is central to productive engagement in scientific practices in that it mediates investigation of the natural world and the construction of scientific explanations. In the *Framework* "engaging in argument from evidence," is defined as follows:

> The study of science and engineering should produce a sense of the process of argument necessary for advancing and defending a new idea or an explanation of a phenomenon and the norms for conducting such arguments. In that spirit, students should argue for the explanations they construct, defend their interpretations of the associated data, and advocate for the designs they propose[5] (p. 73).

When school science emphasizes the central practices of science—developing and using models, developing explanations, and engaging in argument from evidence—the rest of the scientific practices come into play in authentic and generative ways.

Nevertheless, the dilemma remains that most teachers of science have not experienced learning science in this way – engaging in scientific discourse practices to investigate and explain phenomena. For those who have had the personal experience of engaging in scientific practices to learn science more deeply, recall that it is necessary but insufficient for developing the PCK required to engage students in productive participation aligned with the contemporary vision of science learning. As noted previously, research on the nature, sources and development of science teachers' PCK is clear that subject matter knowledge must undergo significant transformation to be useful in supporting the learning of others.[9,10] This is likely similar when considering experience with scientific practices, such as arguing from evidence.[13]

It is also important to consider that reform in science education is not taking place in a vacuum, independent of other subject areas. In her *Educational Researcher* article, Okhee Lee highlights how "argument" has been used to address connections between *Common Core State Standards (CCSS) for English Language Arts* (ELA) and *Next Generation Science Standards* (NGSS) in the United State, especially in early grades.[29] Resources have been developed and used in professional development contexts, especially for elementary teachers, for the purpose of encouraging and advancing integration across subject areas (see the Venn diagram in NRC, p. 12).[4] Lee's critical examination of the content standards and associated literature reveals fundamental contradictions in argumentation as disciplinary norm, as well as developmental progressions in terms of when children are ready to engage in such practices. Of particular interest in her thesis is the past application of *opinion, persuasion, and argument* in the ELA literature (p. 92). She concurs with *Literacy for Science: Exploring the Intersection of the Next Generation Science Standards and Common Core for ELA Standards: A Workshop Summary*,[4] which concluded that "scholars across disciplines have yet to arrive at a common understanding of how ELA/literacy and science functionally interrelate with respect to argument" (p. 93).

Why should we care about the intersections and tensions surrounding ELA and *NGSS*? Science education has a long history of taking the backseat to literacy and mathematics instruction in elementary grades. As Lee points out, *NGSS* makes strategic and intentional connections between science and these subject areas.[29] Nevertheless, classroom teachers are ultimately responsible for enacting new standards. From an epistemic perspective, what counts as argument differs across disciplines. This leaves teachers to resolve complications of subject matter integration (if they are even aware of the disciplinary norms at play). For elementary teachers, in particular, it is critical that professional learning opportunities disrupt the notion that just because multiple reform documents refer to argument/argumentation, that epistemic distinctions are benign. Thus, designing and implementing effective and continuous professional learning opportunities for science teachers that address discourse practices, such as argumentation, is increasingly complex.

7.4 Argumentation as a Conceptual, Epistemic, and Social Practice

It is against the backdrop of reform that many of the intervention studies on teachers' CPD in argumentation have been integrated into the larger landscape of discourse practices for sensemaking in science. Put another way, reform initiatives are framing argumentation in ways that have a profound influence on science learning and what is necessary for teachers to know and be able to do to support that learning. Several contributions to the literature characterize these shifts – attention to domain-specificity in argumentation and epistemic discourse; scrutiny of argumentation as primary or sole scientific discourse genre; and science learning as social practice that takes place in a community of learners.

Erduran and others have problematized the nature of epistemic discourse by emphasizing the role of domain-specificity.[14] Erduran argued that students will come to understand, for example, what a chemist does only when they experience the modes of thinking and acting that are unique to chemistry as a distinct domain of science.[30] For example, students will be able to understand the periodicity in the properties of elements from the Periodic Law according to their atomic numbers. But *how* the Periodic Law came into being, or *why* this knowledge about chemical properties of elements counts as scientific knowledge, are critical questions and are very much a part of "doing science." Teaching about how this information came to be considered scientific knowledge should be an integral part of teaching science. Erduran stated, "When students are embedded in contexts where they can develop, evaluate, and revise chemical knowledge, they will act and think in ways that chemists act and think" (p. 249).[30] In this respect, argumentation can serve as a springboard for students to engage in practices of chemists by giving them opportunities to develop their ideas, investigate problems, and resolve conflicts surrounding validity of chemistry knowledge. While Erduran focused on argumentation, her primary goal for learning was to promote the investigation of phenomena in chemistry that are epistemically aligned and grounded within the discipline. In her work, argumentation is a lever to confront the traditional and prevalent approach of teaching science by providing procedural and declarative knowledge. Rather, the aim is to equip students to participate in holistic scientific inquiry that reflects the ways in which disciplinary experts, chemists in this case, solve problems.[30]

In terms of extending thinking about the nature and purpose of epistemic discourse in science learning, Mortimer and Scott explored interactive and dialogic communication in the service of "meaning making" in science.[31] Similarly, Kelly and McDonald applied a critical lens to argumentation as *the* epistemic discourse of science and explored potential contributions from a pedagogical standpoint.[32] They presented vignettes of scientific discourse from a chemistry class to emphasize the importance of scientific sensemaking and highlighted features that are distinct from argumentation. The authors asserted through their work that arguing to learn is essential for developing deep understanding of scientific knowledge, but that argumentation alone is insufficient for this purpose. Further, Kelly and McDonald expressed concern that a narrow focus on argumentation can be detrimental to other productive pedagogical practices for responsive science teaching, such as supporting students' thinking, developing their ideas about scientists' work, and developing norms of participation in classroom communities for science learning.[32] Their position took into account the productive discourse of scientists. They stated, "...the final form of argumentation, with explicitly stated evidence tailored to a professional audience, is just one discourse of science and is a particularly structured and formal genre of science discourse" (p. 269).[32] The authors' thesis was that emulating scientific practices in school science is difficult given that there are multiple ways in which productive scientific discourse and practices take place in scientific communities. Moreover, these discourses vary by audience (*e.g.*, scientific editors, colleagues), media of communication (*e.g.*, scientific journals, conference proceedings), and purposes (*e.g.*, thinking aloud to solve a problem, supporting a peer, defending a piece of information).

In recent years, research has shifted attention from a focus on individual students and their ability to argue from evidence to the ways in which classroom communities collectively construct explanations for phenomena by negotiating knowledge claims from evidence.[33,34] Recall that the aims of school science are distinct from, yet consistent with, that of scientists' science. Explicit attention to these distinctions may be useful in working with teachers as they retool their knowledge and pedagogical approaches for supporting students' productive participation in science. It seems particularly important in light of Lee's[29] caution about the superficial appearance of subject area complementarities around argumentation (ELA and science) and the negative impact of failing to attend to epistemic alignment.

Shifting instructional practices and assessment to discipline-based inquiry and engaging students in productive discourse and practices, such as argumentation, for the purpose of conceptual, epistemic and social aspects of learning science is a tall order.[35] To be clear, teachers are not to blame in this scenario – all are rendered novice in some way as we attend more closely to epistemic practices in school science. While the *Framework* and *NGSS* provide a vision for students' learning, where do we turn for "images of the possible"[36] of instructional practices that support this vision? How do we take teachers where they are and support their coherent development toward productive discourse and epistemic practices? What do or should next generation professional learning opportunities for science teachers look like?

As an example, consider the classroom vignette that follows. The second-grade teacher, Ms. Windmere, leads a whole class discussion aimed at identifying patterns in evidence that students have been collecting about the relationship between introducing heat energy into a system and melting objects. In the days leading up to this discussion, children worked in small groups to make predictions, record observations, and share initial explanations for the melting of objects or substances (*e.g.*, crayons, chocolate, butter, ice). Ms. Windmere sat with small groups, assisted children with safety precautions, asked productive questions, and modeled writing in a science notebook using a small whiteboard. She encouraged students to make close observations, prompting them to notice that as the objects changed they took the shape of their container. After the melting investigation, the teacher removed the aluminum pans with each item from the burner and covered them with a box. The class then gathered on the carpet as a whole group to discuss their observations, identify patterns in the data, and make informed predictions about what would happen to the substances when removed from the heat source.

Teacher: So what did you find out from our investigations about adding heat?

Student 1: Umm...Say I was using butter and it melted. It would melt faster than say I was using cold.

Teacher: So what makes it melt?

Student 1: Heat.

Teacher: Hmm...Can someone restate what Student 1 said? If I have butter and I add heat, what happens?

Students go through each example material and what happened in the presence of heat. Melting is used consistently by the children and the teacher to describe the change.

Teacher: So who can make a statement that talks about what happens using the words solids and liquids and heat?

Student 2: If you put solids on the stove they will probably turn into liquids.

The teacher repeats Student 2's statement, replacing the word stove with heat. She asks everyone who agrees to put their finger on their nose. It appears that all of the children place their fingers on their noses.

Teacher: Hmm...Someone with their finger on their nose, tell me why you think that. What is your evidence? Student 3, what did you see today that tells you that?

Student 3: So pretty much every material that we tried... all of them umm melted.

Teacher: So did we try all of the materials on the whole planet?

Class: No! [choral]

Teacher: But the materials we did try...

Student 3: Melted!

Teacher: [Walks over to the KLEWS[‡] chart.[37]] Could we say that all solids change to liquids when we add heat?

Class: No! [choral]

Teacher: Can we rephrase it because we didn't try all of them? What could we say instead?

Student 1: Some.

Teacher: So could we say that some solids turn into liquids when we add heat? [Choral responses mixed.] I hear some yes and some no. Someone on the yes side, tell me what's your evidence for that?

When students share their observations/data, they come to agreement on the above statement. The teacher writes this statement as a claim on the KLEWS chart – Some solids turn into liquids when we add heat.

Teacher: Let's go back to the question. Student 4 said he knew this because he said crayons make a change; they turned into a liquid.

Teacher writes student observations in the evidence column of the KLEWS chart.

Teacher: What do you think will happen to our objects next? Now that we have removed them from the heat source?

Teacher records students' predictions and the discussion ends with children excited about making more observations the following day.

[‡]KLEWS is a tool for mapping series of claims, evidence, reasoning, and testable questions across a unit of instruction. The chart text is negotiated in whole group discussion that includes argumentation and negotiation, and results in a scientific explanation of a phenomenon.

Note that Grade 2 students in the United States are typically 7 years old at the beginning of the school year when this discussion was observed.

The whole class discussion is interesting for several reasons. First, the teacher had students share their results with purpose – to look for patterns associated with adding heat to the system. The way in which Ms. Windmere talked about adding and removing the source of heat demonstrated her knowledge of the underlying content and established a productive way for students to think about the system in terms of cause and effect. Second, the teacher's communicative approach was not one of seeking "right answers." In traditional teaching, classroom discourse often follows a variation of the Initiation-Response-Evaluation (I-R-E) pattern.[38,39] In an I-R-E discussion the teacher *initiates* dialogue with a question and a student *responds* to the question with an answer. This interaction is followed with an *evaluation* by the teacher regarding whether the response is correct of not. I-R-E often leads to a closed interaction that engages one student at a time as opposed to whole class, dialogic interactions. Conversely, Ms. Windmere kept the conversation open-ended and orchestrated it in a way that included a variety of students. She intentionally planned instruction such that that patterns would emerge from the data, and reframed questions in support of making those patterns visible to students. Her use of a driving question and KLEWS[37] chart suggests a Claims–Evidence–Reasoning approach framed her instructional moves.[26] Finally, Ms. Windmere skillfully guided the discussion to a powerful conclusion in which children were asked to use their developing explanation for melting to predict the results of a subsequent investigation – *What happens to the objects when the heat source is removed?*

While elements of argumentation are present in the discussion (*e.g.*, agree/disagree), the conversation is a rich context for sensemaking in other ways. Ms. Windmere was thoughtfully eliciting students' ideas and adjusting her next instructional moves responsively. Students were required to use evidence to find patterns, as well as support their claims about melting with evidence. Peers were learning the norms of productive participation in science talks with guidance from the teacher. Taken together, these interactions reflect epistemic and social practices that support sensemaking. Children were learning about science concepts while engaging in the practices of science.

So how do we move from this "image of the possible" in a second-grade classroom to teacher professional learning—with both preservice and practicing teachers—in support of students' meaningful science learning?

7.5 Science Teachers' Learning Needs

Now that argumentation in the context of reform and the existing literature about teachers' CPD has been reviewed, it is pertinent to consider the demands of ambitious discourse practices in the context of what we know about science teachers' learning needs. In 2015, the National Academies of

Science, Engineering & Medicine (NASEM) Board on Science Education, published the report, *Science Teachers Learning: Enhancing Opportunities, Creating Supportive Contexts*, in collaboration with the Teaching Advisory Council and with funding from the Merk Company Foundation.[4] The charge to the consensus committee was to review the research and specify teachers' learning needs in a time of reform, summarize the current state of learning opportunities and supports, identify characteristics of effective learning opportunities, consider the role of school or district context in promoting (or inhibiting) the uptake of new practices by teachers, and develop guidance for schools/districts regarding how to best support teacher learning.

From the review of the research the report concludes that as our understanding of how best to teach in ways that align with the *Framework* and *NGSS* evolves, significant changes in teaching practices also will need to occur. In summary, most teachers have not experienced the kinds of scientific and engineering practices described in reform documents. In many cases, teachers self-select into professional development opportunities based on criteria of convenience and personal interest, rather than planning for coherent experiences that build capacity. Teachers' learning needs are influenced by their preparation, school and community context, and the grades they teach. These factors need significant consideration when designing meaningful professional learning opportunities. There is limited research associated with online professional development, but the power of tailored and networked learning is thought to hold promise. Science teachers' learning occurs in a variety of settings—inside and outside of schools—and through numerous structures (*e.g.*, professional learning communities, professional development sessions, workshops, webinars, Twitter chats). There is a need to further explore and value job-embedded learning. Administrators' understanding of reform and support for all teachers, including those who are new to the profession, is essential to building capacity for the work in a larger system of practices and policies.

Of these conclusions, it is salient to consider that supporting teachers learning needs without an agreed upon, coherent set of criteria for CPD is impossible. Thus, one of the significant contributions of the report is a consensus model for features of effective professional learning opportunities for teachers of science. Decades of research across a variety of subject areas has yielded empirically-based insights regarding teacher professional learning. While these studies vary in design, implementation, and methods, results were synthesized to propose five core features that should guide research and practice on teacher learning.[40] In the report, these are referred to as "the consensus model for effective professional development" (p. 118).

- content focus – learning opportunities for teachers that focus on subject matter content and how students learn that content;
- active learning – can take a number of forms, including observing expert teachers, followed by interactive feedback and discussion, reviewing student work, or leading discussions;

- coherence – consistency with other learning experiences and with school, district, and state policy;
- sufficient duration – both the total number of hours and the span of time over which the hours take place; and
- collective participation – participation of teachers from the same school, grade, or department.

It is of great interest that outcomes associated with professional development models informed by these features are mixed. Several recent reviews of research drawn upon in the science teacher learning report suggest that the professional development activities described in the studies included all or most of the consensus model features.[41,42] Nevertheless, it is difficult to tease apart the possible reasons for incongruent findings given the challenges of studying each component of the model, attending to varying contexts, and coping with limited descriptions of the nature of science content and instructional practices associated with implementation aligned to professional development features. Furthermore, very few studies attempted or were able to "connect the dots" between the effects of CPD on teachers' knowledge, their instructional practices, and their students' learning outcomes.

Of course, it is likely that the field has failed to attend to all of the features that may play a role in teachers' learning and development. Take for example emerging scholarship on teacher identity. In the edited volume, *Studying Science Teacher Identity*, Avraamidou argues that "the construct of identity offers a powerful and multidimensional lens to study science teacher learning and development" (p. 1).[43] While the field is focused on questions of epistemological practices, Avraamidou and colleagues draw on situated learning and communities of practice to address ontological underpinnings – that learning changes us and in so doing acts on identity.[43,44] Not only does identity work to connect individuals to the world, it addresses the role of context and attends to learning as sociocultural.[45] Could identity and life history studies be a missing link in attempting to understand teacher learning across the professional continuum? It seems especially plausible at the primary school level where teachers are prepared as generalists and often do not identity with being "science teachers." Indeed, there is much work to be done regarding research and practice associated with teacher professional learning as the field moves forward.

7.6 Promising Approaches for Science Teacher Learning

In any discussion of professional learning opportunities for teachers of science, not only must teachers' learning needs be intentionally considered, but also there must be clear and justified framing of the repertoire of practices they need to develop. It has been argued that while reform documents and associated standards provide a powerful vision of students' learning, the field has only emerging agreement on core practices for science teaching.[4,46] Put another way, the new standards are intended to be a guide for what students should know and be able to do at the end of instruction; they are not intended to be a prescription for teaching. The distinction can be a subtle one for educators and the public.

For some time we have known that coherent frameworks for teacher preparation allow us to foreground valued pedagogical practices, and limit emphasis on others, with more likelihood of featured practices being taken up.[13,47] Darling-Hammond and colleagues assert, "Repeated experiences with a set of conceptual ideas, along with repeated opportunities to practice skills and modes of analysis, support deeper learning and the development of expertise" (p. 393).[48] While various frameworks for science teacher education have been described in the literature (*e.g.*, explicit teaching of nature of science), a common conceptual framework and associated pedagogical practices in science education have been ill-defined.

In their AERA review of research on science teaching, Windschitl and Calabrese-Barton argue "Supports for continual advancement of teaching be predicated on a clear vision of practice that improves learning – one that does not yet exist" (p. 1104).[49] Put another way, even in a time of reform, science education suffers from a lack of vision and agreement about what counts as highly effective teaching practices specific to science. Without consensus, what counts as "effective" is shaped and reshaped by a wide range of stakeholders and policymakers. In turn, this instability interferes with large-scale attempts to implement coherent, consequential, and equitable science learning experiences for students. The authors make a case for core teaching practices in science. In general, core teaching practices share commonalities that include supporting student work that is central to the discipline and scaffolding learning that increases in sophistication over time through the act of teaching.[49] Additionally, core practices should focus on key priorities of equitable teaching and contribute to the larger picture of supporting students' learning outcomes.

Windschitl and Calabrese-Barton unpack several candidate practices embedded in the larger context of learning outcomes specific to science.[49] One outcome explored by the authors is supporting students in constructing, interrogating, revising and using evidence-based explanations. Two core practices necessary for achieving this outcome are "prompting reasoning about gaps and contradictions in explanatory models" (p. 1141), and "supporting evidence-based argument" (p. 1142). Once again, argumentation as epistemic and social discourse genre is viewed as central to changing classroom practices in powerful ways. To better understand the nature and potential impact of core practices, four examples are described below.

At the University of Washington, Windschitl and colleagues initiated their research and practice in secondary science teacher preparation with a goal of articulating and implementing a core set of science teaching practices.[46] The research team aimed to better delineate the aspects of science teaching broadly defined as *science inquiry, hands-on,* and/or *reform-based teaching*. Their approach was to nominate a set of candidate practices and design a suite of associated tools to support science teacher candidates' abilities to engage students in equitable and rigorous science instruction. A related objective of their work was to be able to implement and study ambitious and equitable science teaching practices across sites and teacher preparation programs.

According to Windschitl and colleagues, classrooms are working communities in which the role of the teacher is to facilitate sophisticated, rich and complex forms of academic conversations among students.[46] One of the challenges they uncovered, however, was that beginning science teachers have difficulty initiating and maintaining epistemic talk that is productive to learning. Thus, the research team identified a collection of discourse-rich practices that create and support patterned instructional moves to establish norms of participation for talk and investigations among members of the science learning community. These discourse practices include eliciting student ideas, supporting ongoing changes in thinking, pressing for evidence-based explanations, and initial planning for engagement with important science ideas. To enable preservice teachers to use these discourse practices effectively and establish ambitious science teaching repertoires in their classroom, the researchers provide tools intended to shape instructional practices. Windschitl and colleagues emphasized that tools coupled with coherent and extended teacher preparation programs and/or professional development opportunities help teachers develop the ability to work towards epistemic goals.

Similarly, Davis and her colleagues at The University of Michigan redesigned their elementary teacher education program to be practice-based.[50,51] That is, learning opportunities are intended to engage novice educators with representations, decompositions, and approximations of professional practice deemed powerful in doing the work of teaching. Davis and Boerst described the design challenges associated with reconceptualizing the program in their commissioned paper for Teaching Works.[50] They argued that the barriers faced by elementary teacher candidates are substantial. In becoming a well-started beginner, preservice elementary teachers not only need to develop a rich knowledge of their disciplines and epistemic practices, but also construct a repertoire of instructional practices to support students' learning – not just for science, but for all subject areas. While pedagogies of practice are often highly resourced and require deep expertise on the part of teacher educators,[52] research findings associated with this approach are encouraging.

Take for instance research on the teaching practice of sensemaking discussions in science using rehearsals.[52] The pedagogy of rehearsals is considered an approximation of practice,[51] and focuses on supporting novices to engage with the complex and contingent work of teaching. Rehearsals differ from micro-teaching in several ways, but importantly in the use of "pauses" during which in-the-moment feedback and problem-solving can occur. "During these pauses, the moves of the teacher (*i.e.*, the candidate in the role of teacher) become objects of inquiry"[53] (p. 2). The research group identified their core practice to be "facilitating sensemaking discussions," which was adapted from the Core Practice Consortium to include the following components: framing the discussion, publicly representing data and/or student ideas, facilitating the discussion and closing the discussion (p. 3). The study investigated teacher educators and how and why they used pauses to support teacher learning during rehearsals of sensemaking discussions. The design is unique in that it involved three science teacher educators at three different program sites, all of whom focused on rehearsals in their work with preservice teachers. Findings from the study highlight the complexity and nuance associated with the ways in which pauses were leveraged for novice teacher learning and provided insights into science teacher educators' decision-making. This work exemplifies the need to develop an agreed upon repertoire of core practices for science teaching that can inform the design of programs of research and practice across contexts and teacher education programs.

While the aforementioned examples focus on models of practice-based preservice teacher education, the next two examples emphasize the professional learning of practicing teachers. Roth and colleagues present a robust and innovative model for CPD—*Science Teacher Learning from Lesson Analysis (STeLLA)*—that not only examined the impact of professional development on teachers' knowledge and practices, but also examined student outcomes to understand the impact of teacher learning opportunities.[54] They addressed this gap by designing a professional development model based on enhancing teachers' science content knowledge and their PCK about students' thinking, while aligning their practices with coherent science content storylines. Roth and colleagues employed video-based analysis of practice as a central feature of their model, which guided teachers to analyze their instruction by examining artifacts of practice, student work, formative assessments, lesson plans, and lesson videos. Even though the general focus of the professional development was not argumentation *per se*, artifacts of instruction included video of classroom discourse that contributed to students' epistemic practices and sensemaking in science. The power of professional learning experiences that involve sharing artifacts of learning and teaching and examining them critically and collaboratively has been described as a common anchor that necessarily focuses teachers' attention on content, student thinking, and learning.[55]

In recent years, Zembal-Saul's research and practice has shifted from supporting argumentation practices in the context of teacher preparation to engaging in research and practice focused on experienced elementary teachers' learning and development in support of children's engagement in epistemic discourse practices.[13,56] The model she has developed extends her prior work to incorporate argumentation into a larger schema for sensemaking in science. Zembal-Saul and colleagues are committed to school–university partnerships in which preservice and in-service teachers study and refine their own instructional practices through assessment of student artifacts. A foundational aspect of her work with teachers engages them as science learners who participate in epistemic discourse and practices to investigate and explain natural phenomena. These experiences are closely linked with video-based cases (of the same or similar instructional sequences when possible) to demonstrate the capabilities of young learners to participate productively in these kinds of discourse-rich science learning experiences. Teachers then engage in cycles of co-planning, co-teaching, and analysis of instruction and student artifacts. Video analysis and related tools are used to support teachers in distinguishing between "doing school" and "doing science." Over time, teachers assume increasing responsibility for planning and adapting instructional materials to better reflect epistemic features of science in their teaching.

As with all design-based research, Zembal-Saul and colleagues focus intentionally on some practices while backgrounding others. In her work specifically, the scientific practice of constructing explanations from evidence is treated as foundational (Figure 7.1). This emphasis aligns well with scientific practices described in the *Framework*, and requires attention to productive participation and discourse-rich learning environments that include argumentation as epistemic talk and social practice.[5] Given that much of Zembal-Saul's work is with K-6 teachers who tend to be unfamiliar and/or uncomfortable with scientific knowledge and practices, she uses the Claims–Evidence–Reasoning framework as a heuristic for teaching about *what counts* as evidence-based explanation in science and how such explanations are

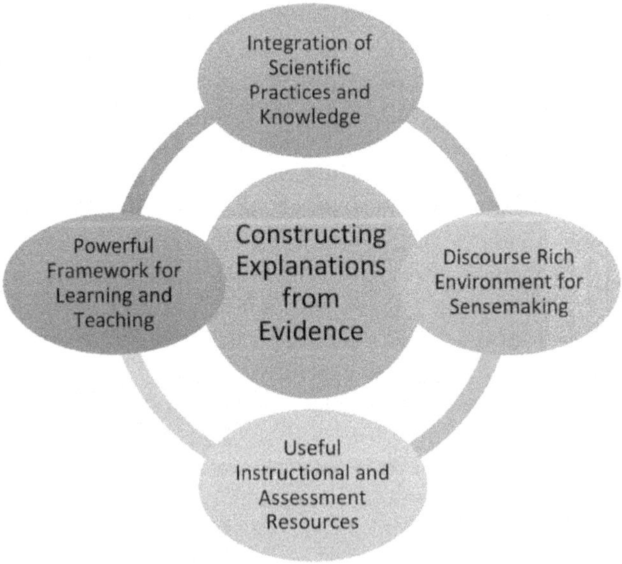

Figure 7.1 Giving priority to evidence and explanation in science teacher professional learning.[74,75]

co-constructed.[20] There are a variety of powerful resources for instruction and assessment that align with the CER framework, including tools for formative assessment (reviewing and responding to student artifacts).

Returning to the key purpose of sharing promising approaches to science teacher development, the goal is to share examples of programs of research and practice that consider science teachers' learning needs, as well as attempt to create coherent frameworks for CPD. Common frameworks, agreed upon core practices, and robust programs of research will allow us to build toward a coherent system of supports, improving the odds of disrupting traditional school science teaching in productive and generative ways.

7.7 Scaling Up Science Teacher Professional Development

It has been asserted multiple times in this chapter that coherence across vision, standards, curriculum, instruction, assessment, and teacher professional learning are crucial to effective change in school science.[4] A potentially weak link in this chain of coherence is the multiple and varied providers of teacher professional learning opportunities – state agencies, curriculum developers, museum personnel and other informal science educators, university faculty, *etc.*[4] The question then becomes how to construct coherent, large-scale systems of supports that increase the potential for the ambitious vision of student learning be realized in school districts, schools and individual classrooms. This priority will require new partnerships, approaches and tools, such as *Networked Improvement Communities* (NICs).[57] NICs involve agencies and organizations, including researchers and practitioners, who collaborate to address particular problems of practice, as well as co-design, implement, and share possible solutions. *ACESSE: Advancing Coherent and Equitable Systems of Science Education* (Penuel *et al.*, in press), is a nationally funded project that leverages the principles and practices of improvement science, NICs in particular, to bring together science education leaders from 13 states (members of the Council of State Science Supervisors) with researchers from the University of Colorado Boulder and the University of Washington for the purpose of improving coherence within and across state systems.[58] Most of the participating states have adopted *NGSS*. This allows for the project to explore whether and how a network of science education leaders can develop a shared understanding of equitable science teaching and formative assessment, as well as how that coherence informs teachers' professional learning opportunities.

The stated goal of ACESSE is "...to enable a network of teams to "get better at getting better;" that is, to learn from their efforts to implement the vision of the *Framework* in ways that can advance the goals of coherence and equity" (p. 7).[58] ACESSE defines coherce in two ways that are consistent with national reform[59,60] – vertical and horizontal. Vertical coherence considers the extent to which stakeholders at all levels of the system share a common vision for equitable and consequential student learning in science. Do leaders in science at state agencies espouse a vision that is consistent with their peers at the national level, as well as school districts within their states? As important as this is to shaping the direction and activities of educational improvement, including teacher professional learning, this kind of alignment has been rare in the past.[61] Horizontal coherence informs what teachers do in their classrooms and includes curriculum frameworks and resources, instructional practices, summative assessments, opportunities for continued professional learning, and more. As with vertical coherence, horizontal coherence also is uncommon. A lack of both kinds of coherence contributes to tensions on the part of teachers and administrators, skepticism about school improvement initiatives, and frustration resulting from mixed messages about expectations. Each participating state is systematically investigating teachers' and relevant stakeholders' views of science learning (*e.g.*, practical measures, surveys, focus groups) and identifying and utilizing "key influencers" (p. 8) from multiple constituencies in their state.[58]

Formative assessment is a core practice that the project is using to address improvement in systems of science education because of its potential to enhance student learning. "[F]ormative assessment is a good leverage point for coherence for another reason: it sits at the intersection of curriculum, instruction, and standards, and to get better at it, teachers need professional development" (p. 8).[58] From that perspective, ACESSE team members co-design, implement, revise and share tools and resources across state systems of science education. One of the more intriguing resources is a series of professional development modules (available free and online[62]) that can be used by state leaders and others responsible for teachers' professional learning. Given the alignment with the *Framework* and important role of scientific practices, including arguing from evidence, the professional development modules place emphasis on formative assessment as eliciting, analyzing, and leveraging students' ideas, their ways of knowing, and their multiple ways of learning – sensemaking discourse is central.

While the professional development modules being co-designed through ACESSE are openly available online, the professional development itself is not provided in an online format. The *Science Teachers Learning* report highlights the possibilities (and uncertainties) associated with online professional development for science teachers.[4] Again, the potential for coherence is an attractive and compelling feature of such an approach. *NGSX: Next Generation Science Exemplar, Learning System for Science Educators* is such an initiative. The development team for NGSX places emphasis on scientific practices, such as explanation-building, modeling, and arguing from evidence. NGSX was designed based on contemporary research that highlights "professional learning systems" that support science teachers' learning in, from and for instructional practices.[63,64] Four features of NGSX are of particular interest: (1) alignment with shifts supported in the *Framework*; (2) the wide range of resources available on the web platform, including videos of classroom practice; (3) the intentional bringing together of teachers, administrators, higher education faculty, and science coaches; and (4) combined online and face-to-face sessions (*i.e.*, hybrid model) with trained and knowledgeable facilitators.[64] NGSX has been adopted at a variety of system levels, including states and school districts.

It is important to note that in terms of teachers' professional learning, the NGSX professional development approach is considered to be "practice-based" in ways that are consistent with the aforementioned studies of preservice teacher education in science.[46,53] Additionally, NGSX design principles align with recommendations for science teachers' professional learning outlined in the NRC report.[4] The five design principles that guide NGSX as a coherent system for scaling-up implementation of *NGSS* are[64] (p. 282–284):

- Situate teacher learning in tasks requiring sensemaking of classroom cases
- Focus PD on the high-leverage practices of argumentation, explanation, and modeling
- Help teachers connect what is new about the science, student thinking about science, and pedagogical supports for science
- Organize teacher study groups working to apply the reforms to their own classroom practice
- Develop peer facilitators' expertise in knowledge-building facilitation

Reiser, Michaels, Moon and colleagues report on the effects of implementation of NGSX at the state level. They describe their hybrid model of online and face-to-face instruction with two main pathways – one for teachers that addresses how to bring *NGSS* science into their classroom practice, and another for peer study group facilitators.[64] Facilitators complete both pathways before leading teacher study groups, increasing the potential for coherence. The study traces the development of 24 teacher facilitators in terms of NGSS implementation expertise and leading teacher study groups. Analysis focuses on changes in facilitators' knowledge, beliefs and pedagogical supports associated with next generation science teaching. A major finding of the study was:

> Supporting learners in the discourse-rich science and engineering practices cannot occur by following routines. Teachers need a rich model of the goals, interactions, and epistemological understandings that translate knowledge building in science into grade appropriate classroom interactions. We found teachers' reasoning about science knowledge building with increasing sophistication following the PD [professional development]. Teachers showed better understanding and facility in generating situations in which models are being developed as generative tools for students to construct, argue for, evaluate, and revise explanations. (p. 294)

Imagine the potential for meaningful and lasting change in science education that is possible when teachers' professional learning opportunities address the contemporary vision of student learning in science, administrators and other key influencers work toward a similar view of improvement in science education, and research-based and practice-proven recommendations for building capacity are aligned and intentional. The ACESSE project and NGSX are two initiatives to watch as the hard work of coherent implementation of *NGSS* and associated science teacher CPD continues to unfold.

7.8 Argumentation and Equity

By this point in the manuscript, readers may have noticed that what teachers need to know and be able to do to support their students' meaningful science learning and scientific practices, including argumentation, is ambitious and complex. Nevertheless, there is an urgent need to address equity, which is a consistent theme in contemporary science education reform and teacher professional learning.[5] Rapidly changing demographics in the United States and globally, and ever-increasing culturally and linguistically diverse classrooms and communities, require teachers to view diversity as an asset and develop specialized knowledge and practices for leveraging that diversity in the service of learning for all students. Zohar and others, unfortunately, have documented the reduced expectations teachers have for what they perceive to be "low ability" students, which is also reported for students who live in poverty or are culturally and linguistically different from those who traditionally participate in the scientific enterprise.[4,16,29] It is no small matter that scientific practices, and in particular arguing from evidence, are linguistically demanding, raising questions about equity and access for all students. *TSS* devoted an entire chapter to supporting norms of participation and language development necessary to productively engage in scientific discourse, which differs from everyday life in terms of what counts as evidence and explanation.[3] The authors of the report acknowledge differences in children's cultural, linguistic, and economic backgrounds that influence their prior experiences and conflict with scientific practices. Notwithstanding, the report maintains that all children are capable and bring funds of knowledge that can be built upon to develop proficiency in science; and that teachers need to be able to leverage the local knowledge and cultural practices of every child in their class.[65] This adds a complex cultural and linguistic layer to an already ambitious goal for the development of teacher knowledge and practices in support of students' meaningful science learning.

Lee, Quinn and Valdés highlight inherent challenges of meeting the expectations of *NGSS* given linguistic demands of the scientific practices.[66] The authors point out that the *NGSS* science and engineering practices are highly interrelated and that successful engagement in one of the practices opens up opportunities for productive engagement in others. For example, arguing from evidence, requires students to develop models to clarify their thinking about phenomena and to construct model-based explanations using evidence. This requires that science teachers not only become fluent with the practices themselves and make corresponding shifts in their pedagogical practices, but also that they develop criteria about how to use opportunities to support student learning regardless of their dominant language proficiency. Lee and Buxton identified empirically-based strategies for the integration of science and language learning, grouping them into five categories: content area literacy strategies; language support strategies; discourse strategies; home language support; and home culture connections.[67] Their work is valuable to consider when designing next generation professional learning opportunities for science teachers.

Gonzalez-Howard and McNeill carried out a study to investigate factors impacting English learners' engagement in learning scientific argumentation.[68] One of the important findings of their study was that students who worked in smaller groups could leverage resources from both English and their first language for engaging in successful science discourse through argumentation. Another study by Swanson and colleagues explored strategies used by teachers to engage their English learners in understanding Chemistry through discourse-intensive scientific and engineering practices.[69] Consistent with the recommendations from Lee and Buxton,[67] Swanson and colleagues found that intentional use of teaching strategies, such as primary language support, deliberately providing scaffolds, and small group instruction were effective in creating rich discourse environments with small groups of students.[69] However, the researchers discovered that argumentation was not as successful during whole-class discussion. This line of research further suggested the important need to address teacher professional learning in science and language development in order to increase productive participation among students for whom the primary language is different from that of the learning community and/or teacher.

Given the rapidly changing demographics in the United States, Zembal-Saul and her colleagues proposed[§] a design-based professional development model for integrating language development and science in a nascent immigrant destination in the northeastern part of the country.[56] Based on census data from 2000 to 2013, the Latinx population in the focus community increased from 5% to 40%. Many families whose first language is Spanish have moved into the area because of an abundance of low wage positions associated with packaging and distribution industries in the region. Almost one quarter of the students in the district are designated English learners (ELs). The project, *Science 20/20: Bringing ELs into Focus through Community–School–University Partnership*, seeks to improve interactions among elementary school teachers (both practicing and preservice)

[§]Credit grant: National Professional Development Grant, Office of English Language Acquisition, Department of Education, 2016–2021, Award T365Z160311.

and administrators, afterschool teachers at a local community center, and families of the children shared across these spaces. The goal is to co-design culturally and linguistically rich learning environments and instructional/assessment tools that provide ELs with opportunities to participate fully in science and engineering practices as a context for language development (reading, writing, speaking, and listening).

The *Science 20/20* conceptual framework is based on three major tenets. First, *Knowledge Building* in science and English language development (ELD) exhibit inherent tensions that the project aims to address. For instance, knowledge building in science rests on epistemic discourse and practices, such as modeling and argumentation. Academic language in science is co-constructed through the sensemaking processes of negotiating, interrogating, and revising evidence-based explanations for natural phenomena. At odds with knowledge building in science, ELD differs epistemically in terms of what counts as argument and evidence (see Section 7.3). Moreover, "best practices" for ELD promote pre-teaching academic vocabulary, which is often interpreted as writing definitions for terminology at the beginning of a unit of instruction.[70] Not only does *Science 20/20* confront these epistemic discrepancies, but it also introduces teachers to practices that leverage students' funds of knowledge in addition to their academic prior knowledge and experiences.[71,72]

Another pillar of *Science 20/20* is a focus on *Productive Participation* in sensemaking, which includes engaging directly with phenomena, participating in rich sensemaking discourse, including scientific argumentation, and using multiple scientific representations (*e.g.*, graphs, models).[47] Contemporary thinking about ELD asserts ELs' participation be invited and scaffolded. For example, as opposed to didactic interactions in which final form science is presented, elementary teachers learn talk moves that invite students to engage in the negotiation of explanations from evidence; in terms of ELD, translanguaging is promoted.[73] Across science and ELD, visuals and small group interactions play an important role in student learning. Finally, *Informal Formative Assessment* serves as the third key tenet of the *Science 20/20* conceptual framework. Providing multiple opportunities for students to demonstrate what they are learning and how they are reasoning through drawing, redrawing, and verbally explaining models, while accepting "kid talk" and gesturing, is essential to this aspect of the work. In addition, teachers are coached to provide non-evaluative feedback aimed at advancing learning. This aligns with ELD recommendations to attend to language proficiencies in reading, writing, speaking and listening when planning and implementing instruction.

Science 20/20 is a five-year project that is currently in its infancy. The centerpiece of professional learning in the project is practitioner inquiry. The systematic examination of one's own instructional practices and student learning is further supported by additional job-embedded approaches, such as professional learning communities at participating schools, coaching by university partners and experienced teachers, and co-teaching. More formal modes of continuous professional development include intensive workshops in summer, winter, and spring. Each year, a new cohort is added to the project, and those who complete at least one round of professional development are eligible to host preservice teachers in their classrooms, strengthening the bond between university preservice teacher education and high quality, coherent field experiences (Practical Digest).

While the goal is to provide equitable access to high quality educational experiences for every child at the research site through community–school–university partnership, the implications of the work may be more far reaching. At the January 2018 meeting, *The Role of Science Education in a Changing World*, 25 science educators and scholars from 15 countries gathered at the Lorentz Center in the Netherlands. Supporting students' meaningful and equitable science learning in multilingual classrooms in an age of mass migration resulting from war, famine, climate change, poverty, and the like, was identified as a common global challenge. We have much to learn from one another as we undertake this important initiative.

7.9 Summary

In the current context of reform in science education it is evident that teachers' professional learning in argumentation is unlikely to occur outside of the larger focus on scientific practices and discourse. What do we gain and/or lose by taking this approach? The foregrounding of sensemaking discourse in science learning and teaching is an important pivot away from traditional science instruction, which has favored memorization and verification of final form explanations. Productive participation in scientific practices requires students to engage with phenomena, design and conduct investigations to gather data in the service of co-constructing causal explanations through argumentation and sensemaking discourses, revise those explanations in light of new evidence, and use them to understand related phenomena. Not only are scientific practices ambitious for students, but they also are challenging for the teachers who support them and who have likely not experienced this kind of science learning themselves. While the conceptual and social aspects of scientific practices may be evident, epistemic features are more elusive. What happens when teachers, through no fault of their own, fail to recognize surface level alignment with literacy practices and/or other instructional work? Does the goal of integrating across subject areas foster efforts that work against understanding how to participate in a scientific knowledge building community? Additionally, as Erduran cautions, how do we address the disciplinary distinctions inherent in argumentation as epistemic discourse and other scientific practices?[15]

Throughout the chapter the need for a coherent framework for CPD and core science teaching practices aligned with the vision for students' science learning is advanced. This is not an argument in favor of uniformity, but rather one by which innovation can be shared and studied across contexts in ways that build capacity for change at all levels of the system. Considering science teacher education as something that should be crafted intentionally across the professional continuum has major implications for those who teach science courses, science teacher educators, professional development providers, district and state leaders, and others. Practical Digest is one example of how this kind of coherence and capacity building might be accomplished.

As the many aspects of science teachers' learning needs are considered and agreement on social and situated theories of learning is advanced it is not surprising that there is an emerging consensus model for features of effective professional development. However, even with consensus about features of continuous professional learning, the nature of those features and how they are designed, implemented, and studied are much less clear. Science teacher educators and scholars have a responsibility to the field to better articulate the design-based aspects of their work. It is not merely about identifying features of the consensus model used in professional learning opportunities that matter, but also the nature of those features and how they are implemented, as well as evidence-based revisions to design principles over time. Only with these rich "images of the possible" for CPD can we gain deeper insights into the theories of action employed by varied approaches to supporting teacher learning.

Finally, the need to craft equitable systems of science education is real and urgent. In the United States, the teaching profession still does not reflect the socioeconomic, racial, cultural, and linguistic diversity of students (and often their communities). Implicit biases and deficit perspectives prevent equity and access to productive participation in meaningful science learning and scientific practices. Attention to equity in all aspects of the work of initial teacher preparation and continuous professional development of science teachers is a fundamental responsibility for all those who work with science teachers.

Practical Digest

The purpose of this addendum is to advance considerations for designing continuous professional development (CPD) opportunities for science teachers that are coherent and attend to scientific discourse and practices. It is also intended to challenge traditional ideas about "noviceness" and the context(s) of teacher learning opportunities. The considerations presented here are based on theories of learning, conceptual frameworks, and empirical studies presented in the associated chapter, supporting the assertion that the development of specialized knowledge and practices for teaching are highly contextualized.

Take, for example, the following comparison that reflects how many science teacher education programs in the United States are structured alongside more contemporary perspectives (Table 7.1).

The shift from traditional to more contemporary perspectives is grounded in the notion that teaching science in ways that are aligned to

Table 7.1 Shifts from traditional to contemporary perspectives on teacher education.

	Traditional	Contemporary
Science Content Courses	Elementary = disconnected, large lecture, introductory courses Secondary = equivalent of subject matter major; lecture and verification laboratory	Specialized courses: • Reflect the content teachers are likely to encounter in their professional lives • Engage preservice teachers in scientific discourse and practices to construct, interrogate and apply explanations for natural phenomena • Connect in meaningful ways to teaching applications and/or contexts
Science Teaching Methods Course(s)	Disconnected from field experiences; Focus on "how to" teach science in general (not discipline specific)	Develop framework(s) to support students' productive participation in epistemic practices and meaning making in science; Engage in decomposition and rehearsal of core pedagogical practices for ambitious and equitable science teaching; Adopt an inquiry stance toward teaching and learning
Field Experiences	Disconnected from teaching methods courses; Mentored by teachers who reinforce traditional practices; University supervision is generic v. discipline specific	Highly integrated with teaching methods; Mentored by practicing science teachers who are engaged in CPD; Supported by university faculty who have deep knowledge of classrooms, students, science content and epistemic practices
Induction Years	Socialized into profession by experienced science teachers who are likely to hold traditional perspectives of science teaching	Ongoing support *via* a network of science educators across the professional continuum who are engaged in CPD focused on rigorous and equitable science teaching; Opportunities to engage in practitioner inquiry and make learning public

the new vision for students' science learning requires that teacher candidates first experience this kind of learning themselves. That is, rather than large lectures and verification laboratories, future teachers should participate productively in epistemic practices of science to investigate and explain natural phenomena. One must experience what it is like to come to understand big ideas in science and the nature of science before s/he can be receptive to imagining the possibilities for designing opportunities for students' science learning.

Likewise, there are more powerful alternatives to science teaching methods courses intent on providing a litany of "how to's" (*e.g.*, activities, steps, tricks) for conveying scientific knowledge. Although it is impossible for future teachers to learn all of the specialized knowledge and practices they will need for their professional careers through their teacher preparation programs, science teaching methods coursework can support the development of conceptual frameworks that foreground student thinking and equitable participation in scientific discourse and practices. Additionally, methods courses can provide opportunities to unpack, rehearse, and analyze core pedagogical practices in science. Given the situative nature of learning, attempting to learn methods of science teaching apart from opportunities to investigate and apply pedagogical practices in classroom contexts is problematic. Moreover, supports for these early experiences in classroom settings are best facilitated by mentors and university faculty who are fluent in contemporary science teaching practices, model an inquiry stance toward teaching and learning, and are skilled at discipline specific coaching. Maintaining these supports throughout program field experiences and into early years of teaching is rare, but necessary in order to avoid socialization of beginning teachers into more traditional mindsets.

Given the complexity of issues surrounding the preparation of science teachers and the multiple junctures at which the process can break down, it is likely more productive to think in terms of systems of professional learning that span the career continuum and are focused on building capacity for consequential and equitable science teaching. Note the appeal of such an approach given that even experienced teachers can be rendered novice when trying to teach science in ways in which they have never experienced learning it. While experienced teachers bring essential understanding of students, science content, school contexts, and more, they may still be new to supporting student engagement in scientific discourse and practices, such as argumentation.

School–university partnership work as discussed in the chapter is utilized as an example of CPD that integrates multiple aspects of the systems in which teacher learning takes place (Figure 7.2). This example is not intended as an alternative to the contemporary perspectives described above, but rather builds upon those considerations in an attempt to strengthen coherence and capacity.

Figure 7.2 Coherent and continuous professional learning *via* school–university partnership.

By building robust partnerships based on simultaneous renewal and a commitment to practitioner inquiry, the boundaries between expert and novice are blurred in productive ways. All member of the partnership are supported in contributing to and learning from interactions with one another, investigating science learning and teaching in classroom contexts, and making the intellectual work of teaching public. These kinds of partnerships require the development of trusting relationships, acceptance of multiple perspectives and experiences, a growth mindset on the part of everyone involved, and a core commitment to robust, consequential, and equitable learning opportunities for all students. While these collaborative relationships require a significant commitment of time, energy, and human resources to develop and maintain, the benefits in terms of coherent and continuous professional learning are invaluable.

Acknowledgements

The authors would like to thank our partners in this work, especially the teachers and students who have welcomed us into their classrooms and allowed us to learn with and from them. We also would like to note that work associated with *TESSA* and *Science 20/20* have involved numerous colleagues over many years. We are changed forever by their contributions to our thinking and practices. Finally, we would like to acknowledge the sources of funding for the work (*TESSA*: Contract grant sponsor: National Science Foundation. Contract grant number: NSF REC 0237922. *Science 20/20*: Contract grant sponsor: US Department of Education OELA NPD; Contract grant number: T365Z160311.). Through this support, we continue to press the boundaries of research and practice associated with science teachers' learning and development.

References

1. S. Erduran, S. Simon and J. Osborne, *Sci. Educ.*, 2004, **88**, 915–933.
2. L. K. Berland and K. L. McNeill, *Sci. Educ.*, 2010, **94**, 765–793.

3. *Taking Science to School: Learning and Teaching Science in Grades K-8*, ed. R. A. Duschl, H. Schweingruber and A. W. Shouse, National Academies Press, 2007.
4. Committee on Strengthening Science Education through a Teacher Learning Continuum, Board on Science Education, Division of Behavioral and Social Sciences and Education, Teacher Advisory Council and National Academies of Sciences, Engineering, and Medicine, *Science Teachers' Learning: Enhancing Opportunities, Creating Supportive Contexts*, National Academies Press, Washington, D.C., 2015.
5. Committee on a Conceptual Framework for New K-12 Science Education Standards, Board on Science Education, Division of Behavioral and Social Sciences and Education and National Research Council, A Framework for K-12 *Science Education: Practices, Crosscutting Concepts, and Core Ideas*, National Academies Press, Washington, D.C., 2012.
6. *Next Generation Science Standards: For States, By States*, The National Academies Press, Washington, D.C., 2013.
7. L. S. Shulman, *Educ. Res.*, 1986, **15**, 4–14.
8. S. M. Wilson, L. S. Shulman and E. R. Richert, *Exploring Teachers' Thinking.*
9. P. Grossman, *The Making of a Teacher: Teacher Knowledge and Teacher education*, Teachers College Press, Teachers College, Columbia University, 1990.
10. S. Magnusson, J. Krajcik and H. Borko, in *Examining Pedagogical Content Knowledge*, ed. J. Gess-Newsome and N. G. Lederman, Kluwer Academic Publishers, Dordrecht, 2002, vol. 6, pp. 95–132.
11. J. Schwab, in *Science, Curriculum and Liberal Education*, ed. J. Schwab, University of Chicago Press, Chicago, 1978.
12. D. C. Smith, in *Examining Pedagogical Content Knowledge*, ed. N. Lederman and J. Gess-Newsome, Kluwer, Dordrecht, The Netherlands, 1999.
13. C. Zembal-Saul, *Sci. Educ.*, 2009, **93**, 687–719.
14. *Argumentation in science education: perspectives from classroom-based research*, ed. S. Erduran and M. P. Aleixandre-Jiménez, Springer, Dordrecht, 2008.
15. S. Erduran, in *Argumentation in Science Education*, ed. S. Erduran and M. P. Jiménez-Alexandre, Springer, 2007, pp. 47–69.
16. A. Zohar, in *Argumentation in Science Education*, ed. S. Erduran and M. P. Jiménez-Alexandre, Springer, 2007, pp. 245–268.
17. R. A. Duschl, in *Improving Science Education: The Contribution of Research*, ed. R. Miller, J. Leech and J. Osborne, Open University Press, Philadelphia, PA, 2000, pp. 187–206.
18. J. L. Lemke, *Talking Science: Language, Learning, and Values*, Ablex Pub. Corp, Norwood, N.J, 1990.
19. K. L. McNEILL and A. M. Knight, *Sci. Educ.*, 2013, **97**, 936–972.
20. K. L. McNeill and J. Krajcik, *J. Res. Sci. Teach.*, 2008, **45**, 53–78.
21. S. E. Toulmin, *The Uses of Argument: Updated Edition*, Cambridge University Press, Cambridge, 2nd edn, 2003.
22. K. L. McNeill, *J. Res. Sci. Teach.*, 2011, **48**, 793–823.

23. L. A. Haefner and C. Zembal-Saul, *Int. J. Sci. Educ.*, 2004, **26**, 1653–1674.
24. L. A. Haefner, P. M. Friedrichsen and C. Zembal-Saul, *Am. Biol. Teach.*, 2006, **68**, 206–212.
25. B. Badiali, C. Zembal-Saul, K. Dewitt and D. Stoicovy, in *Placing Practitioner Knowledge at the Center of Teacher Education: Rethinking the Policy and Practice of the Education Doctorate*, Information Age Publishing., Greenwich, CT, 2012.
26. C. Zembal-Saul, K. L. McNEILL and K. Hershberger, *What's Your Evidence?: Engaging K-5 Children in Constructing Explanations in Science*, Pearson Higher, 2013.
27. K. E. Metz, *Rev. Educ. Res.*, 1995, **65**, 93–127.
28. J. Maloney and S. Simon, *Int. J. Sci. Educ.*, 2006, **28**, 1817–1841.
29. O. Lee, *Educ. Res.*, 2017, **46**, 90–102.
30. S. Erduran, *Found. Chem.*, 2007, **9**, 247–263.
31. E. F. Mortimer and P. H. Scott, *Meaning Making in Secondary Science Classrooms*, McGraw-Hill Education, 2003.
32. S. P. McDonald and G. J. Kelly, in *Perspectives on Scientific Argumentation: Theory, Practice and Research*, ed. S. K. Khine, Springer, 2012, pp. 265–281.
33. E. Manz, *J. Res. Sci. Teach.*, 2016, **53**, 1113–1140.
34. A. R. Cavagnetto, *Rev. Educ. Res.*, 2010, **80**, 336–371.
35. R. Duschl, *Rev. Res. Educ.*, 2008, **32**, 268–291.
36. L. S. Shulman, *Handbook of Teaching and Policy*, 1983.
37. K. Hershberger and C. Zembal-Saul, *Sci. Child.*, 2015, 66–71.
38. P. H. Scott, E. F. Mortimer and O. G. Aguiar, *Sci. Educ.*, 2006, **90**, 605–631.
39. H. Mehan, *Learning Lessons: Social Organization in the Classroom*, Harvard University Press, 1979.
40. L. M. Desimone, *Educ. Res.*, 2009, **38**, 181–199.
41. J. A. Luft and P. W. Hewson, *Handbook of Research on Science Education.*
42. J. H. van Driel and A. Berry, *Educ. Res.*, 2012, **41**, 26–28.
43. L. Avraamidou and C. Zembal-Saul, *J. Res. Sci. Teach.*, 2005, **42**, 965–986.
44. E. Wegner, *Communities of Practice: Learning, Meaning, and Identity*, Cambridge University Press, 1998.
45. J. P. Gee, *Rev. Res. Educ.*, 2000, **25**, 99–125.
46. M. Windschitl, J. Thompson, M. Braaten and D. Stroupe, *Sci. Educ.*, 2012, **96**, 878–903.
47. C. Zembal-Saul, J. Krajcik and P. Blumenfeld, *J. Res. Sci. Teach.*, 2002, **39**, 443–463.
48. L. Darling-Hammond, K. Hammerness, P. Grossman, F. Rust, and L. Shulman, in *The design of teacher education programs*, 2005, pp. 390–441.
49. M. Windschitl and A. C. Barton, *Handbook of Research on Teaching*, 2016, 1099–1158.
50. E. A. Davis and T. Boerst, *TeachingWorks, University of Michigan School of Education*, 2014.
51. P. Grossman and M. McDonald, *Am. Educ. Res. J.*, 2008, **45**, 184–205.
52. M. Lampert, M. L. Franke, E. Kazemi, H. Ghousseini, A. C. Turrou, H. Beasley, A. Cunard and K. Crowe, *J. Teach. Educ.*, 2013, **64**, 226–243.

53. E. A. Davis, M. Kloser, A. Wells, M. Windschitl, J. Carlson and J.-C. Marino, *J. Sci. Teach. Educ.*, 2017, **28**, 275–293.
54. K. J. Roth, H. E. Garnier, C. Chen, M. Lemmens, K. Schwille and N. I. Z. Wickler, *J. Res. Sci. Teach.*, 2011, **48**, 117–148.
55. D. L. Ball and D. K. Cohen, *Teaching as the learning profession: Handbook of policy and practice*, 1999, **1**, 3–22.
56. C. Zembal-Saul and M. Hopkins, US Department of Education, Office of English Language Acquisition, 2016.
57. A. Bryk, L. Gomez, A. Grunow and P. G. LeMahieu, *Learning to Improve: How America's Schools Can Get Better at Getting Better*, Harvard Education Press, 2015.
58. B. Penuel, P. Bell, T. Neill, S. Shaw, M. Hopkins and C. C. Farrell, *AASA J. Scholarship Pract.*.
59. *Knowing what students know*, National Academies Press, Washington, D.C., 2001.
60. *Systems for state science assessment*, National Academies Press, Washington, D.C., 2006.
61. *Literacy for Science: Exploring the Intersection of the Next Generation Science Standards and Common Core for ELA Standards: A Workshop Summary*, National Academies Press, Washington, D.C., 2014.
62. STEM Teaching Tools, http://stemteachingtools.org, (accessed 6 May 2018).
63. M. Lampert, *J. Teach. Educ.*, 2010, **61**, 21–34.
64. B. J. Reiser, S. Michaels, J. Moon, T. Bell, E. Dyer, K. D. Edwards, T. A. W. McGill, M. Novak and A. Park, *J. Teach. Educ.*, 2017, **68**, 280–298.
65. N. S. Nasir, A. S. Rosebery, B. Warren and C. D. Lee, in *The Cambridge Handbook of: The Learning Sciences*, ed. R. K. Sawyer, Cambridge University Press, New York, NY, US, 2006, pp. 489–504.
66. O. Lee, H. Quinn and G. Valdés, *Educ. Res.*, 2013, 0013189×13480524.
67. O. Lee and C. A. Buxton, *Theory Into Practice*, 2013, **52**, 110–117.
68. M. González-Howard and K. L. McNeill, *J. Res. Sci. Teach.*, 2016, **53**, 527–553.
69. L. H. Swanson, J. A. Bianchini and J. S. Lee, *J. Res. Sci. Teach.*, 2014, **51**, 31–64.
70. S. G. Herrera and K. G. Murry, *Mastering ESL/EFL methods: Differentiated instruction for culturally and linguistically diverse (CLD) students*, Pearson, 2016.
71. K. D. Gutiérrez, P. Baquedano-López and C. Tejada, *Mind, Culture Activity*, 1999, **6**, 286–303.
72. E. B. Moje, K. M. Ciechanowski, K. Kramer, L. Ellis, R. Carrillo and T. Collazo, *Read. Res. Q.*, 2004, **39**, 38–70.
73. O. Garcia, S. I. Johnson and K. Seltzer, *The translanguaging classroom: Leveraging student bilingualism for learning*, Caslon, Philadelphia, PA, 2017.
74. C. Zembal-Saul, in *Presentation at the Area Meeting of NSTA-PSTA*, Philadelphia, PA, 2015.
75. C. Zembal-Saul, Michigan Formative Assessment Academy, 2018.

CHAPTER 8

The Development of Argumentation Skills in the Chemistry Laboratory

AVI HOFSTEIN,*[a,b] DVORA KATCHEVITCH[a] AND RACHEL MAMLOK-NAAMAN[a]

[a] The Weizmann Institute of Science, Department of Science Teaching, Rehovot 76100, Israel; [b] The Arabic Academic College, Haifa 33145, Israel
*Email: avi.hofstein@weizmann.ac.il

8.1 Introduction: The History of the Laboratory in Science Education

Laboratory activities have long had a distinctive and central role in science curricula in general and chemistry learning in particular, as a means of making sense of the natural world and providing the learners with opportunities to experience both hand-on as well as minds-on activities. Since the nineteenth century, when schools began to teach science systematically, the laboratory has become a distinctive feature of science education. Throughout the chapter we use the terms *practical work*, which is common in the context of the UK and Germany, and *laboratory work*, which is common in the USA, interchangeably. A precise definition is difficult as these in school practice embrace an array of activities, but generally they refer to experiences in school settings in which students interact with equipment and materials or secondary sources of data to observe and understand the natural world.[1]

Advances in Chemistry Education Series No. 2
Argumentation in Chemistry Education: Research, Policy and Practice
Edited by Sibel Erduran

Published by the Royal Society of Chemistry, www.rsc.org

With the reform in science education in the1960s in many countries, the ideal became to engage students with investigations, discoveries, inquiry, and problem-solving activities. In other words, the laboratory became the core of the science learning process. The *National Research Council*[2] defines such learning activities (*e.g.* inquiry) as: "the diverse ways in which scientists study the natural world and propose explanations based on the evidence derived from their work. Scientific inquiry also refers to the activities through which students develop knowledge and understanding of scientific ideas, as well as an understanding of how scientists study the natural world."

For many years, science educators have suggested (mainly based on research studies) that many benefits accrue from engaging students in science laboratory activities.[3–5] Tobin, for example, wrote that: "Laboratory activities appeal as a way of allowing students to learn with understanding and at the same time engage in the process of constructing knowledge by doing science."

In curricular-type projects developed during the 1960s the laboratory was intended to be a place for inquiring into, developing, the testing of theories as well as providing students with the opportunity to 'practice being a scientist.' Many research studies[4,6,7] were conducted with the goal in mind to explore the effectiveness of the laboratory for attaining the many objectives (both cognitive as well as affective) that had been suggested in the science education literature.

The traditional list of objectives includes:

- Understanding of scientific concepts
- Interest and motivation
- Attitude towards science
- Scientific practical skills and problem-solving abilities
- Scientific habits of mind
- Understanding the nature of science (NOS)
- The opportunity to *do* science

Over the years, hundreds of papers and essays were published with the goal being to explore and investigate the uniqueness of the science laboratory in general and its educational effectiveness in particular. In addition, it has been widely believed that the laboratory provides the only place in school where certain kinds of skills, abilities, and understanding can be developed.[7] In other words, they suggested, the laboratory provides a unique mode of instruction, learning, and assessment. The laboratory provides a unique learning environment.

Precisely what kind of objectives and aims will be attained in the laboratory dependent on a wide range of factors? We suggest that, among others, these will include the teacher's goals, expectations, subject and pedagogical content knowledge as well as the degree of relevance to the topic, the students' abilities and interests, and many other logistical and

Table 8.1 Suggested goals for laboratory activity (Bennett, 2003 pp. 78–79).

Suggested Goals for Laboratory Activity (Bennett, 2003 pp. 78–79)
To encourage accurate observation and description
To make scientific phenomena more real
To enhance understanding of scientific ideas
To arouse and maintain interest (particularly in younger pupils)
To promote a scientific method of thought

economic considerations related to the school settings and facilities (see Table 8.1).

It should be noted that some of these goals, such as "*enhancing learning of scientific concepts*" coincide with the broad goals for science education that are not necessarily laboratory based. The teacher should be in the position to judge whether the laboratory is the most effective learning environment for attaining a certain objective while teaching a certain topic. Teachers should be aware that there has been a great deal of discussion and numerous research studies about which goals are in fact better achieved through laboratory instruction than through other instructional (pedagogical) approaches.[4,6] The many research studies and essays that were cited in the reviews of Hofstein and Lunetta criticized the tradition of conducting experiments without clear purposes and goals. In addition, they revealed a significant mismatch between teachers' goals for learning in the science laboratory and those that were originally defined by curriculum developers and the science education milieu.

8.2 Learning in and from the Science Laboratory

The main goal of this section of the chapter is to argue and demonstrate that the laboratory in science education is a unique learning environment,[4,7,8] such that if designed in an articulated and purposeful manner with clear goals in mind it has potential to enhance some of the more important learning skills such as learning by inquiry, metacognition, and argumentation.

Laboratory activities had a distinctive and central role in the science curriculum, and science educators have suggested that many benefits accrue from engaging students in science laboratory activities.[3,7–10] More specifically, they have suggested that, when properly developed, inquiry-centred laboratories have the potential to enhance students' meaningful learning, conceptual understanding, and their understanding of the nature of science. Inquiry-type laboratories are central to learning science, since students are involved in the process of conceiving problems and scientific questions, formulating hypotheses, designing experiments, gathering and analysing data, and drawing conclusions about scientific problems or phenomena.

At the beginning of the twenty-first century we are entering a new era of reform in science education. Both the content and pedagogy of science learning and teaching are being scrutinized, and new standards intended to shape and rejuvenate science education are emerging.[9,11] In general, one of the characteristics of this reform is the change in the goals articulated for science teaching and learning; namely, that science education should be targeted to all students (*i.e.* attaining scientific literacy for all students) and should be extended beyond the preparation of science-oriented students for academic careers in the sciences. This is in fact a call for also rethink the goals for the learning in and from laboratory work. There are several buzz words that characterize current reform. Among these are student centred learning, learning by the inquiry method, and development of high learning skills such as argumentation, metacognition and asking relevant questions (relevant to the experimental situation). Inquiry in the context of science learning in general and inquiry in the science laboratory in particular are among the important components of this reform.[4,10–12] Bybee[11] wrote that inquiry in terms of skills and abilities includes the following components identifying and posing scientifically oriented questions, forming hypotheses, designing and conducting scientific investigations, formulating and revising scientific explanations, and communicating and defending scientific arguments. It is suggested that many of these abilities and skills are in alignment with those that characterize inquiry-type laboratory work (practical work to include project-based learning), an activity that puts the student in the centre of the learning process.[13,22] Tobin[3] based his review on constructivist ideas of providing students with experiences that would enable meaningful learning in the science laboratory. He wrote that:

> *"Laboratory activities appeal as a way of allowing students to learn with understanding and, at the same time, engage in the process of constructing knowledge by doing science" (p. 405).*

In order to attain this goal, Tobin suggested that students should be provided with opportunities to reflect on findings, clarify understandings and misunderstandings with peers, and consult a range of resources that include teachers, books, websites, and other learning materials in the laboratory. His review reported that such opportunities rarely exist since teachers, in the laboratory, are so often preoccupied with technical and managerial activities.

Based on their review of the literature regarding the laboratory Lazarowitz and Tamir[7] joined the long list of writers claiming that the potential of the laboratory as a medium for teaching and learning science is enormous. They wrote that the laboratory is the *only* place in school where certain kinds of skills and understanding can be developed. They are among those who have suggested that one of the complicating factors associated with research on the effectiveness of the school laboratory is that often the goals articulated for learning in the laboratory have been experiences in the laboratory, and related assessment practices have remained relatively unchanged. In addition, these goals are almost synonymous with those articulated for learning science more generally. Hofstein and Lunetta[4] suggested that much practical work is purposeless and often the explicit objectives of the practical work do not coincide with the purpose of practical experiences. They also claim that many practical tasks have too many different teaching and learning objectives to focus on during instruction.

Gunstone[14] wrote that helping students develop scientific ideas from practical experiences is a very complex process and that students generally did not have sufficient time or encouragement to express their interpretations and beliefs and to reflect on central ideas in the laboratory. Research on learning in the school laboratory makes clear that to understand their laboratory experiences, students must manipulate *ideas* as well as materials in the school laboratory and they must be helped to contrast (and align) their findings and ideas with the concepts of the contemporary scientific community. Manipulating materials in the laboratory is *not* sufficient for learning contemporary scientific concepts. This accounts for the failure of "cookbook" laboratory activities and relatively "unguided" discovery activities to promote desired scientific understanding. Several studies suggested that while laboratory investigations offer excellent settings in which students can make sense of phenomena and in which teachers can better understand their students' thinking, laboratory inquiry alone is *not* sufficient to enable students to construct the complex conceptual understandings of the contemporary scientific community.[10] In the laboratory, students should be encouraged to articulate and share their ideas, to help them perceive discrepancies among their ideas, those of their classmates, and those of the scientific community.

At the end of the twentieth century there was increasing understanding from the cognitive sciences that learning is contextualized and that learners construct knowledge by solving genuine, meaningful problems. The school science laboratory can offer students opportunities to have some control of their activities, enhance their perception of sense of *ownership* and *motivation*.[15] It can be an environment particularly well suited for providing a meaningful context for learning, determining and challenging students' deeply held ideas about natural phenomena, and constructing and reconstructing their ideas. Though a complex process, meaningful learning in the laboratory *can* occur if students are given sufficient time and opportunities to interact, reflect, explain, and modify their ideas. Engaging in *metacognitive* behaviours of this kind enables students to elaborate and to apply their ideas; the process can promote conceptual understanding as well as the development of problem-solving skills. The challenge is to help learners take control of their own learning in the search for understanding while providing opportunities that encourage them to ask questions, suggest hypotheses, and design investigations, "minds-on as well as hands-on."[14]

8.3 New Era New Goals: Goals for Learning in and from the Science Laboratory in the Twenty-first Century

The biggest challenge for practical work, historically and today, is to change the practice of "manipulating equipment not ideas." Unfortunately, the typical laboratory experience in school science is a "hands-on" but not a "minds-on" activity. We are operating in an era in which high-order learning skills are seen as important as the content of science.[16] Higher order thinking and learning skills and activities in the context of learning science are considered to be complex, non-algorithmic, and involve applications of multiple criteria instead of memorizing facts. These activities include asking research questions, solving authentic problems, argumentation, metacognitive skills, drawing conclusions, making comparisons, dealing with controversies, and taking a stand.[17]

Gunstone[14] claimed that meaningful learning in the laboratory occurs when students are given ample opportunities for interaction and reflection in order to initiate discussion. It is suggested that some of these skills could be developed as part of inquiry-based science laboratories. Many of these abilities and skills are in alignment with those that characterize inquiry-based chemistry laboratory work, an activity that places the student at the centre of the learning process.[18] Researchers claim that learning in the laboratory might provide a constructivist environment that fosters higher order thinking, metacognitive and argumentative skills.[19] In this chapter we shall elaborate on two of these variables—the development of argumentative skills and the ability to ask high-level and relevant questions—in the context of the science laboratory.

8.4 Argumentation in the Chemistry Laboratory

Several researchers who focused on the issue of argumentation suggested that the inquiry-type laboratory in science education can provide opportunities for students to develop argumentation skills.[16] However, only a few research studies were conducted with the goal in mind of accepting or rejecting this assumption. For example, Rickey and Stacy[20] found that students who participated in guided inquiry-type laboratories were better at evaluating evidence obtained from their research.

Two recent studies reported in the literature regarding the nature of the experiments as a platform for evoking argumentation both in quantitatively (number of arguments) and qualitatively (level of arguments). Kind *et al.*[21] in the UK investigated the quality of argumentation among 12- to 13-year-old students in the UK in the context of a secondary school physical science program. Their study explored the development of students' argumentation who undertook three different designs of laboratory-based tasks. The tasks described in their paper involved the students carrying out the following: collecting and making sense of data; collecting data for addressing conflicting hypothesis; and paper-based discussions in the pre-collected data phase about an experiment. Their finding showed that the paper-based task (the third one in the above tasks list) generated larger numbers of arguments in a unit of time compared to the two other above-mentioned tasks. In addition, they found that in order to encourage the development of high-level and authentic argumentation there is a need to change the practice that generally exists in the science laboratories in the UK. They suggested that more rigorous and longitudinal research is needed in order to explore the potential of the science laboratory as a platform for development of students' ability to argue effectively and in an articulated way.

The second study was conducted in Israel in the context of 12 years of research and development of inquiry-type laboratories in the context of upper secondary school in grades 10–12 (for more details about the philosophy and rationale of the project see, Hofstein, Shore and, Kipnis[22]). When Driver *et al.*[23] presented their introduction to *argumentation* in the context of learning science they pointed towards the relevance for practical work. They saw argumentation as correcting the misinterpretation of scientific method that has dominated much science teaching in general and practical work in particular. Rather than focusing on the stepwise series of actions carried out by scientists in experiments, they claimed, focus should be directed towards the *epistemic practice* involved when developing and evaluating scientific knowledge. We sense two overlapping learning aims. First, that students should understand the scientific standards and their guiding epistemologies, and next, that they should be able to apply these standards in their own argumentation.

We may find many ways of approaching research of students' epistemological understanding and argumentation skills. Onc contribution comes from psychologists who identify scientific argumentation as the key element of scientific thinking who work from the perspective that certain reasoning skills related to argumentation are domain general. People who are good at scientific argumentation are: (a) able to think *about* a scientific theory, rather than just think with it; (b) able to encode and think about evidence in a similar way, and by this discern evidence from the theory; and (c) able to put aside their personal opinions about what is "right" and rather weigh the theoretical claim against the evidence.

Several research studies indicate that the development of students' argumentation skills and science epistemologies is rather complicated. Students, for example, may hold some beliefs about professional science and very different beliefs about their own practices with inquiry at school; *i.e.* have one set of *formal* epistemologies and another set of *personal* epistemologies.[16] Hofstein and Kind wrote that the overall picture has been that students become good at doing specific types of routine experiments, and solve these from school-based strategies rather than a general understanding of formal scientific epistemologies. The study focused on the process in which students constructed arguments in the chemistry laboratory while conducting different types of inquiry-type experiments. Each experiment consisted of three phases, as demonstrated in Table 8.2.

A way forward to understand how practical work may contribute towards development of students' epistemological understanding and argumentation skills may be to look closer at the "teaching ecology" of the laboratory. It is strongly argued that bringing argumentation into science classrooms requires the enactment of contexts that transform them into knowledge producing communities, which encourage dialogic discourse and various forms of cognitive, social, and cultural interactions among learners.

Table 8.2 Phases in the inquiry-type experiments.

Phases in an inquiry experiment combined with an example of an experiment	Abilities and skills
Phase 1: Pre-inquiry	
• Insert the two solids, A and B, into the plastic bag and mix them by shaking.	• Conducting an experiment using teacher's instructions.
• Pour 10 ml of water into the small glass.	
• Put the glass with the water inside the bag (be careful to avoid any contact between the water and the solids).	
• Put a thermometer inside the bag to measure the temperature of the solids.	
• Tie the bag carefully at its upper part (the thermometer is in the bag).	
• Turn over the glass and let the water completely wet the solids.	
• Record all your observations and answer the enclosed questionnaire.	
Phase 2: Inquiry	
• Plan an experiment to investigate the question.	• Observing and recording observations.
• Present a plan to conduct an experiment.	
• Ask the teacher to provide you with equipment and materials to conduct the experiment.	• Asking questions and hypothesizing.
• Conduct the experiment that you proposed.	• Planning an inquiry experiment.
• Observe and clearly note your observations.	• Conducting the planned experiment and recording observations.
Phase 3: Post-inquiry	
• Discuss with your group whether your hypothesis was accepted or you must reject it.	• Analysing the results, asking further questions, and presenting the results in a scientific way.

The right ecology inviting this practice is created through the social and physical environment,[24] the laboratory tasks[4] and the organization principles used by the teacher.[25] A reconsideration of all these factors is therefore needed to for the science laboratory to contribute meaningfully and effectively towards the new learning goals.

Building an argument has significant social importance for students, in addition to their learning scientific concepts and high-order learning skills. While students are engaged in activities in which they are provided with opportunities to develop argumentative skills, they learn how to conduct a meaningful conversation with their peers. Needless to say, these skills are useful for overcoming life's challenges and are not used solely in the context of science learning. Toulmin's model[26] (See Figure 8.1), consisting of the components of an argument, has been used to characterize argumentation in science lessons and is implicit in using a coding system of others.[24,27,28]

In Israel, Katchevitch, Hofstein, and Mamlok-Naaman[29] implemented Toulmin's framework to investigate the epistemic and argumentative operations adopted by students. During the discourse, the students suggested different explanations for the various phenomena that they observed during the experimental procedure and then analysed the data and presented arguments. The discourse conducted during *inquiry-type* experiments was found to be rich in arguments, whereas that during *confirmatory-type* experiments was found to be sparse in arguments (For a comparison of inquiry and confirmatory-type chemistry experiments see Table 8.3 and an example in Appendix 8.1).

In addition, it was found, that the arguments, which were developed during the discourse of an inquiry-type experiment, were generated during the following stages of the inquiry process: hypothesis building analysis of

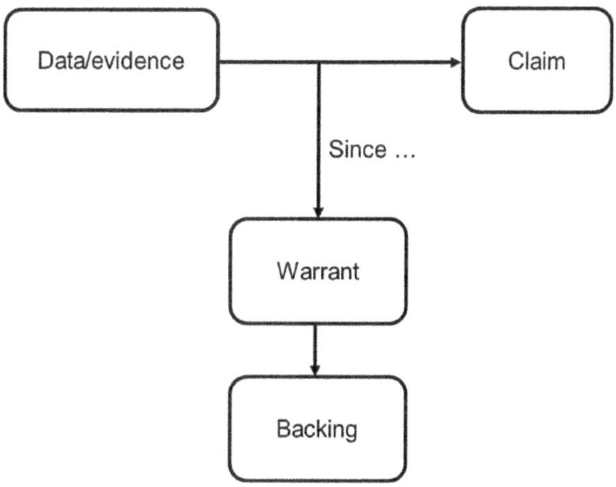

Figure 8.1 Toulmin's model for the components of an argument.

Table 8.3 Skills used in two types of experiments.

Learning skills that are involved during the experiment	Confirmatory-type experiment	Open-ended-type inquiry experiment
Conducting an experiment according to the teacher's instructions	✓	✓
Asking questions		✓
Formulating research questions		✓
Constructing a rational hypothesis		✓
Designing an appropriate inquiry experiment		✓
Conducting the experiment that was planned by the students		✓
Organizing the results	✓	✓
Analysing the results	✓	✓
Drawing conclusions	✓	✓
Summarizing the experiment's procedure	✓	✓

Table 8.4 Level of arguments in the inquiry chemistry laboratory (based on: Erduran *et al.* 2004).

The components of the Argument	Symbol	The level of the Argument
Claim	C	1
Claim + Data or **Claim** + scientific basis **W**arrant	CD/CW	2
Claim + Data + **W**arrant or **Claim** + Data + **R**ebuttal or **Claim** + **W**arrant + Rebuttal	CDW/CDR/CWR	3
Claim + Data + **W**arrant + **B**acking	CDWB	4
Rebuttal that includes **C**laim + **D**ata + **W**arrant	CDWR	5

the results, and drawing appropriate conclusions. On the other hand, confirmatory-type experiment revealed small number of arguments. In addition, the arguments that were posed in the confirmatory-type experiments were low-level in their characteristic (for the various levels of the arguments see Table 8.4, and an example in Appendix 8.2).

Based on detailed analysis of the discourse that was conducted in the chemistry laboratory, the researchers concluded that the open-ended inquiry experiments stimulate and encourage the construction of arguments, especially the stages of hypotheses definition, analysis of the results and reaching conclusions. Some arguments were raised by individuals and some by the group. Both types of arguments consisted of explanations and scientific evidence, which linked the claims to the evidence. Therefore, it is suggested that the mentioned learning environments of open-ended inquiry experiments are effective platforms for raising chemistry-based arguments.

Table 8.5 The criteria for sorting complex/simple inquiry-type experiments.

The type of experiment	Alignment with the concept or topic	Including concepts beyond the curriculum
Complex open-ended inquiry	No	Yes
Simple open-ended inquiry	Yes	No

More recently, these authors investigated the issue of the complexity of the chemistry inquiry experiments and posing arguments. The criteria for sorting complex from simple inquiry-type experiments is presented in Table 8.5. Analysis of the discourse clearly showed that the more complex the experiment the more arguments were posed by the groups of students. In addition, the higher the complexity of the experiments the higher the level of derived arguments.

The research study conducted by Katchevich *et al.*[29,30] focused on the process in which students constructed arguments in the chemistry laboratory while conducting different types of experiments. It was found that *inquiry-type* experiments have the potential to serve as an effective platform for formulating arguments, owing to the special features of this learning environment. The discourse conducted during inquiry-type experiments was found to be rich in arguments, whereas that during *confirmatory-type* experiments was found to be sparse in arguments. In addition, it was found that the arguments, which were developed during the discourse of an inquiry-type experiment, were generated during the following stages of the inquiry process: hypothesis-building analysis of the results, and drawing appropriate conclusions. On the other hand, confirmatory-type experiments revealed small numbers of arguments. In addition, the arguments that were posed in the confirmatory-type experiments were low-level in their characteristics.

To sum up, the open-ended inquiry experiments stimulate and encourage the construction of arguments, especially the stages of hypotheses definition, analysis of the results and the drawing of conclusions. Some arguments were raised by individuals and some by the group. Both types of arguments consisted of explanations and scientific evidence, which linked the claims to the evidence. Therefore, it is suggested that the mentioned learning environments of open-ended inquiry experiments is a platform for raising arguments. In this study we wanted to point out the main factors that stimulate raising argumentation in open-ended inquiry experiments, as well as to characterize situations in which argumentation develops a significant discourse.

One of the key obstacles for conducting chemistry experiments that enhance high-level arguments are the teachers. Clearly, such experiments put the teacher under a lot of pressure and constraints. Inquiry-type experimentation is big challenge for teachers who were used to conducting simpler traditional-type laboratory activities.[23] Keys *et al.*[31] reported on an experimental group who performed the Science Writing Heuristic (SWH) approach. This approach engages students intensively in argumentation within scientific inquiry. The second group, a control group, consisted of students who experienced a traditional approach. They found that the experimental group (those who experienced the SWH) gained significantly higher scores on "argumentation," and understanding of chemistry concepts that underlined the experiments. They concluded that this approach was found to be effective for the preparation of future chemistry teachers, especially in enhancement of their chemistry content knowledge.

8.5 Challenges in Implementing Argumentation in the Chemistry Laboratory

There is a fair agreement that the laboratory provides students unique experiences. As early as the 1960s Kerr[32] suggested that practical work is a unique learning environment in terms of its intellectual and practical abilities and skills involves abilities. A comprehensive study of the literature conducted by Woolnough[33] enabled the taxonomy of aims and objectives of laboratory work to be established. He suggested that these can be structured under the following main headings: understanding of concepts; acquiring habits and capacities; gaining practical skills (including planning and design of a practical exercise, performance, organization, analysis and interpretation of data and application to new situations); appreciating the nature of science; and developing attitudes. Whether these aims and objectives are attained depends very much on the teachers' instructional goals and whether the students are provided with genuine opportunities to be involved in the necessary experiences. In addition, the attainment of these goals is often monitored and regulated by the context in which the laboratory exercise is taking place, the students' characteristics (abilities and motivational pattern), and by the laboratory manual that very often dictates the type of activity that will be conducted, *e.g.* inquiry-type activities as opposed to confirmatory-type experiment, the degree of open-endedness of an activity (see for example, Herron[34]). In addition, very often, the types of activities that are conducted are based on logistical constraints such as the availability of equipment and materials and in recent years also the availability of technological devices such as computers and probes to be used in *Computer-Based-Laboratories (CBL)*. Thus, it is suggested, that if researchers' goal is to obtain a reliable, valid and comprehensive picture regarding the use of and practice in the science laboratory, information must be gathered on the following three distinct features:

- The teacher's behaviour in the chemistry laboratory
- The student's behaviour in the science laboratory, and also
- The type of activity or laboratory exercise in which the student is engaged

It is suggested these factors play an important role in controlling the students' learning in the science laboratory.

8.6 Teachers' Behaviour in the Chemistry Laboratory

There is no doubt that the teacher is the key factor in realizing the potential of the laboratory. One of the important lessons learned during the years of massive curriculum development (in the 1960s) was that the teacher plays an extremely important role in what students learn. The teacher provides organization and an environment that affects whether or not students reach certain instructional goals. For example, if a teacher's goal is to teach observational skills and not just facts that can be observed, this goal should be apparent in the things the teacher says and/or does. Tobin and Gallagher[35] found that science teachers rarely, if ever, exhibit behaviour that encourages students to think about the nature of scientific inquiry and the meaning and purposes for their particular investigation during laboratory activities.

Several studies have reported that very often, teachers involved students principally in relatively low-level, routine activities in laboratories and that teacher–student interactions focused principally on low-level procedural questions and answers. Often, teachers do not perceive that laboratory activities can serve as a principal means of enabling students to construct meaningful knowledge of science, and they do not engage students in laboratory activities in ways that are likely to promote the development of science concepts. They may not perceive that they can manage laboratory activities in ways that are consistent with contemporary professional standards. In addition, many teachers do not perceive that helping students understand how scientific knowledge is developed and used in a scientific community is an especially important goal of laboratory activities for their students.

8.7 Students' Behaviour in the Chemistry Laboratory

In general, there has been only limited efforts to engage students with diverse abilities, experiences, and needs in sharing their ideas and in collaborative inquiry. Tobin[3] wrote that the difficulty of tailoring laboratory activities to the needs of diverse students caused some teachers to avoid laboratory investigations, particularly when working with students having low motivation and skill. The notion that instructional procedures in science education should be matched to learners' characteristics to maximize the effectiveness of teaching and learning has been widely accepted in the science education scholarly literature, if not in school practice, for many years. In the past 20 years special attention has been given to assessing and developing students' conceptual understanding and other cognitive variables.

Simultaneously, less attention has been given to examining variables that influence students' interests and motivation. Hofstein and Kempa[36] postulated that a relationship exists between a student's motivational pattern and characteristics (reasons for learning) and his or her preference for certain instructional techniques in the science classroom or laboratory. Kempa and Diaz[37] probed this relationship. Their study revealed a number of strong relationships between motivational traits and instructional preferences. They found that students they characterized as *conscientious* preferred more *formal* learning environments while others, more motivated by curiosity, enjoyed learning more open-ended situations such as in inquiry laboratory activities. Doing practical work was appealing to the *conscientious* students, but only when those experiences involved explicit instructions, guidance, and closure. On the other hand, students they characterized as *sociable* displayed a distinct preference for group discussions. Other students whom they characterized as *achievers* preferred more individualized or whole class instructional situations. These relationships and other findings suggested the importance of rethinking and reshaping the work of students in the science laboratory to engage students in ways consistent with their diverse experiences, knowledge, and cognitive preferences, perhaps through small group collaboration and inquiry or occasionally through independent inquiry. This suggestion is highly consistent with teaching Standard A[2]:

> *Teachers of science plan an inquiry-based science program for their students. In doing this, teachers: Select science content and adapt and design curricula to meet the interests, knowledge, understanding, abilities, and experiences of students... and work together as colleagues within and across disciplines and grade levels. (p. 30)*

Contemporary teaching standards place a heavy burden on the science teacher. Inquiry-focused teaching now rests on the constructivist notion that learning is a process in which the student actively constructs her or his own ideas that are linked with other ideas in increasingly complex networks. The constructivist model, when practiced, is a relatively radical departure from traditional teaching and learning practice. Teachers are often not well informed about these new models of learning and their implications for classroom teaching and the curriculum. While excellent examples of teaching can be observed, the classroom behaviours of many teachers continues to suggest the conventional belief that knowledge is directly transmitted to good students and that it is to be remembered as conveyed.

Based on the limited research-based evidence one can claim that for most of the chemistry teachers the process of asking students to make claims and justify these claims does not appear to be a normative practice in science education or chemistry education. The two studies reported in this chapter that were conducted in Israel (Katchevitch, *et al.* 2013,[29] 2014[30]), showed that although there are ways to evoke argumentation this happens in cases in which there is use of high-level inquiry techniques and in cases where the experiment is sophisticated and allows many degrees of freedom regarding the students. Engaging students in classroom discourse provides opportunities for the students to construe ideas, to negotiate relationships among different types of experimental-based evidence or representations, and to practice the making of scientific arguments about science content. Chemistry laboratories can be developed (constructed) to provide students with opportunities to plan an experiment, ask relevant questions, hypothesize, and arrive at conclusions. These can provide a platform for argumentation. However, a study conducted in Israel by Taitelbaum *et al.*[38] provided evidence that many teachers have a hard time instructing students in such a learning environment. Teachers find this and similar approaches difficult, demanding and time consuming. Enhancing chemistry teachers' professional development to cope with instructional techniques takes a long time and requires a lot of scaffolding provided by professional development providers. The teacher needs to be familiar with a variety of skills. The inquiry approach puts a lot of stress on the teachers, and it takes time for them to gain self-efficacy in teaching by this way.[38]

Katchevitch, Mamlok-Naaman, & Hofstein[30] found that when the argumentation task presents a complex phenomenon, which includes concepts that are beyond the curriculum, or alternatively, a full enquiry experiment with a scientific background that links a number of content subjects, the discourse is more meaningful and includes many more arguments. On the other hand, in experiments that are not complex (simple) and that are related directly to the concept studied in the formal curriculum material, generally, students know the answer to the inquiry question raised in advance and, consequently, the hypothesis writing, results analysis, and drawing conclusions stages are not controversial but rather formulate an established argument with a scientific background similar to the findings of Kind *et al.*[21]

Apart from those factors that encourage constructing arguments, the task requirements and their related complexity, Katchevitch, Mamlok-Naaman and Hofstein[30] found additional features in the inquiry activity on which an argumentative discourse developed. They found that when students obtained unexpected results in a preceding experiment, or in the experiment that they are planning, the discourse that develops includes more arguments and even refutations. The unexpected results generate a cognitive conflict among the students, which requires them to re-examine what they already know, ask themselves why this knowledge does not form a sufficient basis for explaining the results and whether they have to expand their knowledge or propose explanations based on another scientific background that they had not thought of previously, or that was unknown to them. The conflict is resolved by the group discourse, which is sometimes guided by the teacher.

An additional feature associated with how an argumentative discourse develops is raising questions during the discourse. In addition to the questions that deal with receiving information, the discourse includes questions that require clarification or questions that open up a discussion.

These questions generate attention from the group's members and, therefore, have a very important function in developing an argumentative discourse. We also found that in complex experiments, the students ask more questions and, consequently, many more arguments arise. This finding correlates with the *Questions and Argumentation Model* proposed by Chin and Osborne.[39] In this model, the investigators perceive questions as a factor that motivates discussions. Sometimes the questions are directed at the questioner himself and, sometimes at his peers in the group. However, the need to provide a reply serves as the catalyst for developing the discourse.

However, it is recommended that when teachers select experiments for their classes, they be aware of the potential of these experiments for constructing arguments. Furthermore, they should be aware of the additional features that are likely to contribute to argumentative discourse, such as raising questions that generate a discussion both by themselves and by the group members.

8.8 Summary

The biggest challenge for practical work, historically and today, is the practice of "manipulating equipment not ideas." The typical laboratory experience in school science is a "hands-on" but not a "minds-on" activity. This problem is related to teachers' fear for losing control in the classroom and giving students more responsibility for their learning. Also to be blamed for the current situation is an assessment practice which does not pay enough attention to higher order thinking and a long tradition of developing fool-proof laboratory tasks which guide students through activities without requiring any deeper reflection. The review in this chapter has demonstrated a relationship between these problems in practical work and the "common sense" ideas about science inquiry as a stepwise method.

It has taken science education research a long time to reveal this practice, analyse its underlying rationales and presented alternatives. The development has required a move away from quantitative research methods, which were not sensitive to students' learning in the laboratory, towards more authentic ways of studying what actually goes on in the laboratory. It has also required a thorough analysis of the nature of science inquiry and what makes someone good at doing it. The alternatives which are prominent today combine socio-cultural perspectives on science and learning, but also link to new aims for school science as an important provider of skills and knowledge for citizenship. At the turn of the century we may claim science education is at a better position than ever before for developing a meaningful and appropriate practice for laboratory work. The situation is most promising because of the results and knowledge which were accumulated and achieved. There are many places to start for new development of laboratory teaching strategies and professional development of teachers. These and other tasks calls for science education researchers still to engage with practical work and help develop this area further.

Practical Digest

Examples of Argumentation in the Chemistry Laboratory

Materials in a Small Bag

Student Instructions

Note: Protective glasses and gloves must be worn!

Pay Strict/Attention Regarding

- **fulfilling the instructions** for carrying out stage A precisely
- **recording** as many **observations** as possible
- clear and organized **reporting on** the **observations**
- **participation** of all the group members in carrying out the various tasks
- using correct and precise **scientific language** throughout the course

Stage A: The Pre-inquiry Experiments

First Part

Put the two solids, A and B, into a bag.
Mix them with a glass stirring rod or by shaking the bag.
Put 10 ml of the colourless liquid into a glass receptacle.
Carefully put the receptacle into the bag (so that the liquid and solids do not come into contact with each other). (An alternative possibility: put the corke receptacle into the bag.)
Close the bag well at the top with a rubber band.
Carefully turn the receptacle over so that the liquid will wet the solids. (If you corked the receptacle, then carefully remove the cork while it is still in the bag.)
From time to time tilt the bag to improve the contact.

Second Part

Repeat the first part, but instead of a colourless liquid, use 10 ml of violet-colour liquid.

Third Part

Repeat the first part, but this time measure the temperature of the solids (with the aid of a thermometer) before and after mixing in the liquid.

Stage B: The Inquiry Step

Section I.

1. **Formulate** 5 varied, relevant questions that arose following the observations that were made.
   - **Choose** one of these questions that you would like to investigate.
   - **Formulate** this question clearly as an inquiry question, and to the extent possible, as a link between two variables.
   - **Clearly formulate** a **hypothesis** that relates to the question that you chose to investigate.
   - **Give reasons** for your hypothesis, based on correct and relevant scientific knowledge.
2. **Plan** an experiment that will check the validity of your hypothesis.
   - **Detail** all the steps of the experiment, including the control stage.
   - **List** the equipment and materials needed on the equipment request form.
   - **Consult** with the teacher and make changes if necessary.
   - **Submit** the list of equipment and materials to the laboratory technician.

Section II.

3. **Get the teacher's approval for the proposed experiment.**
   - Carry out the experiment you proposed after approval by the teacher.
   - **Present** the observations and the results in an organized form (table, diagram, graph, *etc.*)
   - **Analyse and interpret** the results
   - **Examine** the connection between the inquiry question and the conclusions.
4. **In the summarizing group discussion**
   - **Express your opinion** about all the stages of the inquiry (limitations, precision, *etc.*).
   - To the extent necessary, **point out** the changes desirable in the inquiry process.
   - **List** additional questions that arose following the whole process.
   - **Prepare** your group's summary of the experiment for **presentation** before the class.
5. **In the summarizing class discussion**

   Relate to our experiment considering the reports of all the other work groups.
6. Ensure that the report is well organized, aesthetic, and readable!

Summary

This open-ended type experiment may show the difference between the skills which students need when conducting an open-ended experiment *vs.* a confirmatory experiment, as described in the following table.

Skills used in two types of experiments.

Learning skills that are involved during the experiment	Confirmatory-type experiment	Open-ended-type inquiry experiment
Conducting an experiment according to the teacher's instructions	✓	✓
Asking questions		✓
Formulating research questions		✓
Constructing a rational hypothesis		✓
Designing an appropriate inquiry experiment		✓
Conducting the experiment that was planned by the students		✓
Organizing the results	✓	✓
Analysing the results	✓	✓
Drawing conclusions	✓	✓
Summarizing the experiment's procedure	✓	✓

Contact between Liquids

Katchevitch, Mamlok-Naaman, and Hofstein (2014) describe the activity, "*contact between fluids*," which was conducted with 11th grade students at a high school in Israel. The activity was an open-ended inquiry experiment, conducted in small collaborative groups (3–4 students). Below are the student instructions to the activity:

Student Instructions

Note: Protective glasses and gloves must be worn!

General Instructions

- **Read** all the instructions well before beginning the experiment.
- **Check** that you have all the necessary equipment and materials at your disposal in order to conduct the experiment.

Pay Strict Attention Regarding

- **fulfilling the instructions** for carrying out stage A precisely
- **recording** as many **observations** as possible
- **reporting** the **observations** clearly and in a well-organized manner

- **participation** of all the group members in carrying out the various tasks
- using correct and precise **scientific language** throughout the course

Equipment and Materials

A Petri dish
about 30 ml of colour water
about 30 ml ethanol
3 Pasteur pipettes
A bottle of liquid soap

Stage A: The Pre-inquiry Experiments

1. Drip coloured water with a Pasteur pipette into a Petri dish until it covers about half the area of the base of the plate. Be sure that the other regions are dry.
2. Drip ethanol with a new Pasteur pipette into the dry part of the plate until the two fluids meet.
3. Describe all the observations. If necessary, you can add more ethanol.
4. Drip a drop of soap solution into the part where the coloured water meets the ethanol.
5. Describe what is happening

Stage B: The Inquiry Step

Section I.

1. **Formulate** 5 varied, relevant questions that arose following the observations that were made.
    - **Choose** one of the questions that you would like to investigate.
    - **Formulate** this question clearly as an inquiry question, and to the extent possible, as a link between two variables.
    - **Clearly formulate** a **hypothesis** that relates to the question that you chose to investigate.
    - **Give reasons** for your hypothesis, based on correct and relevant scientific knowledge.
2. **Plan** an experiment that will check the validity of your hypothesis.
    - **Detail** all the steps of the experiment, including the control stage.
    - **List** the equipment and materials needed on the equipment request form.
    - **Consult** with the teacher and make changes if necessary.
    - **Submit** the list of equipment and materials to the laboratory technician.

Section II.

3. **Get the teacher's approval for the proposed experiment.**
    - **Carry out** the experiment that you proposed after receiving the teacher's approval.
    - **Present** the observations and the results in an organized form (table, diagram, graph, *etc.*)
    - **Analyse and interpret** the results.
    - **Draw conclusions** as much as possible based on the experimental results and rationalize them.
    - **Examine** the connection between the inquiry question and the conclusions.
4. **In the summarizing group discussion**
    - **Express your opinion** about all the stages of the inquiry (limitations, precision, *etc.*).
    - To the extent necessary, **point out** the changes desirable in the inquiry process.
    - **List** additional questions that arose following the whole process.
    - **Prepare** your group's summary of the experiment for **presentation** before the class.
5. **In the summarizing class discussion**

    Relate to our experiment by considering the reports of all the other work groups.
6. **Ensure** that the report is well organized, aesthetic, and readable.

Enjoy the Work!

In this experiment the students observed the contact between two fluids, namely water and ethanol. Students were asked to observe carefully changes that occur. At the beginning of the contact between the fluids, "struggles" were observed - sharp movements of fluids. Finally, a drop of soap in the "struggle" region, after a few seconds, resulted in subduing it and the fluids mixed. Prior to the contact between the fluids, the students observed how water and ethanol spread differently on the surface.

As mentioned above, following the observations, the students were asked to pose questions regarding the phenomena, to select the inquiry-type question for further investigations, suggest a hypothesis, and plan an experiment that would investigate the hypothesis. One of the groups proposed studying the following research question: *How does the ethanol/water relationship affect the water's surface tension?*

The group's discourse was audiotaped, the discourse was analysed using the components of an argument namely: claims, evidence, scientific explanations and rebuttal. Table 8.6 is an example for one of the discourse segments.

Table 8.7 presents the criteria according to which the students were assessed.

Table 8.6 An example for one of the discourse segments.

The components	Symbol	level	Examples of arguments at different levels
Claim	C	1	Nurit: The more powder there is the faster the raisins move, and over time [claim].
Claim + Data or **Claim** + **W**arrant	CD CW	2	Nira: The more reactants that there are in the system, the greater the concentration of solution B, more products will be obtained, more gas will be generated, more bubbles will be created, and more raisins will rise [claim + explanation].
Claim + Data + **Warrant or**	CDW	3	Moriah: As we increased the concentration of the solution, there was a greater amount of sediment [evidence].
Claim + Data + **Rebuttal or**	CDR		Gil: The more we increased the concentration of the solution, the more the quantity of the products increased. We found this by analysing the quantity of the solid [claim + evidence].
Claim + **W**arrant + Rebuttal	CWR		Moriah: Because the reaction has more reactants, there are more collisions between the particles of the reactants and consequently, there are more fertile collisions [explanations]. Gil: And then more of the product that forms the solid that we obtained is created and the solution obtained is more turbid [continued explanation combined with evidence].
Claim + Data + **W**arrant + Backing	CDWB	4	Noam: I want to state that a higher temperature will result in a more frequent occurrence of the reaction [claim]. [He draws a graph] there is an increase in ΔH since this is an endothermic process [evidence]. Alon: There is an increase in ΔS as gas is generated; thus, this is a descending graph [evidence + claim]. Noam: At a higher temperature ΔG is more negative and the reaction will be more spontaneous, according to the graph [he points to the graph that was drawn in the report]. Alon: The spontaneity will be expressed in a broader dispersion of the gas and, as a result, the gas spreads more, because it has greater energy. Ohad: The greater dispersion of the Iodine will be expressed in a greater area that crystallized on the large test-tube [explanation + backing].
Rebuttal that includes Claim + **Data** + **W**arrant	CDWR	5	Yarden: In the first system, there was no reaction at all [claim] Bennie: Not so! There was a reaction, but not like in the other systems. Insufficient gas was generated in order to raise the raisins [refutation based on evidence + explanation].

Table 8.7 Criteria for assessing the hypotheses.

Criteria
The students write hypotheses regarding the research question they chose.
The students explain hypotheses regarding the research question they chose.
The students base their hypotheses on scientific and relevant knowledge.

References

1. E. Hegarty-Hazel in *The Student Laboratory and the Science Curriculum*, ed. E. Hegarty-Hazel, Routledge, American Association for the Advancement of Science, London, 1990, p. 3.
2. National Research Council, National Science Education Standards, 1996, Retrieved May 29, 2006, from: http://www.nap.edu/readingroom/books/nses/html/index.html.
3. K. G. Tobin, *Sch. Sci. Math.*, 1990, **90**, 403.
4. A. Hofstein and V. N. Lunetta, *Sci. Educ.*, 2004, **88**, 28.
5. I. Abrahams and R. Millar, *Int. J. Sci. Educ.*, 2008, **30**, 1945.
6. A. Hofstein and V. N. Lunetta, *Rev. Educ. Res.*, 1982, **52**(2), 201.
7. R. Lazarowitz and P. Tamir in *Handbook of Research on Science Teaching and Learning*, ed. D. L. Gabel, Macmillan, New York, 1994, p. 94.
8. V. N. Lunetta, A. Hofstein and M. P. Clogh, in *Handbook of Research on Science Education*, ed. S. K. Abell and N. G. Lederman, Lawrence Erlbaum, Mahwah, New Jersey, 2007, p. 393.
9. Y. J. Dori, I. Sasson, Z. Kaberman and O. Herscovitz, *Chem. Educ.*, 2004, **9**, 1–39.
10. V. N. Lunetta, in *International Handbook of Science Education*, ed. B. J. Fraser and K. G. Tobin Kluwer, Dordrecht, 1998, p. 169.
11. R. Bybee, in *Quirying into Inquiry Learning and Teaching Science*, ed. J. Minstrel and E. H. Van Zee, AAAS, Washington, DC, 2000, p. 20.
12. G. M. Sere, *Sci. Educ.*, 2001, **86**, 624.
13. A. Hofstein, O. Navon, M. Kipnis and R. Mamlok-Naaman, *J. Res. Sci. Teach.*, 2005, **42**, 791.
14. R. F. Gunstone, in *Practical Science*, ed. B. E. Woolnough, Open University Press, Milton Keynes, 1991, p. 67.
15. A. H. Johnstone and A. Al-Shuaili, *Higher Educ. Chem. (RSC)*, 2001, **5**(2), 42.
16. A. Hofstein and P. Kind in *International Handbook on Science Education*, ed. B. Fraser, K. Tobin and K. McRobbie, Springer, Dordrecht, 2012, p. 189.
17. R. T. White and I. J. Mitchell, *Stud. Sci. Educ.*, 1994, **23**, 21.
18. W. A. Sandoval, *Sci. Educ.*, 2005, **89**(4), 634.
19. P. M. Kind, *School Sci. Rev.*, 2003, **85**, 83.
20. D. Rickey and A. M. Stacy, *J. Chem. Ed.*, 2000, **77**, 915.
21. P. M. Kind, V. Kind, A. Hofstein and J. Wilson, *Int. J. Sci. Educ.*, 2011, **33**, 2527.

22. A. Hofstein, R. Shore and M. Kipnis, *Int. J. Sci. Educ.*, 2004, **26**, 47.
23. R. Driver, P. Newton and J. Osborne, *Sci. Educ.*, 2000, **84**, 287.
24. M. P. Jimenez-Aleeixandre, A. B. Rodriguez and R. A. Duschl, *Sci. Educ.*, 2000, **84**, 757.
25. R. A. Duschl and J. Osborne, *Stud. Sci. Educ.*, 2000, **38**, 39–72.
26. S. Toulmin, *The Uses of Argument*, Cambridge University Press, Cambridge, 1958.
27. S. Erduran, S. Simon and J. Osborne, *Sci. Educ.*, 2004, **88**, 915.
28. Y. Ozdem, H. Ertepinar, J. Cakirgolu and S. Erduan, *Int. J. Sci. Educ.*, 2013, **35**, 2559.
29. D. Katchevitch, A. Hofstein and R. Mamlok-Naaman, *Res Sci. Educ.*, 2013, **43**(1), 317.
30. D. Katchevitch, R. Mamlok-Naaman and A. Hofstein, *Sisyphus J. Educ.*, 2014, **2**(2), 74.
31. C. W. Keys, B. Hand, V. Prain and S. Collins, *J. Res. Sci. Teach.*, 1999, **36**, 1065.
32. J. F. Kerr, *Practical Work in School Science*, Leicester University Press, Leicester, 1963.
33. B. E. Woolnough, *Practical Science*, Milton Keynes, Open University Press, 1991.
34. M. D. Herron, *School Rev.*, 1971, **79**, 171.
35. K. Tobin and J. J. Gallagher, *J. Curric. Stud.*, 1987, **19**(6), 549.
36. A. Hofstein and R. F. Kempa, *Eur. J. Sci. Educ.*, 1986, 7, 221.
37. R. F. Kempa and M. Diaz, *Int. J. Sci. Educ.*, 1990, **12**, 195.
38. D. Taitelbaum, R. Mamlok-Naaman, M. Carmeli and A. Hofstein, *Int. J. Sci. Educ.*, 2008, **30**, 593–617.
39. C. Chin and J. Osborne, *Stud. Sci. Educ.*, 2008, **44**, 1.

CHAPTER 9

Argumentation within Societal Contexts in Chemistry Education

ANDRI CHRISTODOULOU* AND MARCUS GRACE

University of Southampton, Southampton Education School, University Road, Southampton SO17 1BJ, UK
*Email: a.christodoulou@soton.ac.uk

9.1 Introduction

> "The British public's love for takeaway coffee is ruining the environment."

This is one of many headlines in the news recently in England as part of an on-going discussion in the public domain about plastic waste, its management and recyclability. Although widely thought of as 'easy to recycle,' disposable coffee cups are lined with plastic polyethylene in order to make the cups more durable and waterproof, which makes their recycling process more complex than plain paper. What is more, it has recently been acknowledged that there are only a few recycling centres in the UK able to process such disposable cups. It is reported that the UK throws away nearly 2.5 billion of these disposable cups every year.[1,2] As a result of this issue, discussions about whether a 'paper cup tax' should be added to the price of beverages using such cups have started to take place or whether more

Advances in Chemistry Education Series No. 2
Argumentation in Chemistry Education: Research, Policy and Practice
Edited by Sibel Erduran

Published by the Royal Society of Chemistry, www.rsc.org

funding should be made available for researching the development of more sustainable products using bioplastic materials. Essentially, the British public is now asked to consider what should be done about this problem, and take personal action in order to safeguard the environment. Should the public pay more for the same product? Should more specialised facilities be created? What would the environmental and financial consequences of recycling such composite materials be? What alternatives are there to use, and how environmentally sustainable and cost-efficient are these alternatives?

These questions address some of the dimensions that individuals need to take into account when considering issues that have a basis in science, but also affect society, also known as socioscientific issues (SSIs).[3] In such situations, scientific knowledge, social norms, economic considerations and environmental values among others are at play. Scientific knowledge about the materials that such cups are made of, what they consist of and the ways in which they are processed in order to be manufactured and/or recycled would ideally put one in a better position to make an informed decision and act responsibly; thus individuals need to be scientifically literate in order to be able to make such informed decisions and take action.[4,5] Scientific literacy is considered as both the ability to engage in critical evaluation and discussions about scientific issues as these present themselves in everyday life—that is, literacy of science—and, as literacy within science, that is the understanding of scientific content.[6,7] SSI-based science education can be seen as a vehicle for promoting and engaging with both forms of scientific literacy.

Within the field of chemistry education, contexts based on issues such as the one about disposable plastic cups are often used at various levels (*e.g.* personal, societal, professional and technological) so as to make links between the subject matter targeted and its wider implications and applications, in an attempt to make chemistry more meaningful and interesting for students.[8,9] Contexts can be used to illustrate concepts or to apply conceptual knowledge, after it has been taught. They can also be used as starting points, to motivate students and to contextualise the subject matter addressed in an authentic manner (*e.g.* raising authentic questions with students that they can then investigate through different types of inquiry).[10] It is therefore argued that the students' own interests should be taken into account within chemistry education to ensure personal relevance.[11] However, recent reports from the Royal Society of Chemistry[12,13] and research studies[14] suggest that fewer students in England choose to follow a science career, especially in chemistry-related areas. A survey of 2,104 individuals aged 16 or over conducted by the Royal Society of Chemistry on public attitudes to chemistry has found that although the British public overall acknowledged the societal (59%) and financial (72%) value of chemistry, they also held a neutral attitude towards chemistry (51%) suggesting they do not view chemistry as personally relevant to them. Additionally, context-based chemistry education has been found to have a greater effect on the affective domain compared to conceptual understanding or the combination of the two.[15]

Zeidler *et al.*[5] argue that in order to explore students' effective engagement with SSIs we need to go beyond the Science-Technology-Society agenda that dominated work in the field in previous years and that aimed to contextualise science by making links to society and everyday life. The nature of SSIs as controversial, ill-structured issues is an important dimension of SSI-based education and what distinguishes it from other context-based approaches to science education. Marks and Eilks[16] argue that introducing a socioscientific dimension to the teaching of chemistry allows for reflection on scientific questions and a consideration of economic, social and ecological dimensions of these questions. SSIs are not simply hooks for teachers to use in order to get their students' attention in a lesson and to provide a context at the start or end of a lesson. SSI-based education goes beyond that; it aims to not only contextualise the subject knowledge for learners who learn specific concepts through SSI contexts, but also to address the SSI itself, and to allow learners to consider the issue in a critical and evaluative manner, deepening their understanding of its complexity and the levels of controversy that might be involved.[17,18] SSI-based education can engage learners in critical discourse allowing them to make connections between the science and the moral, cultural, social dimensions of the issues discussed[5,19] and as a consequence puts them in a better position to engage in informed decision-making about such issues as they arise in everyday life.

In this chapter, we discuss the role of SSI-based education and the role that argumentation plays within this in the context of chemistry education with two particular aims in mind; the first is to discuss the importance of formulating and developing socioscientific reasoning, and the second is the promotion and engagement of young people with active citizenship as part of science education in general and chemistry education in particular. To do so first we present the relevant areas of Working Scientifically and Chemistry content knowledge that are currently included in the National Curriculum for England, and that could provide the basis for teaching chemistry through socioscientific argumentation. We then review and discuss models suggested that could be adopted and adapted within chemistry education, since as discussed later many of the studies currently exploring teaching and learning science through socioscientific argumentation, are not placed within the subject of chemistry, and when they do they are mainly interdisciplinary, making the role of chemical concepts implicit in the consideration of SSIs. We then present a pedagogical approach, socioscientific inquiry-based learning (SSIBL),[20] that incorporates socioscientific reasoning and citizenship education within a wider framework of responsible research and innovation education and discuss how the SSIBL framework and the concept of 'Green and Sustainable Chemistry' could be used within chemistry education.

9.2 Socioscientific Issues in the Chemistry Curriculum in England

The Science programme of study for the National Curriculum in England for students embarking on their end of compulsory schooling national examinations (the GCSE) includes several statements endorsing the consideration of SSIs, such as ensuring students have an 'appreciation of the relevance of science to their everyday lives' and 'the role of science in understanding the causes of and solutions to some of the challenges facing society'.[21] Particularly relevant in the 'Working Scientifically' section are aspects of scientific thinking that should be applied in all three main science subjects taught (Biology, Chemistry, Physics) and which include:

- appreciating the power and limitations of science and considering ethical issues which may arise
- explaining everyday and technological applications of science
- evaluating associated personal, social, economic and environmental implications; and making decisions based on the evaluation of evidence and arguments
- evaluating risks both in practical science and the wider societal context, including perception of risk
- recognising the importance of peer review of results and of communication of results to a range of audiences

Certain issues are highlighted in the chemistry curriculum. At Key Stage 3 (11–14 year olds) students should be taught about 'the production of carbon dioxide by human activity and the impact on climate' and at KS4 (14–16 year olds), SSIs are most prominent in two sections as outlined below.[21]

9.2.1 Chemical and Allied Industries

- life cycle assessment and recycling to assess environmental impacts associated with all the stages of a product's life
- the viability of recycling of certain materials
- carbon compounds, both as fuels and feedstock, and the competing demands for limited resources
- fractional distillation of crude oil and cracking to make more useful materials
- extraction and purification of metals related to the position of carbon in a reactivity series

9.2.2 Earth and Atmospheric Science

- evidence for composition and evolution of the Earth's atmosphere since its formation

- evidence, and uncertainties in evidence, for additional anthropogenic causes of climate change
- potential effects of, and mitigation of, increased levels of carbon dioxide and methane on the Earth's climate
- common atmospheric pollutants: sulphur dioxide, oxides of nitrogen, particulates and their sources
- the Earth's water resources and obtaining potable water

The disposable plastic cups issue mentioned earlier, the recent ban on using plastic bags, recycling and waste management as well as the different processes used in mining for metals, climate change and air quality are all examples which SSI-based chemistry education could utilise in order to teach the relevant chemistry content knowledge included in the National Curriculum as well as the relevant controversies associated with these issues. In the following section, we first briefly discuss argumentation and second, how it can be embedded within SSI-based education drawing on research studies that have used socioscientific argumentation within chemistry contexts. The Practical Digest provides some further concrete examples and pedagogical strategies for using SSIs in chemistry education.

9.3 Argumentation and Socioscientific Issues

Argumentation is an approach to science learning and instruction that can be used to advance the cognitive, social and epistemic dimensions of science.[22–26] A key component of the argumentative process is the constant consideration of evidence that is required in order to engage in the construction and critique of knowledge claims[27] and the exploration of multiple perspectives.[28] Pieces of evidence need to be evaluated for their ability to provide support to a particular claim and the process of evaluation needs to be grounded on epistemic criteria. Proposing an argument requires one to be able to evaluate the evidence used to support their claim, and at the same time, evaluate their proposed argument against alternatives, in order to be able to account for counter-arguments. Within the field of science education, the process of argumentation can take the form of social, dialogic interactions within classroom discourse.[24,29] Engaging in argumentation practices is also a way to model for the students how authentic, scientific issues are considered and debated by practising scientists in a multidisciplinary environment.[30] Dialogic argumentation is a form of informal reasoning where learners need to coordinate theory and evidence[31] in order to put forward claims supported by sufficient and appropriate information and reasoning.

Socioscientific issues are issues grounded in science that impact on society in different ways[3] and are characterised by a degree of uncertainty and the presence of differing perspectives, which creates space for dialogic interaction and argumentation to take place.[5] Research in the area of SSI-based education has produced evidence that indicates that utilising SSIs to teach science can contribute to developing conceptual understanding,[32–34] decision-making and thinking skills needed in everyday life[16,35,36] and students' motivation and engagement with the content taught.[37,38] Furthermore, using SSIs as contexts for science teaching and learning, allows for student voice to become an integral part of the learning process, which has been found to be critical in efforts to address students' declining interest and positive attitudes towards science.[39,40]

A widely-cited study that provides evidence of the positive influence of argument-based activities relating to SSI to students' understanding of science and their development of argumentation skills is presented by Zohar and Nemet,[33] although within the context of biology. These authors devised an instructional unit which included ten moral dilemmas about human genetics that two classes of 9th graders discussed over a period of 12 lessons. The participants of this study also received explicit instruction on constructing arguments. Comparisons between the intervention and a control group (who did not discuss the dilemmas) of the amount of content knowledge that students used to discuss the same dilemma (cystic fibrosis) at the start and end of the study, showed statistically significant differences in the post-test in terms of (a) the frequency of students that made use of relevant content knowledge, which was higher for the intervention group; (b) the frequency of students that used content knowledge correctly, which was also higher in the intervention group; and, (c) the higher frequency with which students in the control group made unsuccessful attempts to make use of appropriate content knowledge. Students' argumentation skills were also assessed through their written work and audiotaped discussions of two groups of students. The results indicate that although students were able to form simple arguments, counter-arguments and provide rebuttals at the start of the study, their ability to do so at the end of the study had increased, as demonstrated through the increased use of justifications and explicit conclusions made.

Zohar and Nemet's study demonstrates how SSIs can be used to contextualise scientific content and enable students to use that content to support their arguments about specific SSIs. Simon and Amos[41] point out that few studies on argumentation or small group discussion and SSIs are reported specifically in the context of chemistry education. Practice-oriented studies identified in the literature describing instructional units within chemistry education that incorporate SSIs include Stolz *et al.*'s use of the issue of doping as an SSI context for learning about analytical chemistry,[37] Marks and Eilks' study[16] which contextualised the use of synthetic musk fragrances and their impact on the environment and humans by discussing the use of shower gels that include such substances, and Burmeister and Eilks' discussion of how to teach chemical concepts related to plastics through an SSI approach as part of Education for Sustainable Development[42] (see Eilks, Marks and Stuckey[17] for a summary of such contexts within chemistry education). Christenson *et al.* report that out of four topics provided to the secondary school students in their study, the most popular to write about was consumerism and global warming, both SSIs that require understanding of chemistry subject knowledge (*i.e.* materials, and the greenhouse effect).[43] However, they also report that students used less knowledge when writing about these two contexts compared to students that chose to write about nuclear power and GMOs, which might indicate that students might find it challenging to link chemical content knowledge they know and understand to relevant everyday contexts in which this knowledge arises and is relevant.

Within a chemistry context with students in post-compulsory education, Becker *et al.* researched the 'sociochemical norms' utilised during collaborative argumentation of university physical chemistry students when studying the particulate nature of matter through an inquiry-based approach.[44] Sociochemical norms are defined by Becker *et al.* as 'criteria that regulate classroom discourse that are particular to the study of chemistry' such as 'what counts as appropriate justifications in chemistry, how different types of representations should be interpreted, and what counts as a "good" explanation in chemistry'. They analysed the classroom discourse of a group of four students as they engaged in inquiry-based learning activities on physical chemistry over a period of five weeks. Among their findings is the fact that students referred to particulate-level ideas and evidence as justifications about chemical and physical properties, contrary to the Christenson *et al.* findings.

Seung *et al.*[45] report a positive impact on both understanding of the nature and use of chemistry process skills and in the quality of written argumentation by university students that engaged in authentic, inquiry-based learning activities that focused on developing the students' understanding of how chemistry knowledge is constructed and used. Further evidence suggests that when university students engage in argument-based inquiry within chemistry education, improved conceptual understanding is achieved, and that engaging in collaborative oral argumentation as a group has a positive influence on the quality of individual students' written arguments.[46]

Studies in the field of SSI-based education also demonstrate that when students engage in socioscientific argumentation, students draw on types of evidence and information that go beyond scientific knowledge. For instance, Sadler and Zeidler[47] have found that university students used rationalistic, emotive and intuitive types of reasoning when discussing a SSI about genetic engineering. Christenson *et al.* investigated Swedish secondary school students' use of scientific content knowledge in written socioscientific argumentation in four different contexts related to sustainability (global warming, genetically modified organisms, nuclear power and consumerism).[43] They found that students referred to values more than scientific knowledge in their arguments, in all four contexts. However, across the six different dimensions they explored (Sociology, Environment, Economy, Science, Ethics, Policy; SEE-SEP model) they also found that students draw on the domains of the environment, ecology and science more frequently than the other four domains. This would suggest that although students do use scientific knowledge in exploring SSIs, this might be manifested in relation to any of the domains of values, personal experiences or scientific content knowledge, and not necessarily in one or all of these.

Juntunen and Aksela[48,49] make the case for the use of aspects of Green Chemistry based on a life cycle analysis approach to the teaching of chemistry, in order for chemistry education to be in a position to address the goals of scientific literacy in the 21st century. In particular, Juntunen and Aksela[49] have incorporated the use of SSIs, argumentation, inquiry-based learning and the strategy of life cycle analysis (LCA) in order to explore how eight secondary school students (15 years old) in Finland used different types of arguments (scientific, ethical, ecological, socio-economical) when analysing the life cycle of different products and whether their argumentation skills develop as a result of a LCA-based instructional unit. The student participants, based on their own interests chose a product (*e.g.* a pair of jeans, a DVD disk) and then engaged in guided inquiry designing their own investigations to explore the product's materials, how they were used, recycled and their waste management process. They then took part in a role playing debate about the usefulness and responsible uses of their product, as well as possible actions they could take in relation to sustainable ways of using or manufacturing their product. Students' argumentation skills and types of arguments produced were assessed based on the debate and written arguments students wrote in the form of an essay, before and after the debate. A comparison of the types of arguments found in the students' argumentation before and after the intervention indicated that more students used arguments drawing on ecology and scientific evidence in the debate, and the essay post-argumentation, compared to the pre-argumentation task; ethical arguments were the least used during the debate students had whereas socio-economic arguments were strongly present at all stages of students' argumentation.

One possible explanation for the low numbers of ethical arguments identified in students' written argumentation provided by the authors is that ethics is not commonly considered as part of chemistry lessons in Finnish science education classrooms, and thus students are less likely to consider this dimension of SSIs as relevant in their work. This supports Erduran and Mugaloglu's[50] discussion of the place of ethics in chemistry and their argument that more research is need in exploring the nature of moral and ethical reasoning within chemistry and consequently, chemistry education. Another possible explanation for the low incidence of ethical arguments in the debate is that students might feel more comfortable expressing moral and ethical considerations within individual tasks such as writing an essay compared to social configurations such as having a discussion or debate with others. This suggests that tasks that require socioscientific argumentation and reasoning need to be more prevalent and frequently available to students as part of everyday classroom practices in the form of both collaborative and individual activities within chemistry lessons and science education more generally.

The controversial and ill-structured nature of SSIs makes them suitable as a context in which argumentation activities can be planned and carried out, as controversy creates differing views students can discuss and argue about[51] individually or collectively. Argumentation within socioscientific contexts or socioscientific argumentation is a verbal or written reasoning process that allows learners to engage in critical evaluation and discussions about scientific issues as these present themselves in everyday life. The reasoning that students need to employ whilst engaging in socioscientific argumentation goes beyond the coordination of theory and evidence that is based on scientific knowledge; students are not only required to base their discussions on empirical scientific evidence and consider its quality and efficacy. They need to take into account ethical and moral values, economic, practical, social and personal considerations.[52–58] For instance, in Zohar and Nemet's study,[33] in addition to teaching students how to construct arguments students were also instructed explicitly on the differences between values and knowledge, which can be used as evidence in support of claims. As a consequence, the reasoning that is developed and employed by students in socioscientific argumentation takes the form of socioscientific reasoning (SSR).[58–61]

Socioscientific reasoning (SSR) includes the recognition and consideration of multiple perspectives and the role of different stakeholders on the issue, the realisation of the complexity of the issue under examination, including taking into account the different levels of complexity it encompasses, and the need for further collection, analysis and evaluation of relevant evidence, and scepticism in light of potentially biased information.[61] Simonneaux and Simonneaux[56] add to the discussion about the nature of SSR in their study comparing students' reasoning within the context of a local agricultural issue and a global issue (global warming) by pointing out the need to take into account issues of governance and political dimensions when analysing and evaluating risk and uncertainties, as well as issues of personal and cultural relevance and personal identification with the controversy explored. Cultural and social issues are also part of the framework for functional scientific literacy discussed by Zeidler *et al.*[5] Morin *et al.*[59] synthesised the various dimensions of socioscientific reasoning discussed in the literature and produced a socioscientific reasoning framework for sustainability education (S3R) which includes six dimensions. Morin and collegues[60] further elaborate on this model by designing and implementing four levels of sophistication of S3R that can be used for the assessment of students' socioscientific reasoning competencies (Figure 9.1).

SSR poses different demands on the learner who not only needs to consider opposing claims and counter-evidence within a scientific context, but also consider these claims within different social contexts and from multiple perspectives. Morin *et al.*[60] used the S3R framework to analyse French and Australian university students' online exchanges about an SSI. Their findings suggest that these students' S3R reasoning improved by engaging in 'integrative exploratory talk'. This type of talk expands Mercer's[62] concept of

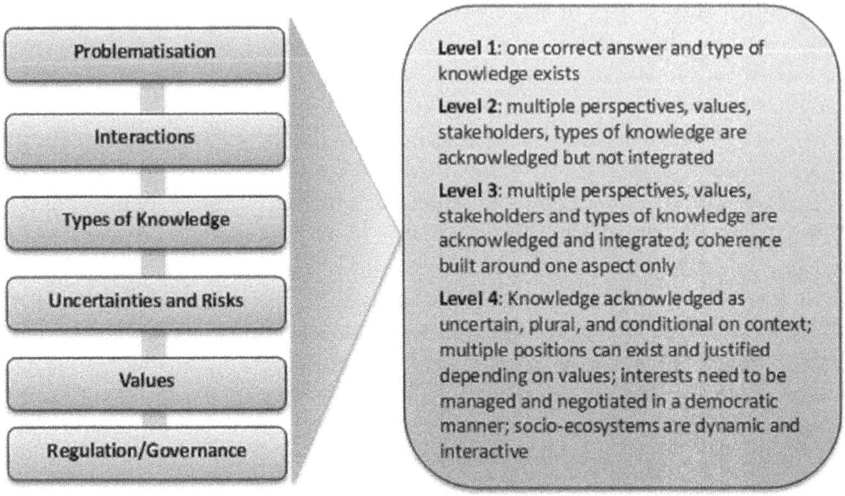

Figure 9.1 The six dimensions of socioscientific sustainability reasoning (S3R) and its four levels of sophistication.[59,60]

'exploratory talk' where students negotiate and examine critically each other's ideas in the process of constructing meaning and understanding. Integrative exploratory talk builds on previous contributions and integrates ideas within group interactions in a way that the arguments constructed are richer and more complex. In a more recent study, Zangori *et al.*[38] have investigated 50 high school students' model-based reasoning about carbon cycling and climate change using an SSI-based pedagogical approach. They report that the use of SSI to frame the content knowledge addressed allowed students to think beyond the content and consider the complexity of the environmental issue explored, as well as provide a context for exploring in-depth the relevant concepts. Therefore, learning environments need to be designed in such a way that allow students to learn more about the context of the issue discussed[51,63] and the levels of complexity that exist.[18] At the same time, the design of appropriate learning environments for SSI-based education need to consider socioscientific argumentation for its potential in promoting students' active citizenship by developing their informal reasoning, decision-making skills and ability to consider multiple types of knowledge and perspectives within societal contexts, issues that are discussed next.

9.4 Active Citizenship through Socioscientific Argumentation

Science as part of SSI-based education should be seen as a human endeavour, and the practice of argumentation can facilitate this realisation by students. Arguing in the context of SSI within science education is a way to humanise science for the students[57,64–66] and to empower them to consider how science and society are interlinked and what their role is within their communities as informed, active citizens. Through dialogue students are provided with opportunities to engage with ill-structured problems about ethical and societal issues related to science such as global warming, the extraction of rare minerals *etc.*, evaluate a situation and reach to conclusions or engage in decision-making about personal action. In this way, students can develop a sense of 'collective responsibility' and 'social consciousness' that allows them to consider issues that overlap between society and science and that are of personal relevance to them or the communities of which they are part.[67]

An approach to promoting active citizenship through socioscientific reasoning and argumentation is to provide students with a decision-making framework. Such frameworks can facilitate students' socioscientific reasoning by allowing them to consider the varying dimensions of SSIs such as knowledge, values and skills[35] and make decisions and formulate perspectives on particular SSIs. Figure 9.2 provides an example of a decision-making framework for use during small group discussions where students are encouraged to come to a joint decision about a real-life, controversial issue.[3,35] The process of reaching consensus can be beneficial for the

Think about the problem. Follow these steps and complete the table as you go.

1. **OPTIONS /ALTERNATIVES**
   - What are the options?
   - Discuss the possible solutions to the problem and list them in a table.
2. **IMPORTANT THINGS TO CONSIDER**
   - How are you going to choose between these options?
   - Discuss the **important** things to consider when you look at each option, and add them to the table.

3. INFORMATION /EVIDENCE

- Do you have enough information about each option?
- Discuss what **science** is involved in the problem.
- Discuss what **extra scientific information** you need to help you make the decision.

4. ADVANTAGES/ DISADVANTAGES

Discuss the advantages and disadvantages of each option, and add them to the table.

5. CHOICE/DECISION

Which option does your group choose?

6. REVIEW/EVALUATION/REFLECTION

What do you think of the decision you have made?
How could you improve the way you made the decision?

OPTIONS (possible solutions) [as many as you think of]	**IMPORTANT THINGS TO CONSIDER** *[for each option]*	**ADVANTAGES** *[for each option]*	**DISADVANTAGES** *[for each option]*
1.			
2.			
3.			

Figure 9.2 The 6-step decision-making framework for small group discussions of socio-scientific issues.[3,35]

students as they explore an issue, since consensus might facilitate the deeper exploration of an issue and promote the use of integrative exploratory talk.[60] For instance, lower secondary school students who were asked to engage in consensus building during an argumentation task about a localised SSI related to energy sources and climate change, performed better than students that were asked to engage in argumentation in order to persuade each other of their position. Learning gains were identified at an individual level based on the quality of the students' written argumentation, and on the quality of the argumentative discourse that took place between pairs of students as they engaged in oral argumentation.[28,68] A framework such as the one presented in Figure 9.2 in combination with a focus on consensus building and deliberation may help students consider scientific, moral, social and other types of information in a balanced manner as it can scaffold the consideration of multiple perspectives and multiple types of knowledge.[69]

One issue to consider in how active citizenship can be promoted through the use of socioscientific argumentation is that provision for *implementing* the decisions taken during collaborative decision-making tasks is not always part of the process of decision-making reported.[35] For example, Simon and Amos[41] explored secondary school students' socioscientific argumentation and decision-making about air quality and have found that although students were able to use different types of knowledge in their arguments (scientific, social environmental, economic) and suggest solutions, they rarely used examples from their everyday lives or attempted to make the issue of air quality personally relevant. Mamlok-Naaman *et al.*[70] used the contexts of alternative fuels and of bioplastics to describe how secondary school students could engage in decision-making within SSI-based chemistry and sustainability education. As part of the decision-making tasks about biodegradable materials students were asked to consider a number of alternative solutions (*e.g.* consuming less, using biodegradable bags) and to eventually make a decision in groups about how to best proceed with decreasing the negative influence that the use of non-degradable plastics have on the environment. Although students were asked to reflect on the consequences of their actions they were not asked explicitly to consider what actions they can take personally to ameliorate the issue discussed and to plan how to implement that action. However, combining decision-making with social responsibility and active citizenship takes the decision-making process a step further and allows students to take action based on their knowledge, beliefs and values.

When using decision-making as part of collaborative argumentation activities, students need to be allowed time to reflect personally and consider their own views in relation to the controversial issue explored; this can put them in a better position to take action as a result of such learning experiences. An approach to active citizenship within science education that puts explicit emphasis on the need to take action is that promoted through the pedagogical framework of SSIBL designed to address responsible research and innovation within educational settings, as discussed in the next section.

9.5 Socioscientific Inquiry-based Learning and Chemistry Education

Responsible Research and Innovation (RRI) has been a major focus by current European funding initiatives, attempting to promote an understanding of the responsibility that technoscientific innovations have to society and the environment and to encourage science *for* and *with* society rather than just science *in* society.[71] A key aspect of RRI is its commitment to prioritising the social, environmental and ethical impacts of technoscientific innovations in society.[72] According to the European Commission, science education is included as one of six key dimensions of RRI (the others being engagement, open access, ethics, gender equality, and governance). Science education's role is to promote and nurture learners' creativity in a way that enables and equips them to deal with societal challenges they will face in a rapidly changing world[73] and to engage practising scientists in communicating their work within educational settings.[74]

One such initiative specifically within science education is the **P**romoting **A**ttainment of **R**esponsible **R**esearch and **I**nnovation in **S**cience **E**ducation (PARRISE) project.[20] The PARRISE project used a design-based approach to develop and implement teacher professional development (TPD) programmes in 18 institutions across 11 countries that aimed to promote the use of SSIBL as a means to attain RRI understanding. One of the key arguments put forward by the PARRISE consortium is that the engagement of students with SSIs on a personal, social and global level deepens students' conceptual understanding of the content taught, and allows them to think more critically about such issues, in a way that results in taking action.[75] The focus of the project was on teacher education and included the development of TPD courses for both pre-service and in-service science teachers.

According to Levinson *et al.*,[20,75] SSIBL can be operationalised in science education through a three-step process illustrated in Figures 9.3 and 9.4. SSIBL inquiries begin by contextualising the science concepts to be taught based on a relevant SSI and then raising authentic questions, which can be the key driver of a lesson or unit (Step 1: ASK). Based on this key authentic question, learners are asked to engage in inquiry-based learning using strategies such as mapping controversies of SSIs, argumentation, debates, analysis of secondary data, practical investigations (Step 2: FIND OUT), and finally as a result of their investigations and learning about the particular SSI, learners are required to take action (Step 3: ACT). Authenticity is a crucial characteristic of the SSIBL framework, as it creates a bridge for personal relevance to emerge through asking questions, and for more complex argumentation to take place. Previous research in the area of inquiry-based learning and argumentation within chemistry education suggests that more complex inquiry experiments that create unexpected results enable students to raise more arguments during their investigations.[77] Unexpected results create potential for raising more arguments as students need to reach consensus about what is happening and why,

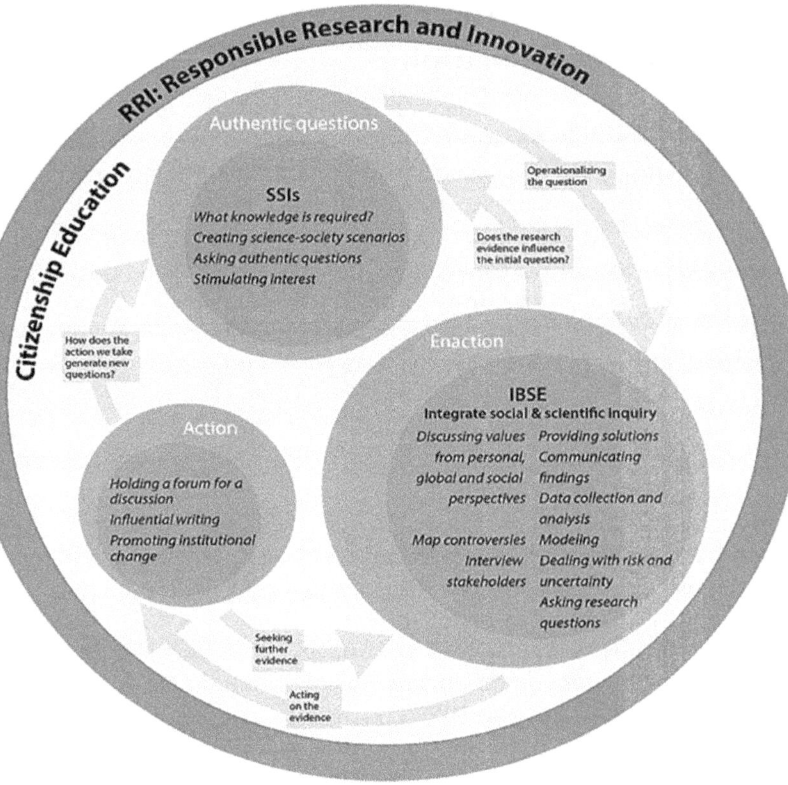

Figure 9.3 The pedagogical framework of Socio-Scientific Inquiry-Based Learning for promoting responsible research and innovation in science education.[20] This figure has been developed by the PARRISE consortium, www.parrise.eu, https://www.parrise.eu/our-approach/.

whereas when results were expected, the arguments raised tended to be confirming these results.

As shown in Figure 9.4 (a practice-oriented version of the SSIBL framework),[76] the tasks students are asked to engage in and consider during this 'ASK, FIND OUT, ACT' process are informed at all stages by social desirability, ethical acceptability and sustainability, which are core features of RRI education. Within the SSIBL framework the potential of SSIs as a way of building character and developing citizenship responsibility[58] is exemplified through the requirement that students take action as a result of their investigations. This increases the sense of social and personal responsibility that students feel in relation to different SSIs and allows them to consider and experience first-hand how their knowledge and actions may have a positive influence on themselves personally, as well as on their local and wider environment.

Social responsibility and socioscientific reasoning can be promoted by engaging students in inquiry-based learning about SSIs. Blonder *et al.*[74]

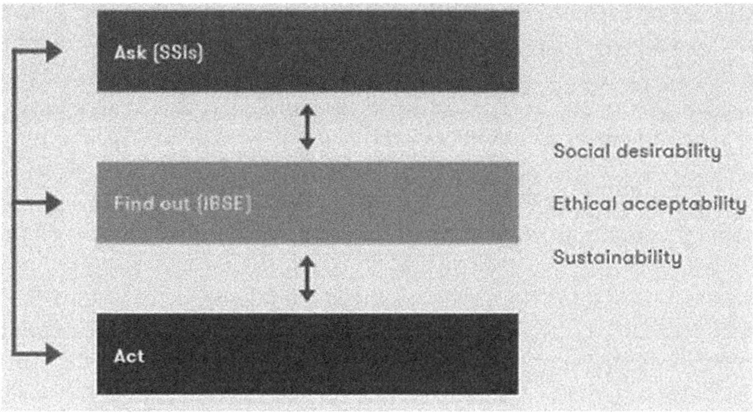

Figure 9.4 The steps followed in the pedagogical framework of Socio-scientific Inquiry-Based Learning (p. 29).[76]
Reproduced from ref. 76 with permission from PARRISE.

argue that the six dimensions of RRI can be used as a guiding framework for implementing SSI-based education within science education. They report on the design of a lesson unit aiming to address RRI aspects through chemistry education by focusing on 'The Story of Lead'. This context combines the use of chemical knowledge with a socioscientific context where students have to engage in evidence-based discussions about the ethical, environmental and social dimensions of using lead (*e.g.* impact on human health). In this study, RRI is an explicit aim of the curriculum whereas the SSIBL approach embeds the dimensions of RRI by bringing together two separate pedagogical approaches within science education (inquiry-based learning and SSI-based education) that have the potential to address the key dimensions of RRI.

As one of the partners in the PARRISE project we have designed a TPD programme to promote the use of SSIBL in secondary science classrooms, which we incorporated into the science teacher preparation programme at our institution. Over a period of three years we have worked with approximately 140 pre-service science teachers (PSTs) exploring how the SSIBL framework can be incorporated into secondary science classrooms. The TPD course designed consisted of up to five workshops and three independent study tasks that aimed to progressively involve PSTs in using SSIBL to teach science.[78] These involved activities that allowed PSTs to make links to their own experiences from teaching science or observing science being taught and then considering and reflecting on these experiences in relation to the different types of inquiry (from open to structured) and the different ways in which links to societal and political/economic/ethical issues could be achieved.

In engaging PSTs with SSIBL we devised three different dilemmas one for each of the three key subjects taught at secondary school level in England (biology, chemistry, physics) and applied the decision-making framework presented in Figure 9.2 in order to engage PSTs in learning and reflecting on the issues discussed. We also engaged PSTs in activities where they had to take on the role of the learner in order to model for them how they could address controversies in their classroom settings (*e.g.* we used a mapping controversies task on a local SSI – the transmission of bovine tuberculosis to cattle from badgers), how they could discuss uncertainty and risk in scientific data (through the climate change dilemma) and how they could run debates on challenging, controversial issues (using the testing drugs on animals scenario). PSTs then took on the role of designer in order to develop lesson plans to teach various topics using the SSIBL approach and the role of the teacher by implementing their resigns, and critically reflecting on their own teaching practices.

The chemistry-based dilemma aimed to address 'evidence, and uncertainties in evidence, for additional anthropogenic causes of climate change' as stipulated in the National Curriculum (see Section 9.2, Part B: Earth and Atmospheric Science). In the next section, we discuss how we approached the teaching of climate change based on SSIBL, and present an example of a SSIBL lesson designed and taught by a PST, which was observed by the first author, in order to discuss some of the challenges and lessons learnt from our work with PSTs on implementing the SSIBL framework within secondary science classrooms.

9.5.1 SSIBL in Action: The Personal, Local and Global Dimensions of Climate Change

Global warming and the potential causes of climate change are a controversial issue due to the way these are presented and discussed in the media and everyday contexts. This is an example of a SSI where the consensus that exists within the scientific community faces resistance in the public domain due to the interplay of dimensions on a political, economic, religious and social level with the scientific domain. From the students' perspective, there are different levels of controversy to consider when planning and teaching using climate change as a socioscientific context. The first level to consider is whether climate change is taking place or not, and the second level to consider is what is causing climate change (whether it is an anthropogenic or a naturally-occurring phenomenon).[78,79] PSTs were also asked to consider how to frame the issue for their students by working in groups to identify key questions and designing lessons addressing personal, local and/or global dimensions, which they then had the option of implementing in their school-based practice.

One lesson observed with a Year 10 class (14–15 year olds) that was designed by one of the PSTs that we worked with included elements of personal, local and global perspectives on the issue. The main aim of the lesson was for students to be able to evaluate the potential impacts of climate change and assess possible actions to reduce impact. To do so, students were asked to work in groups and were given a focus country (*e.g.* Nigeria, Tanzania, Nepal) to do research on, in order to consider the effects of climate change on this as well as on their own country (the UK). Students were asked to present an overview of their findings at the end of the lesson in the form of a debate. Two issues were identified in the implementation of this SSIBL lesson; the first was the way in which students were presented with evidence they could use in support of their ideas and the second the way in which the topic of climate change was framed for the students.

Students were required to engage in socioscientific argumentation, as they were given websites and other relevant resources they could use to draw on information as evidence in order to suggest how countries could be affected based on economic, political and geographical/environmental factors. On the one hand, the challenge posed to this group of students (a high attaining group) was appropriately high based on the information students were asked to explore and use as evidence in their arguments. However, the way that evidence was used and communicated during the debate was not as effective as students did not have any input on how to structure their arguments for them to be of higher sophistication within the particular SSI context (*e.g.* including the consideration of alternatives; including types of knowledge other than science content knowledge; see Figure 9.1). In this case, it was clear from the lesson observation and the subsequent reflective discussion with the training teacher that his pedagogical content knowledge of argumentation in relation to using evidence to support claims[80] was developing and required further support and practice for enhancing his skills in supporting students' oral argumentation skills and groupwork.

The second issue identified through the lesson observation was a tension between the way in which the SSI context was framed by the training teacher and the personal relevance dimension of the SSIBL framework that was operationalised through tasks that focused on taking action. Climate change was framed from a global perspective as an issue located primarily in other countries and placed in the future (*i.e.* students were asked to consider climate change in relation to future generations). This meant that although subsequent activities that required personal and local action were included in the lesson (*e.g.* make a pledge about an action students could take that could have a positive impact in relation to climate change, and a homework task where they were asked to record and estimate their personal energy usage in a day or week, and then outline areas where they could reduce their energy footprint) students had difficulty considering how the issue was of personal relevance to them. As a consequence, when students were asked about actions they could take they discussed the topic as something that is not affecting them directly, but would affect future generations (*e.g.* their grandchildren). This means that the way in which climate change (or other SSIBL contexts) are contextualised and framed in relation to space and time are factors that need to be considered carefully by teachers in order to emphasise for their students the importance of personal and responsible action. Authenticity is an aspect that can be challenging for teachers; what is authentic and relevant for one student might not be the same for others, and teachers might find it time consuming trying to navigate their students towards personal action. One possible solution would be to consider action at the start of the learning process rather than at the end; begin SSIBL inquiries by considering what students think about actions that should be taken for a particular SSI, and what they need to know to be able to take action or to be able to decide which actions are the most appropriate, relevant or effective for students.

Overall, using SSIBL as a pedagogical framework for chemistry education has the potential to create learning environments for students that are contextual in a way that not only creates an interesting and motivating way of learning chemistry, but also it allows them to see the personal relevance of chemistry knowledge to their everyday lives. One way to make the personal relevance and responsible action dimensions of SSIBL more prevalent in chemistry education is to use SSIBL within the context of Green and Sustainable Chemistry Education, an area that is explored in more detail in the next section.

9.6 The Potential of SSIBL for Green and Sustainable Chemistry Education

Green and Sustainable chemistry education is an area of science education that aims to address and educate individuals about environmental issues through the teaching of chemistry. Green Chemistry is defined as 'eliminating the use and production of chemical materials that are hazardous to the environment and human health and as an approach based on finding and planning methods that will prevent the creation of these materials'.[81] As a discipline, Green Chemistry is guided by 12 design principles (Figure 9.5) with the overall aim being to promote sustainable practices within the scientific endeavour of chemists.[82]

Sustainability is a key objective of Green Chemistry as the 12 design principles that guide it provide a framework for designing materials and products that are sustainable, and developing sustainable practices in related industry sectors. The emphasis on the prevention of creating or using hazardous materials requires individuals to take action in order not only to avoid but also to eliminate the use of them, and achieve sustainable living.[83]

1. Waste Prevention	7. Use of Renewable Feedstocks
2. Atom Economy/Efficiency	8. Reduce Derivatives
3. Less Hazardous Chemical Synthesis	9. Catalysis
4. Designing Safer Chemicals (Molecular design)	10. Design for Degradation
	11. Real-time Analysis for Pollution Prevention
5. Safer Solvents and Auxiliaries	12. Inherently Safer Chemistry for Accident Prevention
6. Design for Energy Efficiency	

Figure 9.5 The 12 Design principles of Green Chemistry.[82]

Educational research studies using the principles of Green Chemistry to develop curricula and explore ways of implementing it are now more frequent in the literature[84,85] and are usually combined with educational frameworks that are conducive to the use of socioscientific argumentation and reasoning such as SSI-based education, inquiry-based learning, and context-based learning.[81,86–89] Green and Sustainable Chemistry Education employs strategies such as decision-making that requires chemistry knowledge as well as taking into account Green Chemistry principles and the broader societal (and other) factors that can influence sustainable development.[84] As a result it can develop students' environmental literacy and young people's knowledge of sustainable ways of acting and living. The focus on solutions and action that this approach to chemistry education has means that it is compatible with, and could be promoted through the pedagogical framework of SSIBL.

Examples of products or processes developed within the Green Chemistry discipline could be used as real-life scenarios or case studies creating socioscientific contexts in which students could explore chemical concepts and consider the implications and applications of products and processes routed in Green Chemistry. For instance, the LCA process used by Juntunen and Aksela[49] discussed in Section 9.3 addresses some of the principles of Green and Sustainable Chemistry through socioscientific argumentation that results in taking action. What is more, the emphasis that Green Chemistry has on Sustainability means that the S3R framework (Figure 9.1) can be used as a guiding tool in developing students' socioscientific argumentation and reasoning in Green and Sustainable Chemistry Education. Juntunen[87] discusses how Sustainable Chemistry can be taught using approaches such as inquiry-based learning and SSI-based education. Students can consider the lifespan of products they intent to use in their experiments, or of products they use in their everyday lives and conduct a LCA, the chemistry of healthy food or climate change and consider the implications of their actions to their immediate environment and globally.

Green and Sustainable Chemistry Education through SSIBL has the potential to influence positively students at a personal level and at a local level, including their communities and schools. By engaging students in projects about how they can help their schools produce less waste, or how they can make their school community more energy efficient means that students are learning chemistry and gain an insight to how taking responsible action as a result of their learning can fundamentally bring change to their lives and their communities.

9.7 Conclusions

The aim of this chapter was to present and discuss current research and practice in the area of socioscientific argumentation in chemistry education. We have discussed socioscientific reasoning and collaborative decision-making frameworks that can be utilised to engage students in socioscientific argumentation. The use of SSIBL within a wider Green and Sustainable Chemistry Education approach has been suggested as a fruitful way forward in not only developing students' socioscientific reasoning skills but also their sense of social responsibility and active citizenship. In doing so, some issues have been raised which we outline here in the form of recommendations for further research.

One area for further consideration and research is the need to account for the nature of chemical knowledge in the development of learning environments that promote chemistry education within societal contexts as this is likely to influence the extent to which students are able to make the links between key concepts taught in the chemistry curriculum, and how these can be utilised when learning and acting on SSIs.[43] This is particularly important within secondary science education settings, since older students seem to be in a better position to move between the sub-micro and/or symbolic to the macro level of chemistry knowledge[90] compared to younger students.[44] At the same time, many of the studies currently exploring teaching and learning science through socioscientific argumentation are not placed within the subject of chemistry, and when they do they are mainly interdisciplinary, making the role of chemical concepts implicit in the consideration of SSIs. Therefore, we need to continue researching effective approaches for combining SSI-based education with the effective teaching of chemical subject knowledge at the secondary school level.

Another area to consider is the professional development of both pre-service and in-service chemistry teachers for being in a position to use SSIBL effectively in their practice. SSI-based education and inquiry-based learning both require specific pedagogical content knowledge and orientations, and developing these takes time and effort. Our own work so far indicates the tensions that might arise between the PSTs' knowledge and understanding of the SSIBL framework and their classroom-based implementation when PSTs attempt to utilise SSIBL in secondary science teaching[91] and therefore more research is needed to establish what makes a 'good' SSIBL teacher and what are the particular factors that influence chemistry teachers' willingness and competencies in introducing SSIBL within their teaching.[74]

We have discussed how the SSIBL framework can be used within the wider Green Chemistry and Sustainability Education approach. However, most of the educational research in this area, as well as on SSI-based education within chemistry education has been conducted at university level. Thus, more research is needed in understanding how the design principles of Green Chemistry could be utilised within secondary education in order to help students understand the origins, implications and applications of the chemical knowledge they are taught in school. Burmeister *et al.*[85] argue that the use of controversial issues within chemistry education allows learners to experience how chemical knowledge, societal influences, technology and industry interact in order to create new products, eliminate the use of others and further the scientific endeavours of chemists. Doing so within the SSIBL framework can allow learners to develop their critical and problem-solving skills and participatory decision-making, which are central skills for Education for Sustainable Development.

Finally, we would argue that it is imperative to consider how to achieve balance between the current realities of chemistry teaching and learning at secondary school level, especially in educational systems with high-stakes examinations, and the need to make chemistry education more socially and personally relevant for young people.

Practical Digest

Pedagogical Strategies and Short Scenarios for Classroom-based Implementation of Socioscientific Argumentation within Chemistry Education

Grace[92] summarises approaches readily useable in lessons such as role play, consequence mapping, evaluating media reports, and weighing up probability and risk, in order to engage students in productive discussion and argumentation about socioscientific issues (Table 9.1).

The following are scenarios that could be used to explore more localised SSIs, which are relevant to the everyday lives of the communities involved, related to chemistry subject knowledge.

More Immediate Chemistry-related SSIs

Students are fascinated (sometimes rather morbidly) by the often well-documented historical accounts of how chemicals have affected people adversely. Occupational diseases can be evocative examples, such as the condition known as 'phossy jaw'. This was rife among workers in the matchstick industry in 19th century Britain. The disease was caused by breathing in the vapour of white phosphorus, the active ingredient of matches. The symptoms began with toothache and inflamed gums and this progressed to abscesses and eventually to a rotting of the jaw bones and it sometimes also caused brain damage.[93] The situation developed into an SSI when the growing concerns over employees' health, combined with poor working conditions, resulted in the 'London matchgirls strike' of 1888, with high-level political consequences.[94] Such graphic accounts may serve well to arouse curiosity among students, and can sometimes provide a more useful way of engaging students with the complexities of SSIs than by introducing them to global SSIs, which may initially seem remote from their lives.

- **Occupational diseases caused by chemicals**
  Phossy jaw is an example of an 'occupational disease' from the past, but today there are still many chemical agents that cause

Table 9.1 Pedagogical strategies for engaging students in productive discussion and argumentation about socioscientific issues.[91,92]

Class discussion A very popular approach but: – Danger of students being rushed into making up their minds on the issue, and forming their opinions too soon – Danger that some students dominate the discussion • How do you 'chair' the discussion? • Do you stay neutral? • Do you declare your views? • Do you present an opposing view for balance?	**Consequence mapping ('what if' scenarios)** *E.g.* What if cars in the city are completely banned? What if cars in the city are completely unrestricted? • Then compare maps • Students identify their own values and those of others • Students appreciate that tackling climate change involves considering values as well as scientific evidence
Role play Another very popular approach • Students can explore other peoples' perspectives • But do they really empathise with other peoples' perspectives? • Students may be forced to adopt perspectives opposing their own views	**Evaluating media reports** Get students to answer the following questions: – What are the scientists/researchers claiming? (*i.e.* what are their conclusions?) – What evidence is provided to support their conclusions? – Does the evidence provided convince you that the conclusion is correct? – What extra evidence is needed? How might this be collected?
Probability and risk Promotes discussion about science and society: – How risk statistics are derived – How the risks associated with the issues are presented to the public – Why scientists are unable to state categorically the risks associated with certain issues – How we decide whether or not to carry out risky activities – do we always calculate the risk first?	And to highlight the socially-framed nature of scientific research: – Who did the research? – Who do they work for? – How certain are they about their conclusions? – Do other scientists agree with their conclusions?

occupational diseases contracted by employees as a result of exposure to substances at their place of work. Such agents can be categorised as: *irritants*, which cause reversible inflammation (*e.g.* isopropyl alcohol, propanone (acetone)); *sensitisers*, which produce an allergic response (*e.g.* natural rubber latex proteins, nickel, chromates, methanol (formaldehyde)); *carcinogens*, capable of causing cancer (*e.g.* asbestos, arsenic, beryllium, cadmium, chromium, nickel, polycyclic aromatic hydrocarbons and aromatic amines); and *mutagens*, which can cause changes in the DNA of cells (*e.g.* benzene, lead, vinyl chloride).[95] Although these are often localised issues, collectively they are not insignificant; each year occupational carcinogens cause over 150 000 deaths worldwide.[96]

- **Are natural chemicals safer than artificial ones?**
  This is of course a common misconception. There are many accounts of naturally-occurring substances causing health and environmental debates. Take formaldehyde (methanal) for example. This section demonstrates how one chemical can elicit a range of issues. This is a naturally-occurring chemical, common as an intermediate product of burning carbon compounds (*e.g.* in forest fires, tobacco smoke and car exhaust). It is rapidly broken down by sunlight or by bacteria present in soil or water and does therefore not persist in the environment. Formaldehyde also occurs widely in living organisms as an essential cellular metabolite and is found in the bloodstream of animals, including humans, but it does not accumulate. It is metabolised quickly into formic acid, which in turn is converted into carbon dioxide or excreted in urine. However, formaldehyde is more widely recognised by its use in many industrial processes as a precursor to more complex compounds, and is thus important in producing a huge range of products from facial tissues and crease-resistant fabrics, to foam insulation, paints, adhesives and explosives.

  People often associate formaldehyde with the more familiar term formalin and incorrectly perceive the two as synonymous. Formalin is an aqueous solution of formaldehyde gas with added methanol to prevent it polymerising. Formalin is used as a disinfectant, effective at killing bacteria and fungal spores, and is also used in embalming and preserving human and animal tissues in 'specimen jars', which is perhaps how it is most familiar to the public. This image, coupled with recent international consensus that formaldehyde over certain concentrations is a human carcinogen,[97] defines formaldehyde as a dangerous chemical in the eye of the public, and it is therefore not surprising that several SSIs have developed around it.

  There was public outrage in Indonesia in 2005–6 when it was discovered that formaldehyde was being added to foods to extend shelf life. The 'Indonesia food scare' emerged in Jakarta when the government reported that 60% of noodle shops were serving noodles containing formaldehyde and people were arrested. Similar reports later came from Vietnam and Thailand about formaldehyde in tofu, fish and chicken. However, the amount of the chemical present was never clear, and the media added to the furore using phrases like 'the corpse-preserving chemical' and with emotive opening lines in newspaper columns as in the Sydney Morning Herald (9 January 2006): "It's used for everything from embalming bodies to manufacturing fertilisers and fungus killers. Now the chemical formaldehyde is at the centre of massive food scare in Indonesia."

More recently there has been a scare in western countries about formaldehyde, now widely perceived as a carcinogenic chemical and an embalming fluid, being present in vaccines for children. Another of the many uses of formaldehyde is to inactivate viruses and bacterial toxins (*e.g.* poliovirus, Hepatitis B antigen, diphtheria and tetanus toxins) in the very early stages of the vaccine manufacturing process. Then the formaldehyde is diluted out, but trace amounts may remain in the final vaccine. Parents understandably are concerned about injecting anything into their children, and some campaign groups still insist it could be dangerous. However, scientists and health experts are in agreement that the amount is far too small to cause any harm. There is around 1.1 mg of naturally-occurring formaldehyde in a 2-month-old baby's blood; this is ten times more than the amount found in any vaccine (less than 0.1 mg).

- **The importance of concentration of chemicals**

  It is clear that this issue, as with all SSIs, requires us to consider the importance of levels of chemical concentration and the associated risks. All chemicals which are beneficial (and indeed essential), are also harmful in high enough concentrations, and we need to be aware of these thresholds. There are some very useful teaching resources explaining this, including 'The Dose Makes the Poison',[98] which includes a series of infographics, one of which highlights visually the amount of formaldehyde in vaccines:

  > *"The amount of formaldehyde in a vaccine is so tiny that it doesn't even affect the naturally-occurring levels of formaldehyde in a child's blood."*
  >
  > *"A 200 g pear contains up to 12 000 μg of formaldehyde naturally."*
  >
  > *"Vaccines contain up to 100 μg, or 0.83% of the formaldehyde in a pear."*

  Other thought-provoking infographics in the series include statistics on lethal doses (LD–50) of common everyday substances such as:
  - Toothpaste 33 tubes
  - Espresso coffee 78 shots

- Water 6 litres
- Alcohol (40% ABV) 13 shots

Can We Remove the Controversy?

Some SSIs revolving round particular chemical substances do not necessarily include disagreement over 'thresholds' arguments. One is the concern about people living in granite houses being exposed to radioactive radon. Some granite contains 10–20 ppm uranium, which decays to release radioactive radon gas and dust. The dust is inhaled and trapped in our airways and emits radiation that damages our lungs, increasing the risk of lung cancer. This is a different kind of issue from those above because there is a general consensus now that this is a serious issue – there is little disagreement. Public Health England (PHE) states that radon increases the risk of lung cancer and causes over 1100 deaths from lung cancer each year in the UK.[99] Social factors significantly increasing risk involved include the length of exposure, and interestingly, whether you are a smoker (or even an ex-smoker). PHE recommends that radon levels should be reduced in homes where the average is more than 200 becquerels per metre cubed (200 $Bq\,m^3$), and this recommendation has been endorsed by the UK Government. However, this consensus has not removed controversy from the issue, it has simply refocussed the controversy onto a different aspect. A mapping study revealed that up to 600 000 homes may fall into this high risk category, making millions of people vulnerable. The socially acute question now therefore is what should be done for these people? Remedial action to protect the houses is very costly and expecting people to move home has profound economic and social impacts.

It is worth noting that chemical deficiency can also have a social impact. A well-known example is an iodine-deficient diet leading to endemic goitre, also known as 'Derbyshire neck', a swelling of the neck resulting from enlargement of the thyroid gland and common among the people in Derbyshire for many centuries. Scientific investigations revealed that it was most prevalent in Carboniferous limestone areas of the Derbyshire-Peak District. On limestones, the soil pH environment ensures that most soil iodine is bound as the iodate anion, which is less volatile and therefore not available for absorption into plants, preventing it from entering the human food chain. Goitre declined in the area in the 1930s when the population's diet improved with consumption of produce from outside the region.

However, the UK's iodine deficiency problem is still with us. The country is the seventh most iodine-deficient nation in the world.[100] Recent studies in the UK have revealed iodine deficiency among young and pregnant women, which is of particular concern as demand for iodine increases during pregnancy and lactation and iodine intake is important in pregnancy for foetal brain development. The WHO recommends that iodine deficiency should be corrected through iodisation of household table salt, but in contrast to other countries, availability of iodised salt in supermarkets is low, table salt accounts for only a small percentage of total UK salt consumption, and UK public health campaigns are encouraging people to reduce their salt intake. So, with the building scientific evidence that low maternal iodine concentrations during pregnancy are linked with reduced verbal intelligence quotient (IQ) and reading abilities in children,[101] this is becoming a SSI. At the time of writing, the issue has not yet caught the attention of public health authorities or politicians. There is no government drive to respond to dietary experts' call for educational campaigns to increase awareness of the importance of iodine during pregnancy, and to promote the consumption of iodine-rich foods, such as milk and white fish. In Australia and Denmark, the governments have mandated the use of iodised salt in the production of products such as bread that are consumed by a large proportion of the population, but in the UK a mandatory iodine supplementation policy is so far proving to be too extreme.

Acknowledgements

The PARRISE project was funded by the European Commission (grant agreement 612438).

References

1. Disposable Packaging: Coffee Cups, Second Report of Session 2017–19. House of Commons Environmental Audit Committee, https://publications.parliament.uk/pa/cm201719/cmselect/cmenvaud/657/657.pdf (accessed 23 July 2018).
2. W. Poortinga and L. Whitaker, *Sustainability*, 2018, **10**, 873.
3. M. Ratcliffe and M. Grace, *Science Education for Citizenship: Teaching Socio-scientific Issues*, Open University Press, Maidenhead, 2003.
4. R. Levinson, *Sisyphus-J. Educ.*, 2017, **5**, 25–39.
5. D. L. Zeidler, T. D. Sadler, M. L. Simmons and E. V. Howes, *Sci. Educ.*, 2005, **89**, 357–377.
6. D. A. Roberts, in *Handbook of Research on Science Education*, ed. S. K. Abell and N. G. Lederman, Routledge, New York, 2006, vol. 1, ch. 28, pp. 729–780.
7. D. A. Roberts, in *Exploring the Landscape of Scientific Literacy*, ed. C. Linder, L. Ostman, D. A. Roberts, P.-O. Wickman, G. Ericksen and A. MacKinnon, Routledge, New York, 2011, ch. 3, pp. 11–27.
8. O. de Jong and K. S. Taber, in *Handbook of Research on Science Education*, ed. N. G. Lederman and S. K. Abell, Routledge, New York, 2014, vol.2, ch. 23, pp. 457–480.

9. J. K. Gilbert, *Int. J. Sci. Educ.*, 2006, **28**, 957–976.
10. A. M. Bulte, H. B. Westbroek, O. de Jong and A. Pilot, *Int. J. Sci. Educ.*, 2006, **28**, 1063–1086.
11. J. Bennett and J. Holman, in *Chemical Education: Towards Research-based Practice*, ed. J. K. Gilbert, O. De Jong, R. Justi, D. F. Treagust and J. H. Van Driel, Springer, Dordrecht. 2002, ch. 7, pp. 165–184.
12. Inspiring, engaging and expert: the formula for world-class science and chemistry education, http://www.rsc.org/globalassets/04-campaigning-outreach/campaigning/specialist-teachers-report.pdf, (accessed 23 July 2018).
13. Public attitudes to chemistry, http://www.rsc.org/globalassets/04-campaigning-outreach/campaigning/public-attitudes-to-chemistry/public-attitudes-to-chemistry-research-report.pdf, (accessed 23 July 2018).
14. T. Mujtaba, R. Sheldrake, M. J. Reiss and S. Simon, *Int. J. Sci. Educ.*, 2018, **40**, 644–667.
15. M. Karpudewan, W.-M. Roth and Z. Ismail, *Asia-Pacific Edu. Res.*, 2015, **24**, 35–43.
16. R. Marks and I. Eilks, *Chem. Educ. Res. Pract.*, 2010, **11**, 129–141.
17. I. Eilks, R. Marks and M. Stuckey, *Educ. Quím.*, 2018, **29**, 9–20.
18. R. Levinson, *Int. J. Sci. Educ.*, 2006, **28**, 1201–1224.
19. T. D. Sadler, *St. Sci. Educ.*, 2009, **45**, 1–42.
20. Promoting Attainment of Responsible Research and Innovation in Science Education: Our approach. https://www.parrise.eu/our-approach/, (accessed 23 July 2018).
21. National curriculum in England: science programmes of study. Available at: https://www.gov.uk/government/publications/national-curriculum-in-england-science-programmes-of-study/national-curriculum-in-england-science-programmes-of-study (accessed 23 July 2018).
22. A. R. Cavagnetto, *Rev. Educ. Res.*, 2010, **80**, 336–371.
23. R. Duschl, *Rev. Res. Educ.*, 2008, **32**, 268–291.
24. R. Driver, P. Newton and J. Osborne, *Sci. Educ.*, 2000, **84**, 287–312.
25. R. Duschl and J. Osborne, *Stud. Sci. Educ.*, 2002, **38**, 39–72.
26. S. Erduran and M. P. Jimenez-Aleixandre, in *Argumentation in Science Education: Perspectives from Classroom-Based Research*, ed. S. Erduran and M. P. Jiménez-Aleixandre, Springer, 2008, ch. 1, pp. 3–27.
27. M. Ford, *Sci. Educ.*, 2008, **92**, 404–423.
28. M. Garcia-Mila, S. Gilabert, S. Erduran and M. Felton, *Sci. Educ.*, 2013, **97**, 497–523.
29. R. Alexander, *Towards Dialogic Teaching: Rethinking Classroom Talk*, 4th edn, Dialogos, Cambridge, 2008.
30. L. A. Bricker and P. Bell, *Sci. Educ.*, 2008, **92**, 473–498.
31. B. Koslowski, *Theory and Evidence: The Development of Scientific Reasoning*, MIT Press, Cambridge, Massachusetts, 1996.
32. T. D. Sadler, W. L. Romine and M. S. Topçu, *Int. J. Sci. Educ.*, 2016, **38**, 1622–1635.

33. A. Zohar and F. Nemet, *J. Res. Sci. Teach.*, 2002, **39**, 35–62.
34. G. J. Venville and M. Dawson, *J. Res. Sci. Teach.*, 2010, **47**, 952–977.
35. M. Ratcliffe, *Int. J. Sci. Educ.*, 1997, **19**, 167–182.
36. L. Simonneaux, in *Argumentation in Science Education: Perspectives from Classroom-based Research*, ed. S. Erduran and M. P. Jiménez-Aleixandre, Springer, 2008, ch. 9, pp. 179–199.
37. M. Stolz, T. Wittteck, R. Marks and I. Eilks, *EURASIA. J. Math. Sci. Tech. Ed.*, 2013, **9**, 361-370.
38. L. Zangori, A. Peel, A. Kinslow, P. Friedrichsen and T. D. Sadler, *J. Res. Sci. Teach.*, 2017, **45**, 1249–1273.
39. J. K. Gilbert and A. S. Afonso, in *Chemistry Education: Best Practices, Opportunities and Trends*, ed. J. Garcia-Martinez and E. Serrano-Torregrosa, Wiley-VCH, Weinheim, 2015, ch. 6, pp. 123–148.
40. M. Logan and K. Skamp, *Res. Sci. Educ.*, 2008, **38**, 501.
41. S. Simon and R. Amos, in *Socio-scientific Issues in the Classroom: Teaching, Learning and Research*, ed. T. D. Sadler, Springer, Dordrecht, 2011, ch. 10, pp. 167–192.
42. M. Burmeister and I. Eilks, *Chem. Educ. Res. Pract.*, 2012, **13**, 93–102.
43. N. Christenson, S. N. C. Rundgren and H. O. Höglund, *J. Sci. Educ. Tech.*, 2012, **21**, 342–352.
44. N. Becker, C. Rasmussen, G. Sweeney, M. Wawro, M. Towns and R. Cole, *Chem. Educ. Res. Pract.*, 2013, **14**, 81–94.
45. E. Seung, A. Choi and B. Pestel, *EURASIA. J. Math. Sci. Tech. Edn.*, 2016, **12**, 991–1008.
46. J. P. Walker and V. Sampson, *J. Res. Sci. Teach.*, 2013, **50**, 561–596.
47. T. D. Sadler and D. L. Zeidler, *J. Res. Sci. Teach.*, 2005, **42**, 112–138.
48. M. Juntunen and M. Aksela, *CEPS J. Center Educ. Policy Stud. J.*, 2013, **3**, 157–180.
49. M. K. Juntunen and M. K. Aksela, *Chem. Educ. Res. Pract.*, 2014, **15**, 639–649.
50. S. Erduran and E. Z. Mugaloglu, in *International Handbook of Research in History, Philosophy and Science Teaching*, ed. M. R. Matthews, Springer, Dordrecht, 2014, ch. 10, pp. 287-315.
51. T. D. Sadler and L. A. Donnelly, *Int. J. Sci. Educ.*, 2006, **28**, 1463–1488.
52. V. Albe, *Res Sci Educ.*, 2008, **38**, 67–90.
53. S. D. Kolstø, and M. Ratcliffe, in *Argumentation in Science Education: Perspectives from Classroom-based Research*, ed. S. Erduran and M. P. Jiménez-Aleixandre, Springer, 2008, ch. 6, pp. 117–136.
54. T. Patronis, D. Potari and V. Spiliotopoulou, *Int. J. Sci. Educ.*, 1999, **21**, 745–754.
55. T. D. Sadler, in *Socio-scientific Issues in the Classroom: Teaching, Learning and Research*, ed. T. D. Sadler, Springer, Dordrecht, 2011, ch. 1, pp. 1–9.
56. L. Simonneaux and J. Simonneaux, *Cult. Stud. Sci. Educ.*, 2009, **4**, 657–787.

57. D. L. Zeidler and T. D. Sadler, in *Argumentation in Science Education: Perspectives from Classroom-based Research*, ed. S. Erduran and M. P. Jiménez-Aleixandre, Springer, 2008, ch. 10, pp. 201–216.
58. D. L. Zeidler, in *Handbook of Research on Science Education*, ed. N. G. Lederman and S. K. Abell, Routledge, New York, 2014, vol. 2, ch. 34, pp. 697–726.
59. O. Morin, L. Simonneaux, J. Simonneaux, R. Tytler and L. Barraza, *Sci. Educ.*, 2014, **98**, 517–542.
60. O. Morin, L. Simonneaux and R. Tytler, *J. Res. Sci. Teach.*, 2017, **54**, 825–851.
61. T. D. Sadler, S. A. Barab and B. Scott, *Res Sci Educ.*, 2007, **37**, 371–391.
62. N. Mercer, *The Guided Construction of Knowledge: Talks Amongst Teachers and Learners*, Multilingual Matters, Clevedon, England, 1995.
63. M. Hand and R. Levinson, *Educ. Phil. Th.*, 2012, **44**, 614–629.
64. G. S. Aikenhead, *Science Education for Everyday Life: Evidence-based Practice*, Teachers College Press, New York, NY, 2006.
65. J. F. Donnelly, *Sci. Educ.*, 2004, **88**, 762–784.
66. J. F. Donnelly and J. Ryder, *Hist. Educ.*, 2011, **40**, 291–313.
67. W.-M. Roth and A. Calabrese-Barton, *Re-thinking Scientific Literacy*, Routledge Falmer, New York, 2004.
68. M. Felton, M. Garcia-Mila and S. Gilabert, *Inf. Log.*, 2009, **29**, 417–446.
69. O. Acar, L. Turkmen and A. Roychoudhury, *Int. J. Sci. Educ.*, 2010, **32**, 1191–1206.
70. R. Mamlok-Naaman, D. Katchevich, Y. Malka, M. Burmeister, T. Feierabend and I. Eilks, in *World Wide Trends on Green Chemistry*, ed. Z. Vania and L. Mammino, Royal Society of Chemistry, UK, 2015, ch. 4, pp. 45–60.
71. R. Owen, P. Macnaghten and J. Stilgoe, *Sci. Publ. Pol.*, 2012, **39**, 751–760.
72. A report on responsible research and innovation, http://www.diss.unimi.it/extfiles/unimidire/243201/attachment/a-report-on-responsible-research-innovation.pdf, (accessed 23 July 2018).
73. M. Geoghegan-Quinn, *Proceedings of the Science in Dialogue–Towards a European Model for Responsible Research and Innovation*, Denmark Odense, 2012.
74. R. Blonder, E. Zemler and S. Rosenfeld, *Chem. Educ. Res. Pract.*, 2016, **17**, 1145–1155.
75. R. Levinson and PARRISE Consortium, in *Science and Technology Education Promoting Wellbeing for Individuals, Societies, Environments*, ed. J. L. Bencze, Springer, Dordrecht, 2017, ch. 21, pp. 477–502.
76. R. Levinson, M. C. Knippels, F. van Dam, *et al.*, , 2017, Socio-Scientific Inquiry-Based Learning: connecting formal and informal science education with society, https://www.parrise.eu/wp-content/uploads/2018/04/parrise-en-rgb.pdf, (accessed 23 July 2018).

77. D. Katchevich, A. Hofstein and R. Mamlok-Naaman, *Res Sci Educ.*, 2013, **43**, 317–345.
78. PARRISE Good Practices: Teacher Training Materials. https://www.parrise.eu/teacher_training_materials/ (accessed 23 July 2018).
79. R. Amos and A. Christodoulou, *Sch. Sci. Rev.*, 2018, in press.
80. K. L. McNeill, M. González-Howard, R. Katsh-Singer and S. Loper, *J. Res. Sci. Teach.*, 2016, **53**, 261–290.
81. T. Günter, N. Akkuzu and S. Alpat, *Res. Sci. Tech. Educ.*, 2017, **35**, 500–520.
82. P. Anastas and N. Eghbali, *Chem. Soc. Rev.*, 2010, **39**, 301.
83. S. Shamuganathana and M. Karpudewan, *Chem. Educ. Res. Pract.*, 2017, **18**, 386.
84. I. Eilks and F. Rauch, *Chem. Educ. Res. Pract.*, 2012, **13**, 57–58.
85. M. Burmeister, F. Rauch and I. Eilks, *Chem. Educ. Res. Pract.*, 2012, **13**, 59.
86. L. B. Armstrong, M. C. Rivas, M. C. Douskey and A. M. Baranger, *Curr. Opin. Green Sustainable Chem.*, 2018, **13**, 61–67.
87. M. Juntunen, Holistic and inquiry-based teaching of sustainable chemistry, *LUMAT-B Int. J. Math Sci. Technol. Educ.*, 2016, **1**(3).
88. L. Summerton, G. A. Hurst and J. H. Clark, *Current Opinion in Green and Sustainable Chemistry*, 2018, **13**, 56–60.
89. C. W. Keys, B. Hand, V. Prain and S. Collins, *J. Res. Sci. Teach.*, 1999, **36**, 1065.
90. V. Talanquer, *Int. J. Sci. Educ.*, 2011, **33**, 179–195.
91. A. Christodoulou, R. Amos, M. Grace and R. Levinson, Paper presented at the 12th Conference of the European Science Education Research Association (ESERA), Dublin, Ireland, 2017.
92. M. Grace, in *Learning to Teach Science in the Secondary School*, ed. R. Toplis, Routledge, London, 2015, 4th edn., ch. 5.4, pp. 220–237.
93. C. Jacobsen, W. Zemann, J. A. Obwegeser, K. W. Grätz and P. Metzler, *Oral Maxillofac. Surg.*, 2014, **18**, 31–37.
94. L. Raw, *Striking a Light: The Bryant and May Matchwomen and Their Place in Labor History*, Continuum, UK, 2009.
95. J. Wah Lim and D. Koh, *Chemical Agents That Cause Occupational Diseases*, Wiley Blackwell, 2014.
96. T. Driscoll, D. I. Nelson, K. Steenland, J. Leigh, M. Concha-Barrientos, M. Fingerhut and A. Pruss-Üstun, *Am. J. Ind. Med.*, 2005, **48**, 432–445.
97. Formaldehyde. https://ntp.niehs.nih.gov/ntp/roc/content/profiles/formaldehyde.pdf (accessed 23 July 2018).
98. The Dose Makes the Poison. https://www.chemicalsafetyfacts.org/dose-makes-poison-gallery/ (accessed 23 July 2018).
99. Public Health England: The risks to your health from radon. http://www.ukradon.org/information/risks (accessed 23 July 2018).

100. World Health Organisation: Global Database on Iodine Deficiency. http://www.who.int/vmnis/iodine/en/ (accessed 23 July 2018).
101. Iodine deficiency in the UK: grabbing the low-hanging fruit, http://www.btf-thyroid.org/images/documents/lancet_iodine.pdf (accessed 23 July 2018).

CHAPTER 10

Argumentation in Organic Chemistry Education

AYBUKE PABUCCU

Canakkale Onsekiz Mart University, Department of Mathematics and Science Education, Anafartalar Campus, 17100, Canakkale, Turkey
Email: apabuccu@comu.edu.tr

10.1 Why Do Some Students Struggle with Organic Chemistry?

Researchers have pointed out that what organic chemists take for granted in their own problem-solving is not comparable to how students solve problems, and it takes students many years to reach a similar state of reasoning.[1] Cartrette and Bodner (2009) compared the problem-solving ability of organic chemistry graduate students and faculty to investigate the important factors of development of expertise in problem-solving.[2] The researchers determined unsuccessful problem-solvers often used single features or isolated facts instead of multiple features presented in the problem statement. Indeed, the problems in organic chemistry are classified regarding their content (*i.e.*, mathematical, non-mathematical, and mechanistic problems). The use of mechanisms is probably the most challenging part for students in organic chemistry classes.[3] Bhattacharyya and Bodner (2005) investigated the extent to which organic chemistry graduate students made use of the electron-pushing formalism while proposing mechanisms for SN1, SN2, and Diels–Alder reactions.[4] They found that students had limited understanding of the implicit meaning of the curved-arrow formalism. Ferguson and

Advances in Chemistry Education Series No. 2
Argumentation in Chemistry Education: Research, Policy and Practice
Edited by Sibel Erduran

Published by the Royal Society of Chemistry, www.rsc.org

Bodner (2008) also used common mechanistic problems, which required the use of arrows.[5] They determined that students relied on single, unrelated, and erroneous memorized pieces of knowledge and showed little substantial conceptual understanding of the basic concepts (*e.g.*, acid–base chemistry). The inability or unawareness to recall or to apply the appropriate content knowledge was the main barrier that prevented the students from making correct assumptions or proposing reasonable mechanistic steps. That students used terms such as 'electrophile,' 'reduction,' and 'electrophilic addition' as empty envelopes that held no further meaning was also apparent in the study.[5]

Organic chemistry is one of the most visual sciences, considering the generation and interpretation of mainly domain-specific symbols and structural representations that have no counterparts in our daily lives. Lewis structures, stereochemical information, Fischer and Newman projections, and different 2D and 3D representations are just a few examples of how chemists visualize chemical information.[3] Manipulating, translating between, and correctly interpreting these representations are huge challenges to most of the students in a chemistry class,[6] requiring various cognitive skills. The cognitive skills currently under investigation that influence students' performance in organic chemistry classes can be organized into representational competence, spatial ability, and scientific reasoning strategies.[3] Spatial abilities play an important role in students' performance at organic chemistry,[7] however, many organic chemistry students struggle to understand the different spatial arrangements of the molecules and the graphic representations.[8] Strickland *et al.* (2010) investigated the representational competence of organic chemistry graduate students by analysing the relationship between students' understanding of common organic terms—like acid–base, functional group, or electrophile/nucleophile and their verbalization of the corresponding structural representations.[9] They described that even at graduate level students' explanations were often based on very superficial information rather than on process-orientated attributes. This behaviour led them to pay more attention to structural change and limited their interpretation of structural representations. Students could verbally explain the general behaviour of electrophiles or nucleophiles, but they had trouble identifying this behaviour in the mechanisms presented. Besides the question of how students construct and translate between various types of representations and representational conventions, a host of other studies focused on the nature of students' reasoning strategies that may influence their performance in organic chemistry. Kraft *et al.* (2010) undertook a qualitative study to understand what kind of cues organic graduate students used for generating meaning while engaged in mechanistic tasks.[10] Their research focused on the identification of the reasoning strategies that students used. In their analysis, they found that many of the students were using case-based or rule-based reasoning strategies. Only a few were engaged in model-based reasoning, which is the more successful reasoning strategy. Rule- and case-based reasoning were often triggered by single cue associations, an experience in class, or a memorized rule (*i.e.*, 'nucleophiles attack electrophiles' and some molecules are 'good leaving groups'). This allowed students in this study to reproduce memorized sequences of steps or mechanisms without a complete understanding. Kraft *et al.* (2010) discovered that the rules mentioned were often correct but remained factual and were not taken to judge reactivity or to decide between different possible mechanistic steps.[10]

In addition to aforementioned research, some researchers focused on the overall experience in learning in organic chemistry. For instance, Grove and Bretz (2012) carried out an investigation to better understand the difficulties that college students encountered in learning organic chemistry and the reasons why students had the difficulties that they did.[11] The researchers emphasized the central role of the learner in constructing knowledge in their study. Indeed, they believed that the learner must consciously choose to learn meaningfully by seeking connections and not default to rote memorization. They described four groups of students regarding the learning strategies that they employed to learn organic chemistry: Meaningful Learners, Transitional Learners, Unaware Learners, and Indifferent Learners. They characterized students as meaningful learners if they remarked how important it was for them to leaving the course with a meaningful understanding of the material. These students' commitments to learning in a meaningful manner also influenced their perspective on learning.

Although all meaningful learners prepared for exams quite differently from one another, they all focused upon internalizing course material to make key concepts more personally meaningful. For instance, George stated that he frequently created his own study guides to synthesize the important concepts from the textbook and lecture notes while Luna made sure that she could explain key concepts to her study partner in her own words. Afterwards, she and her study partner created reaction and synthesis problems for each other and exchanged them for additional practice. Transitional Learners were described as the students who were not initially focused on meaningful learning at the beginning of their study of organic chemistry. However, they did undergo significant changes in their approach to learning over the year and transitioned to more meaningful approaches. Indeed, these students may occasionally decide to use rote memorization techniques during periods of stress. Moreover, unlike meaningful learners, they were less certain about organic chemistry's relevance to their chosen professions. Lastly, both Unaware and Indifferent Learners were defined as relying upon rote memorization techniques to learn organic chemistry. However, what differentiated these two groups was the recognition by Indifferent Learners that more meaningful approaches did exist, yet they chose not to use them. For instance, one Indifferent Learner admitted that that he does not enjoy organic chemistry, so he does not want to spend more time with it. Thus, he preferred to memorize as many example problems as he could, and he hoped he would see the same or similar problems in exams. The results of this study strongly suggest that students strongly rely on rote memorization and that the biggest factor that hindered meaningful learning was the perceived lack of relevancy of the class itself. Besides, the researchers observed that more meaningful learning techniques and low metacognitive awareness were mutually exclusive. That is, they concluded that metacognition plays a crucial role in helping students to recognize that the more meaningful ways to learn organic chemistry exist.

Another study focusing on the overall performance of undergraduate students in organic chemistry was conducted by Anderson and Bodner (2008).[12] The researchers exemplified the experience of many students who had been successful in general chemistry but struggled in organic chemistry. In a case study about a student named Parker, they gave an overview of the emergent problems during an organic class. In the study, Parker admitted that his trouble with reaction mechanisms did not come from an inability to understand individual reaction mechanisms; it came from a difficulty with understanding the purpose of reaction mechanisms. Parker believed that understanding why reactions occurred was critical because it gave him the ability to apply what he was learning in new contexts. According to the researchers, there was no doubt that the instructor tried to present "the why of organic chemistry reactions" to the students. After all, the important question here is that why Parker still feels that organic chemistry was not about whys. Anderson and Bodner (2008) believed that Parker's inability to attribute useful meaning to chemical symbols contributed greatly to his feeling.[12]

For Parker, the letters, lines, dots, and arrows did not symbolize anything that corresponded to physical reality for him. Besides, Parker's inability to find answers to questions of why led Parker to become an instrumental learner as opposed to relational learner. An instrumental learner would approach organic chemistry by focusing on the rules for reactions. For instance, Parker was searching the textbook for important rules that he then highlighted for ready reference as he solved problems. Eventually, he found himself encumbered with so many rules that he had difficulty remembering them. Moreover, he often applied them inappropriately or in inappropriate contexts. Thus, Parker recognized that his strategy of learning rules was not the best way to learn, but he could find no other approach. An important source of Parker's problems with organic chemistry was the pace of the course, which was so fast that he felt it was impossible for the instructor to devote time to helping students understand the conceptual underpinnings behind why reactions occurred the way they did. Furthermore, Parker had difficulty selecting the appropriate conceptual knowledge to apply to a problem. In his attempt to transfer what he knew from general chemistry to the problems he encountered in the organic course, Parker tended to focus on the properties of individual atoms or elements, rather than look at the structure and properties of a whole molecule. He seemed to view a molecule as a collection of atoms, rather than as a system of electrons. Lastly, it is seen that Parker focused on the products of a chemical reaction instead of taking a process-orientated perspective on mechanistic problem-solving tasks. However, the use of mechanisms requires switching from a mostly product-orientated thinking in general chemistry to a more process-orientated thinking.

As described in the presented studies, the rules and principles learned in the general chemistry class are often not translated appropriately into the organic chemistry classroom. For instance, many domain-specific organic concepts—such as aromaticity and resonance—are not straightforward for students, as the mindset in general chemistry focuses on single atoms or small entities. Besides, students had various misconceptions about some general chemistry concepts ubiquitously used in organic chemistry. For example, Brønsted acid–base knowledge and skills are required in greater than 85% of the more complex organic and biochemical reactions. Recognizing acidity and comparing compounds necessitates a robust understanding of acidic strength and the application of the acid–base concepts in use, the Brønsted–Lowry and Lewis theories. Bhattacharyya (2006) undertook a qualitative study with graduate students to determine the nature of their mental models when applying different acid–base concepts. The most often stated characteristic to describe acidity was 'bond strength,' whereas steric or solvent effects were less often mentioned. Their explanations were often based on one recalled characteristic but did not consider the interplay of multiple aspects of acid–base theory.[13]

Some other studies have taken a closer look to define what kind of intuitive thinking students use to determine acid–base properties[14] and how intuitive thinking influences the quality of their decision-making process. McClary and Talanquer (2011) described various heuristic strategies in their study with undergraduate organic chemistry students while engaged in ranking acids, especially organic molecules.[15] The participants in this qualitative study were science and chemistry majors enrolled in their first-year organic chemistry course. They noticed various heuristics that allowed students to eliminate cues and to focus on one single attribute, for instance the presence of a functional group that is considered to be an acid. Cartrette and Mayo (2011) carried out a qualitative study with organic chemistry majors and investigated how students solved organic problem-solving exercises that required the application of acid–base theory.[16] They reported that although organic chemistry majors, students were able to compare the acidity of organic molecules in terms of resonance, inductive effects and electronegativity, they struggled to apply the concepts while doing their problem-solving exercises. The researchers assumed that a poor understanding of the Lewis acid concept prevented students from drawing solid connections between acid–base concepts and the terms electrophile and nucleophile. Besides, students often considered the relationships between nucleophiles (electrophiles) and bases (acids) only when explicitly prompted to do so.

The aforementioned discussion suggests that students' experiences in organic chemistry are influenced by a complex set of factors. However, it seems that many of the difficulties that students encounter in learning organic chemistry ultimately stem from an over-reliance on rote memorization without using more meaningful techniques.

10.2 Research on Argumentation in Organic Chemistry

Research on argumentation in organic chemistry is scarce. Some researchers focused on the nature of students' reasoning. For instance, De Arellano and Towns (2014) designed a study to investigate students' understanding of alkyl halide reactions in an undergraduate organic chemistry course.[17] They interviewed 22 undergraduate students, ranging from their second to fourth year of study. The researchers used a think-aloud protocol with a set of questions dealing with reactions and mechanisms of alkyl halide molecules. They used Toulmin's model of argumentation to identify students' source and quality of reasoning. Those students who were successful showed an appropriate connection between the property of a reagent, nucleophile, or base and the corresponding reaction type, SN1, or elimination reaction as well as knowledge about intermediate and mechanistic steps of the reactions. However, they found many students seemed to be able to produce the right product even without a substantial understanding. These results provide further evidence that a constant focus on correctly using and applying the basic chemical concept in organic chemistry is crucial to improve the organic chemistry classroom practice. Furthermore, they stated that students who try to rationalize mechanisms instead of reproducing them should be valued for their effort.

Additionally, Christian and Talanquer (2012) studied the use of reasoning strategies that science and engineering majors used in self-initiated study groups in an undergraduate organic chemistry course.[18] They found the same predominant use of rule- and case-based reasoning at the undergraduate level and defined a fourth reasoning mode: symbolic reasoning mode. They used it to classify students' argumentation when they mainly manipulated representations like atoms or bonds on a purely symbolic level without a clear reference to their chemical nature. Students primarily used case- and symbolic reasoning while talking about reactivity and mechanisms and spent much more time on static representations instead of discussing process-orientated mechanistic issues.[18] They further argued that the results on the use of reasoning strategies applied by students studying organic chemistry seemed to be a homemade problem. A huge amount of time in class is spent on learning to construct, use and translate structural representations and to visualize structures, which predominantly require rule- and case-based reasoning. Often exams or in-class assessments are organized in a comparable way and influence this narrowed learning focus. This explains why the main cognitive processes students used in their study were basic cognitive processes, such as remember and apply and why they utilized few higher-level cognitive processes involving an evaluative or critical analysis.

Hand and Choi (2010) investigated students' use of multiple modal representations (such as graphs, drawings, and mathematical or chemical equations) within their written arguments as a consequence of completing a series of investigations of an organic chemistry laboratory course.[19] One hundred and eleven students were involved in using the Science Writing Heuristic approach where they are required to use the argument structure of question, claim, evidence and reflection in completing the written report for their instructor on their laboratory investigations. They discovered a connection between students' quality of arguments given for an explanation and their use of representations in their lab book. That is, students who were able to embed multiple modal representations in evidence made strong reasoned connections to support their claim(s) and construct a cohesive argument. Further, this research revealed that the understanding of a concept or a model can be related to the way the corresponding representations are constructed. Thus, this study suggests there is a need to build support structures pedagogically for the individual to help students understanding the role and function of multiple modal representations in science.

10.3 Example Activities to Support the Incorporation of Argumentation in Organic Chemistry Lessons

Apart from the presented studies above, Erduran and Pabuccu (2012) developed three activities (Alkanes in Stress, Organic Potions, Cheating on the Exam) to promote the incorporation of argumentation in organic chemistry lessons.[20] For all activities, the students were expected to work in small groups to complete writing frames which were designed to structure their written arguments. In this chapter, I will present two of them. First one, named Alkanes in Stress, was about conformational analysis of butane (see Figure 10.1) and that required the engagement of spatial abilities. As seen in Figure 10.1, five possible Newman Projection formulas for the conformations of butane were given in the activity. Then, students were asked to choose which four to correctly place on the graph at erased points. One of the formulas was isobutane and it was included in the activity as a distractor. Because it was not butane, it was expected to be labelled as 'none' to indicate not having a corresponding place on the graph. For the activity, students had to place missing formulas on a graph, and then produce the best explanation to support their choices. Researchers utilized this activity several times in organic chemistry classrooms.[8,21] On the first trial, Erduran and Pabuccu (2015) studied with 46 second-year pre-service science teachers taking an organic chemistry class.[21] Then, 89 pre-service teachers from two classes participated in their study.[8] Each time the groups were observed, and audio recorded to investigate the nature of their argumentation. In the analysis, the researchers focus was the number of rebuttals in the transcripts. Toulmin defined rebuttals as the exceptional circumstances under which the claim would not be true.[22] Both studies determined that it was not easy for pre-service teachers to understand the different spatial arrangements of the molecules, so they spent most of their time discussing them. This implies that spatial abilities play an important role in pre-service teachers' engagement in argumentation. Pabuccu and Erduran, (2017) also found that the

Question 1

The graph below shows the changes that arise from rotation about the C2-C3 bond of butane. Here are the two Newman projection formulas for the conformations of butane. The other four Newman projection formulas for the conformations of butane in the graph were erased and they replaced with the letters A, B, C and D respectively.

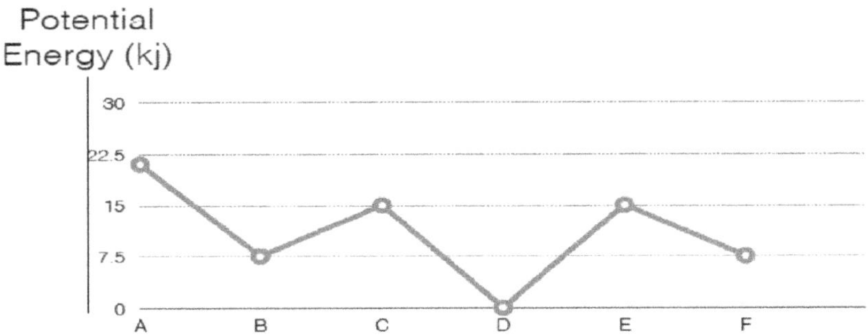

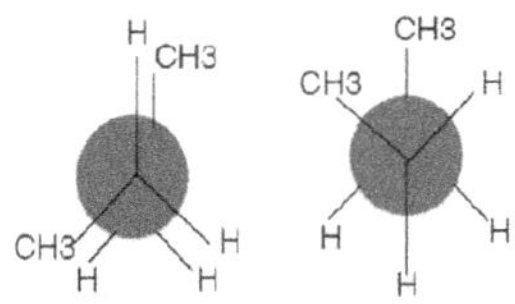

Below are the five different Newman projection formulas. Look at them and try to estimate the erased Newman projection formulas in the graph. Complete the below table and use the evidence from the evidence cards to substantiate your claims.

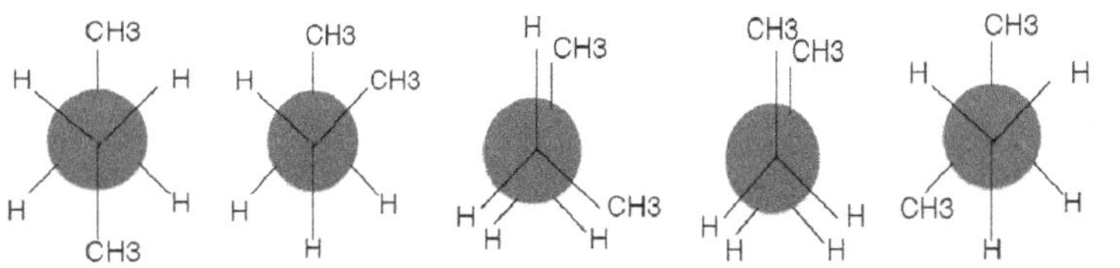

Figure	A/B/C/D/None	Type of Strain	Evidence
I			
II			
III			
IV			
V			

Figure 10.1 Student exercise for Alkanes in Stress activity resources. Reproduced from page 16 of ref. 20.

activity managed to engage pre-service teachers to make multiple rebuttals in their discussions and the discussions ended up improving pre-service teachers' understanding the conformational analysis of butane.[8]

Another activity developed by Erduran and Pabuccu (2012), named Cheating on the Exam (see Practical Digest), gave students an opportunity to construct arguments for the compound given in the reaction classifying as

STUDENT EXERCISE
Scenario In an organic chemistry exam, the teacher found some organic chemistry reactions written on Tom's desk. The teacher warned Tom that she would fail him. Tom objected because he believed that there were no questions in the exam about the reactions written in his desk. Having looked at the Tom's desk, April, another student in the class, was not in agreement in Tom and she said the first question of the exam could be answered with the help of the reaction mechanisms written on the desk. Who was right, Tom or April? Whom do you agree with and why? Try to justify your decision by looking at Tom's desk and the first question of the exam. **1st question of the exam** Decide whether each compound given below is (a) a nucleophile or an electrophile, both or neither (b) Brønsted Acid or Lewis acid, both or neither. *In the reaction of Alcohols with Hydrogen Halides,* Tert-Butyl alcohol, Hydrogen chloride, tert-Butyloxonium ion; tert-Butyl cation and H_2O *In the reactions of Electrophilic addition of HX to alkenes* Alkene and HX *In the Nucleophilic substitution reactions* Methyl bromide and Hydroxide ion *In the reactions between carbocation and Halogen anion* Carbocation and Halogen anion

Figure 10.2 First page of the student exercise for the Cheating on the Exam activity resource.
Reproduced from page 26 of ref. 20.

(a) a nucleophile or an electrophile, and (b) a Brønsted acid or Lewis acid and to use the evidence from the cards to justify their claims (see Figures 10.2–10.4).[20] Besides, students were expected to evaluate the evidence presented on the cards (see Table 10.1) and select them to support their points of view about the compounds' behaviour in the reactions. Since some of the evidence can be ambiguous and could indicate that the

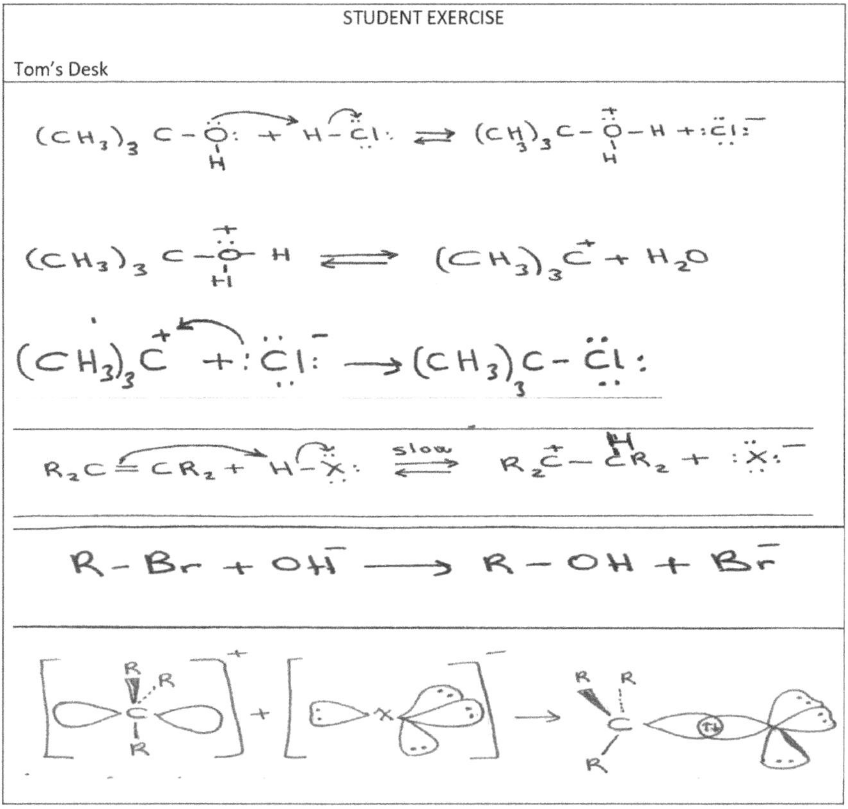

Figure 10.3 Second page of the student exercise for the Cheating on the Exam activity resource.
Reproduced from page 27 of ref. 20.

compound in the reaction is both an electrophile and a nucleophile, the activity provides an opportunity to generate cognitive conflict and argumentation for students. For this activity students need to know the concepts of proton, electron, covalent bond, electrophile, nucleophile and mechanism of organic reactions and the definitions of acids and bases.

For the activity, students were asked to support their classifications with evidence cards. They do not need to use all evidence cards (see Table 10.1). Besides, some cards contain irrelevant or wrong information. Indeed, the activity asks for the ability to distinguish between relevant information from irrelevant information, and to evaluate the accuracy.

I utilized this activity in my organic chemistry classes with 60 second-year pre-service science teachers at a public university in Turkey. The students discussed the questions in small groups. 13 groups were utilized for the activity. Data were collected through writing frames and audio recordings of the group discussions. The groups were observed, and audio recorded to investigate the nature of their argumentation. In the data analysis, many

STUDENT EXERCISE

Nucleophile or Electrophile?

Decide whether each compound given in the first question of the exam whether it is a nucleophile or an electrophile, both or neither. Then write down your evidence to justify your decisions in the table that follows.

Nucleophile	**My evidence**	**Electrophile**	**My evidence**
Both	**My evidence**	**Neither of them**	**My evidence**

Bronsted Acid or Lewis acid?

Decide each compound given in the first question of the exam whether is Bronsted acid or Lewis acid, both or neither. Then write down your evidence to justify your decisions in the table that follows.

Bronsted Acid	**My evidence**	**Lewis acid**	**My evidence**
Both	**My evidence**	**Neither of them**	**My evidence**

Figure 10.4 Third page of the student exercise for the Cheating on the Exam activity resource.
Reproduced from page 28 of ref. 20.

misconceptions about nucleophiles, electrophiles, and acid–base theories were determined in the explanations. Pre-service science teachers had particular difficulties about the Lewis theory, and they confused the Arrhenius and Brønsted–Lowry Theories. Moreover, it was observed that a few groups managed to make multiple rebuttals in their discussions and each of them (higher-level argumentation) ended up changing student's misconception(s). Thus, the results implied that engaging in higher-level argumentation helped them to better understand the organic chemistry topic they were discussing.

Table 10.1 Evidence cards for the Cheating on the Exam activity. Reproduced from page 30 of ref. 20.

No	Cards
1	It gives protons
2	It takes protons
3	It gives H^+ ions into the solution
4	It gives OH^- ions into the solution
5	It is looking for electrons
6	It is looking for the nucleus
7	It has a non-bonding electron pair
8	It is positively charged
9	It is negatively charged
10	It is neutral
11	It has carbon with a sp^2 hybrid orbital
12	Carbon has empty orbitals
13	It is a tertiary carbocation
14	It is an acid–base reaction

10.4 The Effects of Incorporating Argumentation into the Organic Chemistry Classroom

The main aim of the chapter was to present argumentation as a strategy to improve meaningful learning in an organic chemistry class and so weaning students away from rote learning. Often students consider organic chemistry as nothing more than memorization that cannot involve any argumentation. One of the reasons students tend to resort to a memorization-orientated approach to learning was that they do not understand the purpose of reaction mechanisms. They simply do not comprehend why chemicals behaved the way they did. One students from the Anderson and Bodner (2008) study, named Parker, noted:[12]

> *"I don't feel like I know what I'm doing because I don't know why. And I think that's what most kids struggle with is that you don't have any why. You don't have any knowledge of what's really going on, so you're just kind of floundering and memorizing. And you get through it and you say, "it's done, it's behind me." And you don't realize that there's a whole wealth of knowledge that could be tapped into and applied, that's there, but you don't understand the reasons why, so you can never apply it."*

In regular organic chemistry classes, the 'whys' were communicated to the students through structures and mechanisms that utilize lines, letters, curved arrows, that held no physical meaning for some students.[4,12] This allowed students to reproduce memorized sequences of steps or mechanisms without a substantial understanding of the chemistry.[3,10] It is rare to find instructors who devote extensive class time to developing an understanding of the reason why the electron-pushing formalism is used by organic chemists.[12] However, it takes a significant amount of time before line structures and mechanisms truly become symbols for students.[1] Moreover, many studies have suggested that instructors need to spend more time helping students to understand the conceptual underpinnings behind why reactions occur the way they do. In this context, using an argumentation activity such as Cheating on the Exam can be helpful to foster awareness of how and why a reaction is following a certain path. As previously mentioned, for this activity students were expected to evaluate the evidence presented on the cards and select cards to support their points of view about the compounds' behaviour in the reactions (see Figures 10.2–10.4). For instance, to decide whether tert-butyl alcohol is a Brønsted acid or Lewis acid, both or neither, students were expected to review the mechanism of the reaction between alcohols and hydrogen halides. To make the classification correctly, they should be able to use the curved-arrow symbolism properly. Additionally, students needed to know the mechanism of a reaction to assess whether the reaction involved nucleophiles and electrophiles or not.[23] In summary, students were engaged in questioning how and why a reaction is following a certain path as well as the usefulness and relevance of the curved-arrow notation as a tool to explain mechanistic steps. In this way, this activity contributes to students' understanding of the implicit meaning of the curved-arrow formalism, so that the activity contributes to their mechanistic problem-solving ability and so their understanding of acid theories, nucleophiles and electrophiles. Here, the argumentation strategies embedded in the task encourage peer learning that extended beyond argumentation and had influence on conceptual understanding.

Furthermore, the Cheating on the Exam activity can help students to be successful in organic chemistry by switching their mechanistic problem strategy from product-orientated to process-orientated as this activity supports students use of process-orientated thinking. In the typical organic chemistry course, students focus on only the reactants and products, rather than taking a mechanistic approach to the problem.[12] Hence, many successful chemistry students struggle in organic chemistry. Another source of difficulty students encounter in organic chemistry results from their attempt to transfer the rules and principles learned in the general chemistry class into the organic chemistry class. For instance, a poor understanding of the Lewis acid concept prevented students from drawing solid connections between acid–base concepts and the terms electrophile and nucleophile.[16] Because the key concepts in the activity were acid/base theories, nucleophile and electrophile, the Cheating on the Exam activity could be useful for students to self-monitor their learning on these topics and so to actively and regularly reflect upon which strategies worked and which did not.[11] Thus, participation in argumentation can develop students' metacognitive awareness.

The Cheating on the Exam activity also helps students develop a deeper understanding of why they are expected to learn reaction mechanisms. In the typical organic chemistry course, instructors tell the students to use mechanisms to help them learn organic chemistry. But it is also clear that having been told something is not the same as having the student learn what we have told. The Cheating on the Exam activity provides them some problems that cannot be solved by resorting to the memorization of rules, so helps students understand the value of mechanistic reasoning. For instance, often organic chemistry students focused on the rules for reactions and they found themselves encumbered with so many rules that they have difficulty remembering them. Even though they recognized that their strategy of learning rules was not the best way to learn, most of them could find no other approach. Metacognition plays a crucial role in helping students to recognize that more meaningful ways to learn organic chemistry exist. Hence, using an argumentation activity provides an excellent opportunity for teachers to engage students in scientific reasoning, decision making, and reflection – all important aspects of metacognition.[24]

Further, the findings from research on general cognitive skills imply that students should constantly be engaged in the use of strategies to deal appropriately with structural representations in organic chemistry. The act of interpreting common representations should become a large part of the discourse in class. Thus, additional time should be spent in class on the application of these strategies, such as how to visualize molecular structures and how to translate between different molecular representations.[25] Strickland *et al.* (2010) claimed that one reason for the problems encountered could be the missing emphasis on metacognition, *i.e.*, students are not spending enough time in class on the critical analysis of their own constructed structural representations.[9] The argumentation activity developed by Erduran and Pabuccu (2012), named Alkanes in Stress, can be helpful to encourage students to reflect on their own reasoning and decision-making processes.[20] Besides, as previously mentioned, Erduran and Pabuccu (2015) found the activity to be helpful in enhancing pre-service teachers' conceptual understanding regarding organic chemistry concepts.[21] Indeed, the researchers observed pre-service teachers helped each other to understand different spatial arrangements of atoms relative to rotations about bonds during the higher-level argumentation.

10.5 Reflections as a Teacher Educator

I have utilized some argumentation activities (Alkanes in Stress, Organic Potions, and Cheating on the Exam) developed by Erduran and Pabuccu (2012) in my organic chemistry classes with pre-service science teachers.[20] Firstly, I should note that my students were not familiar with argumentation as a strategy for organic chemistry or any kind of science education. Indeed, our normal classroom discourse was predominantly monologic and focused on the facts and the rules. It seemed that there was not much to debate in organic chemistry for me and my students. However, such a discourse strategy does not function well when the goal of instruction is to promote students' understanding of organic chemistry topics and so there is a requirement to wean students away from rote learning. Besides, at the beginning of the semester, when I asked my students, most of my students stated that learning organic chemistry topics required rote learning. Thus, the main reason of my thinking about infusing of argumentation in my organic chemistry classes was to help my students to see there are other ways to memorizing. After the recognition that more meaningful approaches do exist, they can choose to use them.

In my experience, I can tell that my students had a better time when I utilized the Cheating on the Exam activity as compared to traditional organic classes. Before the activity, both my students and I had some concerns about the activity. From my point of view, I was wondering if the activity promotes students' argumentation about organic chemistry topics, because grouping students and asking them to generate an argument or to evaluate the evidence for a given phenomenon will not always result in a beneficial outcome for students.[26] Moreover, personally I believe that it is harder to initiate argumentation in an organic chemistry context than in a scientific or socio-scientific context. It was encouraging to see that the activity managed to engage students to in collaborative scientific argumentation. From my students' point of view, they were suspicious about the fun part of the activity as most of them admitted that that they do not enjoy organic chemistry, so they did not want to spend more time with it. Thus, some of them stated that they preferred to memorize as many example problems as they could instead of talking about organic chemistry. Hopefully, they were still willing to do the activity (without any grade or reward expectation) and they also changed their mind after the implication of the activity. Following the activity, they were eager to apply more argumentation activities before the exam. Surprisingly, most of the students stated that they did not understand how time flew during the activity and they admitted that they found the activity good for preparing for the exam. Some also stated that this activity was helpful because it gave them an opportunity to ask questions that they had never had time, opportunity courage to ask during traditional class time. Besides, some admitted that they recognized many of their misconceptions and they came across new questions that they had never thought of before.

Practical Digest

Teacher outline for the activity named Cheating on the exam[20]

Teacher outline

This activity requires students to use and evaluate evidence presented on cards to argue whether the compounds provided regarding different reaction mechanisms are either (a) a nucleophile or an electrophile, both or neither, (b) a Brønsted acid or Lewis acid, both or neither. The students are asked to work in groups and to conduct presentations following group work.

Aims. The aim of this exercise is to consider the evidence about whether the compound is (a) a nucleophile or an electrophile, both or neither, and (b) a Brønsted acid or Lewis acid, both or neither. Students will use the evidence presented on cards to argue for the appropriate classification of the compound.

Learning goals.

- Students will have the opportunity to construct arguments for the compound given in the reaction classifying either (a) a nucleophile or an electrophile, and (b) a Brønsted acid or Lewis acid and use the evidence from the cards to justify their claims.
- Students will learn to evaluate the evidence presented on the cards and select them to support their points of view about the compounds' behaviour in the reactions. Since some of the evidence can be ambiguous and could indicate that the compound in the reaction is both an electrophile and a nucleophile, the activity provides an opportunity to generate cognitive conflict and argumentation for students.

Teaching points. For this activity students will need to know the concepts of proton, electron, covalent bond, electrophile, nucleophile and mechanism of organic reactions and the definitions of acids and bases.

Teaching sequence.

- Distribute the activity sheet with columns and the evidence cards. Explain that the students will need to place each compound in the column where they think the statement goes. Arrange the students into groups of three or four.
- Now explain that the students will need to classify the compounds and they will need to support their classifications with evidence cards, which will need to be cut out.
- Explain that they not need to use all evidence cards. Warn students about the evidence cards because some evidence cards contain irrelevant or wrong information. The activity asks for the ability to distinguish between relevant information from irrelevant information, and to evaluate the accuracy.
- Allow about 15 minutes for students to work as a group to sort out the cards. While they are working in their groups, go around and probe their reasoning for selecting cards to put in one column over another column. Encourage students to use the textbook to look up any of the evidence that they may not be sure about.

Hold a class plenary at the end. Ask each group's idea about each compound respectively. Ask the students to report on their discussions and indicate the outcome of the group's exploration. Then ask who would like to argue against the presented ideas. Encourage the groups to rebut each other's argument by providing evidence that would counter their position. Ask questions such as "what information you would use from the cards to prove that his argument is not true?"

Student exercise for the activity named Cheating on the Exam[20]

Decide whether each compound given below is (a) a nucleophile or an electrophile, both or neither, and (b) a Brønsted acid or Lewis acid, both or neither.

P.S. Please ask your teachers for the reaction mechanism of each of the following reactions.

In the reactions of alcohols with hydrogen halides
tert-butyl alcohol, hydrogen chloride, tert-butyloxonium ion; tert-butyl cation and H_2O

In the reactions of electrophilic addition of hydrogen halides to alkenes
alkene and HX

In nucleophilic substitution reactions
methyl bromide and hydroxide ion

In the reactions between the carbocation and a halogen anion
carbocation and halogen anion

Nucleophile or electrophile?

Decide whether each compound given in the first question of the exam is a nucleophile or an electrophile, both or neither. Then write down your evidence to justify your decisions in the table that follows.

Nucleophile	My evidence	Electrophile	My evidence
Both	My evidence	Neither of them	My evidence

Brønsted acid or Lewis acid?

Decide whether each compound given in the first question of the exam is a Brønsted acid or a Lewis acid, both or neither. Then write down your evidence to justify your decisions in the table that follows.

Brønsted Acid	My evidence	Lewis acid	My evidence
Both	My evidence	Neither of them	My evidence

Evidence cards

1. It gives protons
2. It takes protons
3. It gives H^+ ions into the solution
4. It gives OH^- ions into the solution
5. It is looking for electrons
6. It is looking for the nucleus
7. It has a non-bonding electron pair
8. It is positively charged
9. It is negatively charged
10. It is neutral
11. It has carbon with a sp^2 hybrid orbital
12. Carbon has empty orbitals
13. It is a tertiary carbocation
14. It is an acid–base reaction

References

1. G. M. Bodner and D. S. Domin, *Univ. Chem. Educ.*, 2000, **4**(1), 24.
2. D. P. Cartrette and G. M. Bodner, *J. Res. Sci. Teach.*, 2009, **47**(6), 643.
3. N. Graulich, *Chem. Educ. Res. Pract.*, 2015, **16**(9), 9.
4. G. Bhattacharyya and G. M. Bodner, *J. Chem. Educ.*, 2005, **82**, 1402.
5. R. Ferguson and G. M. Bodner, *Chem. Educ. Res. Pract.*, 2008, **9**, 102.

6. R. B. Kozma and J. Russell, *J. Res. Sci. Teach.*, 1997, **34**(9), 949.
7. M. Harle and M. Towns, *J. Chem. Educ.*, 2011, **88**, 351.
8. A. Pabuccu and S. Erduran, *Int. J. Sci. Educ.*, 2017, **39**(9), 1154.
9. A. M. Strickland, A. Kraft and G. Bhattacharyya, *Chem. Educ. Res. Pract.*, 2010, **11**(4), 293.
10. A. Kraft, A. M. Strickland and G. Bhattacharyya, *Chem. Educ. Res. Pract.*, 2010, **11**(4), 281.
11. N. P. Grove and S. L. Bretz, *Chem. Educ. Res. Pract.*, 2012, **13**, 201.
12. T. L. Anderson and G. M. Bodner, *Chem. Educ. Res. Pract.*, 2008, **9**, 93.
13. G. Bhattacharyya, *Chem. Educ. Res. Pract.*, 2006, **7**, 240.
14. J. S. B. Evans, *Trends Cognit. Sci.*, 2003, **7**, 454.
15. L. M. McClary and V. Talanquer, *Int. J. Sci. Educ.*, 2011, **33**, 1433.
16. D. P. Cartrette and P. M. Mayo, *Chem. Educ. Res. Pract.*, 2011, **12**, 29.
17. D. C. R. De Arellano and M. Towns, *Chem. Educ. Res. Pract.*, 2014, **15**, 501.
18. K. Christian and V. Talanquer, *Int. J. Sci. Educ.*, 2012, **34**, 2231.
19. B. Hand and A. Choi, *Res. Sci. Educ.*, 2010, **40**, 29.
20. https://www.naturfag.no/binfil/download2.php?tid=1919020 (last accessed December 2018).
21. S. Erduran and A. Pabuccu, *Relevant Chemistry Education – From Theory to Practice*, ed. I. Eilks and A. Hofstein, Sense Publishers, Rotterdam, 2015, ch. 8, pp. 143–161.
22. S. Erduran, S. Simon and J. Osborne, *Sci. Educ.*, 2004, **88**, 915.
23. M. E. Anzovino and S. L. Bretz, *Chem. Educ. Res. Pract.*, 2015, **16**(4), 797.
24. A. Llewellyn, *Educ. Stud. Math.*, 2012, **81**(3), 385.
25. M. Stieff, *Sci. Educ.*, 2011, **95**, 310.
26. V. Sampson and D. B. Clark, *Res. Sci. Educ.*, 2011, **41**(1), 63–97.

CHAPTER 11

Argumentation in Physical Chemistry

MARCY H. TOWNS,*[a] RENÉE S. COLE,[b] ALENA C. MOON[c] AND COURTNEY STANFORD[d]

[a] Purdue University, Department of Chemistry, 560 Oval Drive, West Lafayette, IN 47907, USA; [b] University of Iowa, Department of Chemistry, W331 Chemistry building, Iowa City, Iowa 52242-1294, USA; [c] University of Michigan, Department of Chemistry, 930 N. University, Room 2408, Ann Arbor, MI 48109, USA; [d] Virginia Commonwealth University, Department of Chemistry, 1001 West Main Street, Richmond, Virginia 23284-2006, USA
*Email: mtowns@purdue.edu

11.1 Introduction

Physical chemistry is one of five traditional sub-disciplines of chemistry. In the undergraduate chemistry curriculum, it is frequently taught as a two-semester (or three quarter) sequence of courses that provides breadth and depth in understanding chemical phenomena.[1] Students typically enroll in this sequence during the third or fourth year of their undergraduate curriculum. Physical chemistry has long had a reputation as being a difficult course, largely due to the abstract nature and heavy use of mathematical equations in presenting concepts.[2–8] In order for students to have a meaningful understanding of physical chemistry, it is critical that the mathematical symbolism frequently used in the course has meaning at both the macroscopic and particulate levels. However, there is ample evidence in the literature that translating across representational levels is challenging for students.[9–17]

Advances in Chemistry Education Series No. 2
Argumentation in Chemistry Education: Research, Policy and Practice
Edited by Sibel Erduran

Published by the Royal Society of Chemistry, www.rsc.org

Complicating these difficulties is an overemphasis on symbolic representations in physical chemistry textbooks.[18] Without explicit intervention by the instructor, this creates an environment where students are expected to interpret symbolic information at the particulate and macroscopic levels but are rarely provided with tools to help them make these connections. Addressing these challenges requires approaches to instruction that move beyond the textbook and didactic presentations. The examples provided below illustrate the difficulty of thermodynamic concepts; particularly when it comes to translating between symbolic, macroscopic, and particulate levels. Students received the definition of enthalpy ($H=U+PV$) and a reminder that U is a state function and were prompted to justify why the term PV is a state function. Although this would be a straightforward question for a physical chemist, students struggled to correctly answer the question in multiple case studies, as shown in the excerpted arguments below.

Case B1

Claim: H is a state function but PV is not. (Melissa)
Data: *Cause it's not a delta, it's just a P and a V.* (Melissa)
Rebuttal: *Well if you use this equation* [H = U + PV] *and you rearranged it, it is still dependent on internal energy.* (Beth)

Case G1

Claim: PV a state function (Garrett/Mark)
Data: U is a state function (Mark)
Warrant: *Because it's a measure of energy, which is a state function* (Garrett)
Backing: *The pressure is what it is, it doesn't matter what path the system took to get there. And the same with the volume* (Dr Green)

There are a few features that make these concepts difficult. The first is the concept of path and state function. The values of path functions, like work or heat, can differ for processes with the same initial and final states depending on how the change in the system is achieved. In contrast, state functions only depend on the initial and final state of the system and the value does not change regardless of the path taken to achieve the final state. This is important for many variables in physical chemistry because it means that one can "create" a path for which equations and variables are more accessible. Pressure (P) and volume (V) are measurable, and explicitly represent physical entities. Internal energy (U) and enthalpy (H) are more abstract concepts for students and cannot be measured directly. This prompt explicitly requires students to translate between symbolic and macroscopic levels. However, this argument shows how students may try to reason only with the symbolic level, which results in a shallow understanding of thermodynamics. The instructor intervenes to incorporate some macroscopic reasoning about the nature of pressure and volume. As will be illustrated in this chapter, argumentation provides a useful strategy for addressing these difficulties. Further, the instructor plays an important role in facilitating translation across levels.

As science education evolves, new practices and frameworks emerge. Recent reform efforts have included an emphasis on science practices rather than simply the accumulation of science knowledge.[19–21] The Next Generation Science Standards (NGSS) being incorporated into K-12 education in the United States have supplied an impetus to engage college faculty in discussions about science and engineering practices.[22] The practices are described as behaviors that scientists engage in as they investigate and build models and theories about the natural world. One of these practices is argumentation, which is a characteristic of authentic science settings and communication within science. Facilitating argumentation improves learning content knowledge,[23–27] but instructors must explicitly model and support argumentation practices in order for students to use them.[9] One approach that shows promise is the incorporation of inquiry-based instructional practices, which have been shown to promote argumentation and allow instructors to authentically represent scientific practice as well as prepare students to engage in the discursive practices of scientific inquiry.[25,28,29] POGIL (Process-Oriented Guided Inquiry Learning) and other active learning pedagogies help create learning environments that encourage students to engage in meaningful discourse, including argumentation, to construct knowledge and develop skills such as communication and critical thinking.[30]

Our research, which focused on investigating argumentation in POGIL physical chemistry classrooms, was first reported in 2012.[31] Initially our research sought to adapt a methodology used in research on undergraduate mathematics education to document the collective production of meaning in a classroom. By examining classroom discourse using Toulmin's argumentation scheme[32] to document and analyze students' conceptual progress, we demonstrated the efficacy of this technique in undergraduate chemistry classrooms, specifically in physical chemistry. Through analysis of student discourse in the classroom we gained insights into student reasoning in physical chemistry as well as into the role of materials and instructor facilitation in supporting student argumentation.

11.2 Participants, Settings, and Methods

The data for the studies summarized in this chapter were collected at two different institutions in the Midwestern United States. Institutional review board approval was obtained for these studies and all participants provided informed consent to participate.

11.2.1 Participants and Settings

In 2009 and 2010 the classroom of Dr Black was the setting for our study. Dr Black taught at a public university in the Midwestern United States with approximately 14 000 students. In 2013 and 2014 the classroom of Dr Green was the setting for our study. Dr Green taught at a private college of approximately 1000 students. Physical chemistry at these institutions (as is typical in the United States) is an upper level course taken by students majoring in chemistry or chemistry-related disciplines. It is usual for physical chemistry to have pre-requisites of at least one calculus course, although two is more common, and one or two physics courses. While there were differences in the number of class meetings per week and the time of each class meeting, the overall instructional time for each case study was the same. Most courses involve about 45 hours of face-to-face instruction over the term, although the frequency and duration of each class meeting will vary. An overview of each of the classrooms is provided in Table 11.1.

Both instructors in these studies used POGIL materials and pedagogies. However, there were distinct differences in implementation. Dr Black used a pattern of mini-lecture or introduction to the material, followed by students engaging in small group work on a portion of the POGIL activity and then whole class discussion of that material. Once the whole class discussion of the questions was completed, another cycle of a brief presentation, small group work, and whole discussion would repeat. Dr Green used a pattern of students engaging in small group work to complete a POGIL activity followed by a lecture on that material. There were also differences in group management. Dr Black maintained consistent groups throughout the semester, while group membership changed throughout the semester in Dr Green's class. Although both instructors used the POGIL activities in a manner consistent with the philosophy of POGIL and emphasized small group work, Dr Black spent very little time lecturing while Dr Green employed almost no whole class discussion.

The POGIL approach used by these instructors uses social constructivist theories of learning as a foundation for the creation and revision of curricular materials and facilitation strategies.[33,34] The materials are designed to prompt students in the co-construction and application of knowledge. Students are typically guided through activities *via* a learning cycle of exploration, concept development, and application. In the physical chemistry materials used in case studies B1, B2, and G1, each activity began with a focus question for consideration and discussion. Often the students may not be able to answer the focus question – they need more content knowledge. The students then work through a set of critical thinking questions (CTQs) that build on prior knowledge and help students to construct a model specific to the activity that has predictive power. The final CTQs of an activity require students to explain the concepts they have been exploring and constructing during the activity. The final portion of the activity contains exercises that the students may use to apply concepts they have been constructing.

In case studies B1, B2, and G1, data was collected when students were completing activities in the POGIL workbook that covers thermodynamics.[35] The sequence of activities spans a typical set of concepts in thermodynamics including work, enthalpy, heat capacity, entropy, Gibbs Energy, Helmholtz energy, and the first and third laws of thermodynamics. In case study G2, the

Table 11.1 Comparative overview of classroom demographics for case studies B1, B2, G1, and G2.

<table>
<tr><th></th><th>B1</th><th>B2</th><th>G1</th><th>G2</th></tr>
<tr><td></td><td colspan="2">Instructor: Dr Black</td><td colspan="2">Instructor: Dr Green</td></tr>
<tr><td>Years of POGIL experience</td><td>9 years</td><td>10 years</td><td>7 years</td><td>8 years</td></tr>
<tr><td>Setting</td><td colspan="2">Medium Midwestern University
Physical Chemistry: Thermodynamics
Spencer, Moog & Farrell POGIL Materials (Spencer et al., 2004)</td><td colspan="2">Small Midwestern College
Physical Chemistry I and II</td></tr>
<tr><td></td><td colspan="2"></td><td></td><td>Shepherd & Grushow POGIL Materials (unpublished)</td></tr>
<tr><td>Number of participants</td><td>15 Students 10 Females 5 Males</td><td>18 Students 5 Females 13 Males</td><td>1 Students 3 Females 7 Males</td><td>6 students 4 Females 2 Males</td></tr>
<tr><td>Participant demographics</td><td colspan="2">3rd and 4th years</td><td>2nd, 3rd and 4th years</td><td>3rd and 4th years</td></tr>
<tr><td></td><td>At least one semester of calculus (except one student)</td><td>At least one semester of calculus</td><td>2 Semesters of calculus (except one student)</td><td>two semesters of calculus</td></tr>
<tr><td></td><td colspan="2">Chemistry majors</td><td colspan="2">Science majors</td></tr>
<tr><td>Class time</td><td>Three days a week</td><td>Two days a week</td><td colspan="2">Five days a week</td></tr>
<tr><td></td><td colspan="2">15 weeks</td><td colspan="2">8 weeks</td></tr>
<tr><td></td><td colspan="2">1/3 to 1/2 class small group work, rest whole class discussion</td><td>2/3 small group work and 1/3 lecture</td><td>Mostly small group work with some lecture</td></tr>
</table>

data was collected when students were completing activities covering similar content, but the POGIL activities were from a set of materials being beta-tested as part of an effort to revise the POGIL physical chemistry activities.

11.2.2 Methods

In both classrooms data was collected *via* two cameras. One was focused on the whole classroom and was positioned to capture the professor as he or she interacted with the class or lectured to the class. Another camera was focused on one of the small groups in the class. In each classroom, multiple groups were observed initially to identify a group to focus on for the ten thermodynamics lessons. Groups were selected for more detailed analysis based on having regular sustained interactions. The video of the small group discussion and whole class discourse for each day in each class was transcribed verbatim. Annotations were added to the transcripts to indicate the nature of the classroom activity (small group work, whole class discussion, lecture) as well as which question prompts from the POGIL materials were being discussed. These transcripts were then used as the data source for subsequent analysis.

To characterize student argumentation as well as instructor facilitation and its impact on argumentation, we used multiple frameworks. Toulmin's argumentation scheme allowed us to analyze the discourse and create argumentation logs. The argumentation logs were then analyzed based upon the research questions of interest. Additional frameworks that corresponded to the research questions were often used to guide analysis. A brief description of each of these frameworks is provided below.

11.2.3 Toulmin's Argumentation Scheme

Toulmin's argumentation scheme[32] was used to code the transcripts such that individual arguments were extracted. This scheme was selected because the basic structure of a claim based on data and supported with a warrant (reasoning) forms the foundation of scientific reasoning and discourse. There was also a history of use in other active learning environments.[23,25,28,36]

To generate argumentation logs, the researchers identified components of arguments present in classroom discourse by focusing on the nature and purpose of the utterances of both students and instructors. The first step was for members of the research team to carefully read through transcripts to identify the presence of claims. Usually, these were responses to questions in the POGIL materials. If a claim was accompanied by additional information regarding data or reasoning, it was extracted as an argument. In the excerpt below, the class is having a conversation about the entropy of solids, liquids, and gases.

Case B1, ChemActivity T6: Entropy, Model 1 (whole class discussion)
Instructor: *Which one of those (solid, liquid, gas) has the most entropy?*
Multiple Students: *Gas.*

Instructor: *Why?*
Luke: *It has the least interactions.*
Instructor: *It has the least interactions which means I'm looking at ways to distribute the particles, so what can I, how many ways can I distribute the particles in a gas?*
Beth: *A lot.*

The instructor asked the students which phase has the greatest entropy, and several students responded "gas," From this conversation, a claim was identified: "Gas has the most entropy." To identify the data used as the foundation for the claim, the instructor asked why the students believed this claim to be true. Luke then responded that gas molecules have fewer interactions with other molecules than liquids or solids, which he knew from prior knowledge of gases, liquids, and solids.

After the claim and data had been identified, researchers looked for evidence of a warrant or explanation as to how the data led to the claim. In this example, the instructor provided the reasoning that if gases have the fewest interactions that means there are no restrictions to where the gas molecules can be. To justify the warrant, the instructor asked the students "How many ways can I distribute the particles in a gas?" to help the students recognize that there are more possibilities in how the individual gas particles can be arranged compared to liquids and solids. This type of statement is referred to as a backing. Together all of this information was used to generate one complete argument about the entropy of gases compared to liquids and solids.

Argument from Case B1, ChemActivity T6: Entropy, Model 1 (whole class discussion)
Claim: Gas has the most entropy. (Multiple students)
Data: It has the least interactions. (Luke)
Warrant: I don't really have any restrictions on where I put the gas molecules. (Dr Black)
Backing: There are a lot of ways to distribute the particles. (Dr Black/Beth)

After the transcripts were coded using Toulmin's Scheme, argumentation logs were generated for each class period. The argumentation logs presented each argument in a consistent manner (claim/data/warrant/backing), even though arguments rarely developed in this sequential manner in the actual flow of conversation. Additionally, to condense data when isolating arguments, dialogue was paraphrased if it could retain the same sense of function of a particular statement in an argument. In these classroom settings the data components were often not verbalized by the students, but were implied based on the information given to the students in the ChemActivities. Direct quotes were identified by using italicized text. Each argument was tagged with the ChemActivity and relevant question number that was the focus of discussion. The argumentation logs transformed a daily transcript of over 30 pages into 2–4 pages of arguments that took place in either small group discussion or whole class discussion.

11.2.4 Inquiry-oriented Discursive Moves

The inquiry-oriented discursive moves (IODM) framework provides a method for analyzing discourse that allows one to characterize the instructor's discursive moves and their role in sustaining inquiry-oriented classroom environments.[37] There are four main classifications of moves including questioning or requesting, revoicing, managing, and telling. Each of these moves is parsed into four sub-categories to further define its function. For example, within the *revoicing* move, an instructor may *repeat*, *rephrase*, *expand*, or *report*. *Questioning or requesting* moves are particularly important because they may be used to elicit further reasoning from students through asking *evaluating, clarifying, explaining,* and *justifying* questions. *Telling* moves are used to provide more information or in some cases to respond to student questions. Within these moves instructors may *respond* to questions, *initiate* new lines of discussion, *summarize* information, or *facilitate* discussion across the classroom. Finally, instructors engage in *managing* moves, wherein they *direct* or *arrange* discussion, *motivate* engagement, and *check* on student reasoning and progress.

This framework was used to code the transcripts across all four case studies. Each instance of instructor talk was coded in terms of its function in the classroom. A single turn could include multiple moves.

11.2.5 Johnstone's Triangle

Johnstone's triangle[10,11] was used to characterize the POGIL materials and the components of arguments as relating to the particulate, macroscopic, or symbolic level. While there are some variations in how Johnstone's triangle has been interpreted,[10,38,39] the basic themes of characterizing representational levels and investigating how students translate between them provides a mechanism to probe how students reason about chemical concepts.

11.2.6 Chemical Thinking Learning Progression

Sevian and Talanquer describe chemical thinking as the "development and application of chemical knowledge and practices."[40] The emphasis is on "analyzing synthesizing, and transforming matter for practical purposes." Viewing argumentation through the lens of a chemical thinking learning progression (CTLP) allowed us to evaluate the degree to which students engaged in more authentic science practices during classroom discussions.

11.2.7 Material Analysis

The questions in the inquiry materials themselves were analyzed along several dimensions. These included components of the learning cycle, type of POGIL question, and the nature of the task itself. Given the emphasis of many STEM education reform efforts on promoting scientific practices, including argumentation, the Task Analysis Guide for Science (TAGS)[41] provided a lens to look at the POGIL questions in terms of their cognitive demand and the degree to which the integration of content and practices was required.

11.3 Student Argumentation – Understanding of Content and Reasoning

Analysis of student arguments over time provides a mechanism to document student understanding and how it develops in an active learning classroom environment. Three criteria emerged from our analysis of argumentation logs in small groups and in whole class discussion that identified when ideas began to function "as if shared" across the classroom.[31] The first two had been identified in Rasmussen and Stephan's (2008) earlier research in undergraduate mathematics classrooms, but the third criterion was unique to this study.

- Criterion 1: Warrants or backings are initially present, but then fade.
- Criterion 2: An idea shifts roles in subsequent arguments indicating consolidation of knowledge.
- Criterion 3: Specific ideas are used repeatedly to justify claims on different days.

To identify normative reasoning, we analyzed the argumentation logs for regularities that occurred in reasoning about chemical and physical properties. We found Criterion 3 to be especially useful in this regard. For example, we noted in our 2013 work that "temperature is a measure of kinetic energy of particles" was repeatedly used as data or warrant in argument logs.[42]

Our analysis of the argumentation logs allowed us to discover themes in the data, warrants, backings, and rebuttals across arguments. We found that the students repeatedly used particulate-level ideas to justify their reasoning about physical and chemical properties.[42] Specifically, the students and instructors repeatedly used information about the structure of molecules and the motion and spacing of particles in their data, warrants, and backings for claims about chemical and physical properties. This was an interesting finding because it has been noted in the research literature that students have difficulty using particulate-level reasoning to describe chemical phenomena.[14,15] We also found rebuttals and counter-claims that refuted incorrect use of particulate-level reasoning in the data set. Overall, the interpretation of these argumentation patterns allowed us to establish that particulate-level reasoning had become normative within the classroom as students developed their understanding of heat, energy, enthalpy, heat capacity, and entropy. This insight led us to more thoroughly consider the degree to which different aspects of Johnstone's triangle were incorporated into arguments.[43] We found that the majority of arguments pertained to only one portion of Johnstone's triangle with the majority of arguments using the symbolic level only. This may be reasonable given the amount of reasoning about mathematical equations present in a physical chemistry course. Typically, students would be working with mathematical expressions and balanced chemical equations or carrying out calculations during these discussions.

Reasoning at the macroscopic level was the second-most common with particulate-level arguments occurring the least frequently. While fewer than 4% of the discussions in either the small group or the whole class used all three levels of Johnstone's triangle, students were more able to coordinate across two levels. Macroscopic and symbolic discussions, where students coordinated between symbolic and macroscopic representations, accounted for 25% of the whole class discussions and 27% of the small group discussion. Macroscopic and particulate discussions occurred across 11% of whole class discussions and 4% of the small group discussions. Frequently in these discussions the instructor facilitated further coordination between the levels of Johnstone's triangle by asking for an elaboration of student reasoning. This request for explanation, justification, or elaboration led to the researchers exploring the instructor's discursive moves (described in section 15.5). To describe student reasoning, we used the framework of chemical thinking[40] to identify and delineate the reasoning students used in their arguments. Evaluating the content of the arguments allowed us to describe their causal reasoning, which provided a new dimension to our understanding of arguments in physical chemistry.[44] Arguments were described as descriptive, relational, linear, or multicomponent as shown in Table 11.2.

Across both classrooms in the study, we found that relational reasoning occurred the most frequently in arguments, followed by linear, descriptive, and multicomponent. Relational arguments occurred frequently and had the pattern of relying on a relationship in the absence of causal justification. Linear reasoning was identified in arguments containing mathematical derivations where students moved stepwise through substitutions to achieve a final expression. There were some linear arguments focusing on Hess's Law that were a hybrid between conceptual and mathematical arguments. While there were mathematical aspects to these arguments, students used justifications grounded in the chemical processes to create the argument. In many of the linear arguments, the students or students and instructor included cause and effect reasoning.

Descriptive reasoning did not prompt the students to propose cause and effect mechanisms. Students identified features within the prompt of the

Table 11.2 Features of modes of reasoning in CTLP used to analyze arguments. Reproduced from ref. 44 with permission from the Royal Society of Chemistry.

Mode	Features
Descriptive	• Salient properties are recognized • Explicit properties are verbalized • Phenomenon is instantiation of reality • Reasoning based on experiences from daily life
Relational	• Explicit and implicit properties are highlighted • Spatial and temporal relations are noticed • Phenomenon is effect of single variable (no mechanism)
Linear	• Mechanisms proposed that involve linear cause-effect relationships
Multicomponent	• Mechanism weighs effects of several variables

problem, which in some cases produced superficial arguments and in others revealed confusion on the part of the students. Finally, multicomponent arguments were the least frequent in the dataset, indicating the difficulty of considering the effect of multiple variables. Ultimately, the analysis revealed that linear and multicomponent reasoning was indicative of more verbalization of student reasoning and understanding. These arguments were more complete than the relational and descriptive arguments.

Another reason instructors promote student argumentation is to help develop students' critical thinking skills such as evaluating and considering alternative arguments. The process of interpreting and reflecting on someone else's reasoning is known as decentering.[45,46] Argumentation logs from case studies G1 and B2 were first analyzed to identify arguments that included rebuttals or argument sequences that included counter arguments. These arguments were further analyzed for evidence of decentering by determining whether the rebuttals or counterarguments indicated that the student had interpreted a previous argument made by another student. An example of an argument that provides evidence of decentering is shown below. In this example, Aiden and Dr Black justified that the internal pressure of a system is constant during a reversible process.

Case B2, ChemActivity T2: The First Law of Thermodynamics, CTQ 6a (whole class discussion)

Claim: The pressure of the system is constant for the process. (Aiden)
Data: The system is reversible. (Aiden)
Warrant: So the total energy is constant. (Aiden)
Rebuttal: *That's true, but it doesn't have anything to do with...* (Dr Black)
Warrant: *Because the system is in equilibrium.* (Aiden)
Backing: *That's what "reversible" means is the system is in the equilibrium throughout the process.* If it's in equilibrium, then the internal pressure and my external pressure have to be equivalent. (Dr Black)

Aiden makes the correct claim and cites the correct data but includes an irrelevant warrant. Dr Black notes the irrelevant warrant and responds directly to that, which prompted Aiden to respond with a relevant warrant. Dr Black then expands upon this to tie the whole argument together. This example illustrates an important feature of decentering—targeted rebuttals. Further, this example shows the potential outcomes of decentering. The argument was strengthened with the shift towards a relevant warrant. This new warrant and backing were co-constructed and induced by Dr Black's rebuttal.

However, instances of rebuttals or counterarguments, such as the one below, were found where there was no evidence that students were evaluating previous arguments. In this example, the students were prompted to consider a system for which the total initial and final entropy were equal. In particular, they considered whether the forward and reverse processes would be spontaneous.

Case G1, ChemActivity T6: Entropy, CTQ 4b & 4c (small group work)

Claim: The forward and reverse processes would be spontaneous. (Garrett)

Data: (S_{tot})final = (S_{tot})initial (POGIL Materials)

Warrant: The process is reversible. (Garrett)

Rebuttal Claim: The forward process wasn't spontaneous (Mark)

Rebuttal Data: (S_{tot})final = (S_{tot})initial (POGIL Materials)

Rebuttal Warrant: Because it says for an event to occur spontaneously ΔS total must be positive (Mark)

Rebuttal Backing: *You see the argument that you just told me for 3c… I was thinking because the … and the reverse in this case. It says it reversible but it doesn't say the reverse process because if you think in a reverse process it can… the initial is still going to equal the final change, once you get there* (Stephanie)

Garrett made an incorrect claim, as the fact that the process is reversible means that it is not spontaneous. Ideally, Garrett's peers would have responded to this in order to elicit what Garrett was misunderstanding. However, this was not the case. Mark rebutted by saying Garrett's claim is incorrect and provided reasoning that is pulled directly from the POGIL workbook. Stephanie provided a backing that showed she was considering what Mark said. Neither Mark nor Stephanie addressed what Garrett said, and this is further illustrated by the lack of a consensus argument or further participation by Garrett. When decentering is absent, alternative conceptions have the potential to remain hidden and arguments maintain weaknesses. Since decentering can promote more productive discourse that supports students' conceptual development, it is important for instructors to include facilitation strategies that encourage and model decentering. Our research has also highlighted features of decentering that instructors should monitor and facilitate in the classroom.[48] Some approaches are described in Section 11.5.

11.4 Classroom Argumentation – Influence of Course Materials

When considering argumentation in any classroom environment, the materials and prompts that drive student thinking must also be considered. This led us to consider the links between the prompts in the POGIL materials and the types of arguments produced. Ideally, course materials should provide appropriate scaffolding to help students structure their reasoning and support productive argumentation. Given that the intent of POGIL materials is to provide a basis for small group discussion and development of ideas, an initial step in this analysis was to characterize each of the questions in the POGIL materials in terms of their role in the learning cycle (exploration, concept development, application) and the nature of the question (directed, procedural, convergent, divergent). These categories were selected based on the structure and descriptors used in POGIL writing workshops that provide support for creating POGIL materials,[34,47,48] although the category of procedural questions arose from our analysis of the materials themselves.[49] Questions were also coded in terms of the type of task students were required to perform, such as perform a calculation, complete a derivation, describe a process, make a prediction, *etc.* There were 233 distinct questions (combining both Focus Questions and CTQs) that were present in in the eleven activities from the Spencer, Moog and Farrell POGIL Materials[35] used in three of the case studies. The questions that generated arguments were not consistent across the different case studies, as illustrated in Table 11.3.

Further analysis indicated that many of these differences were due to the differences in facilitation. However, the analysis of argumentation by question type provided some insights into the role of questions on student argumentation. In Dr Black's course, over half the questions in each phase of the learning cycle (exploration, concept development, and application) resulted in arguments. In Dr Green's course, only about 20% of exploration questions resulted in argumentation while the results for concept development and application questions were similar to what was observed in Dr Black's class. Taken together, these results are similar to those found by Kulatunga *et al.* (2014) that exploration questions generate fewer arguments than concept development and application questions.[49] When organized in

Table 11.3 Summary of the number of questions generating arguments in each of the case studies. SGW indicates arguments generated during small group work, while WCD indicates arguments generated during whole class discussion.

Case Study	Total questions	SGW	WCD	Questions in common
B1	126	93	81	48
B2	146	58	132	44
G1	71	71	NA	NA

terms of question type (directed, convergent, divergent, procedural), a similar pattern was observed. Approximately 60% of the directed, convergent, and procedural questions resulted in argumentation in Dr Black's class, while only about 30% of the directed and convergent questions resulted in arguments in Dr Green's class. The reason for the larger number of arguments generated by exploration/directed/convergent questions in Dr Black's class was related to the classroom norms developed, primarily through the use of whole class discussion and group roles.

This analysis also highlighted the difficulty students have in making sense of mathematical equations. Questions that involved students deriving equations or providing mathematical expressions (usually categorized as procedural questions) were the least likely to generate arguments, and the arguments that were generated typically focused on the mechanics of manipulating terms rather than connecting the expression to the chemical concepts being modeled. In the following argumentation logs, the materials gave the definition of work as force multiplied by distance, which was then modified to describe work in chemical contexts as pressure operating through a volume. The model also included a piston cylinder diagram and the equation that $P_{ex} = F/A$ and the sign convention that in chemistry work done on the system is positive while work done by the system is negative. Students were then guided to derive work in terms of pressure and volume. Question 2c prompted students to "Modify your work equation to express the infinitesimal work, δw, in terms of external pressure, P_{ex}, and the infinitesimal change in volume, dV." Question 2d followed that by prompting students to "Integrate this expression to describe the relationship between work, w, and the initial and final volumes of a gaseous sample for a process under constant P_{ex}. The student arguments generated in response to those questions for Case Study G2 are shown below.

Case G2, ChemActivity T.3: Work, CTQ 2c (small group work)
Claim: $w = -PdV$ (Nadine)
Data: $dw = -FdL$ (previous argument)
Data: $P = F/A$ (Briana)
Warrant: *If you divide the volume by the area, you get the L. Right? So if you have force times L, that's going to be the same as pressure times volume. I mean, assuming that the pressure is equal to F over A.* (Briana)

Case G2, ChemActivity T.3: Work, CTQ 2 (small group work)
Claim: $w = -P(V_2\text{-}V_1)$ (Kevin)
Data: $w = -PdV$ (previous argument)
Warrant: *Well, it's if the change in volume is positive, then the answer's going to be negative, and vice versa. It needs to be either. It needs to be opposite the change in volume. So the negative needs to be there.* (Kevin)

Since the materials never ask students to reflect on how the equations are related to the physical processes and whether they make sense, which is also true of most questions in other physical chemistry textbooks, it must be assumed that either the instructor (and textbook author) assumes students make these connections or is only concerned with the mathematics. There are many examples throughout the four case studies of students manipulating equations with no discussion of meaning. Even when students are prompted to interpret the equations, they rarely ascribe the equations to models of physical processes but rather simply "name" the variables.

Our case study data suggests that if a classroom learning environment emphasizes collaborative teamwork, students are more likely to engage in argumentation as they make sense of models and generate ideas. These arguments also illustrate some of the difficulties students have in interpreting diagrams and information that would otherwise remain invisible to instructors. One example is seen as students worked to interpret another model in the same activity highlighted above. In this case, students were provided with a model of a piston and asked to describe the process being depicted. The diagram students were exploring is shown in Figure 11.1 below.

Specifically, students were prompted to "Consider the removal of one of the 1 atm weights from the piston in State A: (a) Is the system still at equilibrium at the moment that the weight is removed? When the weight is removed, what will happen?"

Case B1, ChemActivity T1: Work, CTQ 11a (small group work)

Claim: The system is not at equilibrium when the weight is removed. (Melissa)

Data: The pressure, P_{ex} goes up. (Melissa)

Warrant: P_{ex} in the Diagram in Model 3, page 45, shows that [P_{ex}] goes up. (Melissa, points to diagram)

Rebuttal: It's saying there's less pressure. (Adam/Beth)

Rebuttal Claim: Still in equilibrium. (Melissa)

Rebuttal Data: *But it still shows over here even when it's removed that they're still in equilibrium but just kind of moved up.* (Melissa)

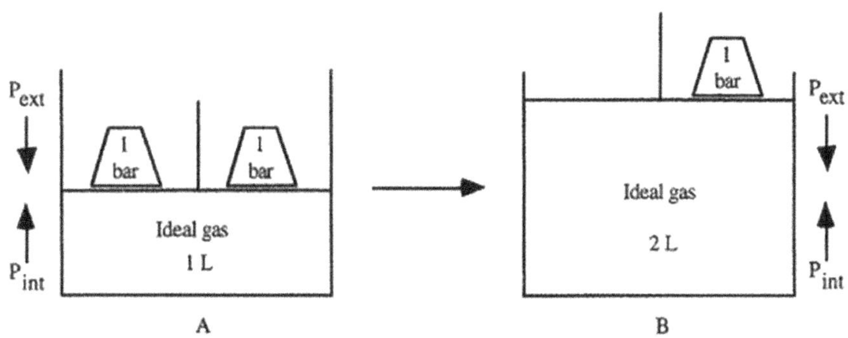

Figure 11.1 Model 3 diagram from ChemActivity T1.
Reproduced from ref. 35 with permission from The POGIL Project.

In this case Melissa seems confused about what was happening in the diagram. Her initial interpretation is that since P_{ex} was written physically higher up on the diagram, then P_{ex} went up. She is attending to a surface feature of the diagram rather than thinking about the physical process and what it is intended to depict. Students continued to negotiate the meaning of the diagram as seen in the subsequent argument.

Case B1, ChemActivity T1: Work, CTQ 11a (small group work)

Claim: They're still in equilibrium but at a different volume. (Melissa/ Beth)

Data: *If you look at the arrows they're the same size.* (Carrie refers to diagram)

Rebuttal Claim: System is not at equilibrium immediately after weight removed. (Beth)

Rebuttal Data: It's not showing in between. (Adam)

Rebuttal Data: From the moment it comes off. (Beth)

Rebuttal Warrant: The moment the weight is removed it changes. (Beth)

Consensus Claim: System not in equilibrium. (Melissa)

Consensus Data: [Internal] pressure will push plunger up to reach equilibrium. (Adam)

Consensus Warrant: $\mathrm{P}_{int} > \mathrm{P}_{ext}$ at moment when weight is removed. (Carrie/Adam)

Based on this argument, Carrie seems be interpreting the model as a force diagram, where the size of the arrows represents the magnitude of the force. Since they are of equal length, she initially determined that $P_{ext} = P_{int}$ for both diagrams. Eventually the students worked through the questions and arrived at an appropriate answer to the prompt. In discussing the student interpretation of the diagram with the authors of the textbook, the textbook authors were surprised at the degree to which students focused on surface features of the diagram that were not intended to be key features for interpretation.

The POGIL activities were also analyzed using the TAGS framework.[53] This analysis revealed that the activities consisted of primarily scripted tasks that were designed to promote the integration of content and practice. This characterization is consistent with the overall POGIL pedagogy of scaffolding student learning of content while developing process skills such as communication, critical thinking, and problem solving. In the physical chemistry materials, the most common practices targeted were "using mathematics and computational thinking" and "constructing explanations." We believed that if a clear link existed between the nature of the question prompt and students' chemical thinking then we could improve the prompts used in the course to encourage linear and multicomponent reasoning. However, the linkage was not easily identified. We found that linear causal arguments frequently were associated with prompts that required derivation of an equation. But relational arguments resulted from all types of prompts. Multicomponent arguments were identified when students were required to make a prediction in many cases.

11.5 Promoting Productive Discourse Through Facilitation

The use of IODM allowed us to compare facilitation across implementations.[50] Looking for similarities and differences across the case studies allowed us to characterize strategies that appear to promote (or inhibit) productive argumentation during classroom discourse. A notable difference in the IODM profiles resulted from Dr Black's use of whole class discussion rather than lecture to elaborate on and extend student understanding developed during small group work. Dr Black and Dr Green both promoted translation across the different representational levels of Johnstone's triangle, but the manner in which this was supported was quite different.

In case study G1, lecture was the dominant form of instructor discourse, with a mix of initiating, summarizing, and responding moves. Dr Green would model arguments incorporating translation across the levels, which was then reflected in student argumentation. In contrast, the most prevalent moves during whole class discussion in Dr Black's classes were *questioning* moves, especially moves that evaluated the class's understanding or required the students to focus their attention on specific content that required further evaluation. Overall, questioning moves accounted for 43% of the instructor's moves.

The second-most prevalent move for Dr Black (27%) was *revoicing,* where the instructor repeated and often expanded upon a student response to probe and drive forward student reasoning. The instructor used *telling* moves less often (23% of the whole class discussion moves). These moves allowed the instructor to directly respond to student questions or to evaluate a student response and redirect the discussion if required. These moves also served to summarize ideas, suggest next steps, or to point out information that had yet to be considered. Managing moves were used much less frequently (7%) and served to keep the class on track.

This analysis revealed a set of discursive moves used in combination to orchestrate class discussion and to support the class's coordination of levels of the chemistry triplet. The sequence of moves involved *questioning, revoicing, and expanding,* abbreviated QRE. This set of moves was used by the instructor to scaffold student thinking, and evidence strongly suggests that the instructor used this move sequence to coordinate levels of the chemistry triplet and to improve student understanding. Frequently QRE moves elicited student reasoning and prompted students to connect the symbolic level to other levels of representation. This required the instructor to be aware of the level of students' reasoning and to have targets (particulate level or macroscopic level for example) that she wished to drive them towards. Additionally, the use of revoicing was critical in acknowledging student ideas and synthesizing them into a more coherent knowledge structure. Simply questioning or questioning and telling lacks the elicitation and recognition of student contributions that revoicing allows.

The findings from this research demonstrated that it is critical that faculty encourage classroom discussion where student ideas are elicited and meaning across levels of representation is explicitly negotiated. Simply presenting equations and defining the terms is not enough explanation and scaffolding for students to understand the meaning of the equation or a calculated value and its connection to other levels of representation. Argumentation where claims are supported with data and reasoning became a critical feature of meaningful learning.

The use of IODM also allowed us to compare facilitation for a single instructor across implementations to document how changes in facilitation influence student argumentation in physical chemistry. When we did this for the B1 and B2 implementations, we found key differences in facilitation, accompanied by interesting differences in argumentation. The analysis revealed that in both iterations, the instructor engaged in primarily *telling* and *questioning* moves. However, within these categories, there were differences between iterations. In the *telling* category, we observed a slight increase in *responding* and *summarizing* moves from B1 to B2. In the questioning category, there was a noteworthy decrease in the number of *evaluating* moves, accompanied by an increase in number of *justifying, explaining, and clarifying* moves. In the *revoicing* category, we observed a decrease in number of *repeating* moves and an increase in the number of *reporting* moves. In the *managing* category, there was no change between iterations.

Changes in the discursive moves used by the instructor were accompanied by two important changes in the resulting argumentation. In contrast to the B1 implementation, the B2 implementation included more warrants overall and more student-generated warrants. Further, the B2 implementation included a relative increase in student–student co-constructed arguments and single student constructed arguments with a decrease in instructor–student co-constructed arguments. The following argument sequences illustrates both of these differences (*i.e.*, increase in warrants and student contributions). Both argument sequences are in response to the same prompt, in which students were asked to predict the final temperature of one mole samples of Ne and N_2 in which equal amounts of heat are inputted. Numbers indicate the order in which the arguments occurred in class.

Case B1, ChemActivity T4: Heat Capacity, CTQ 2 (whole class discussion)

1. *Initial argument*

    Claim: Ne and N_2 will have different temps even if the same amount of energy is added [Student 1]

    Data: N_2 is bigger [Student 1]

    Warrant: Bigger molecules take more heat to move [Student 1]

    Rebuttal: The bigger the molecule is, does not have to do with the amount of energy [Instructor]

Claim: Nitrogen can rotate and Ne can't [Student 1]
Data: Nitrogen has a bond [Instructor]

2. *Initial argument*
   Claim: N_2 has a lower temp than Ne [Instructor]
   Data: It can vibrate and rotate (Ne can't) [Instructor and students]
   Warrant: Some of the heat energy can be put into rotational and vibrational motion [Instructor]
   Backing: Only the translational energy is kinetic energy [Instructor]

3. *Initial argument*
   Claim: Neon has no rotational energy or vibrational energy [Instructor]
   Data: The bond (N_2 has one and Ne doesn't) [Student 2]
   Warrant: Because it looks the same however it moves [Student 1]
   Warrant: You have to have a bond in order to vibrate [Instructor]
   Backing: An axis of rotation tells that there's a difference when it's been rotated, and I need two things interacting in order to vibrate [Instructor]

Case B2, ChemActivity T4: Heat Capacity, CTQ 2 (whole class discussion)

1. *Initial argument*
   Claim: the N_2 will be hotter? [Student 1]
   Data: Kinetic energy equals $\frac{1}{2}$ mv^2 and N_2 weighs more than Ne [Student 1]
   Warrant: *If it weighs more, it's going to have a slower velocity. And since the term's v squared, and in that sense, we figure that the velocity would have more weight than the temperature* [Student 1]

2. *Alternate argument*
   Claim: Ne is hotter [Student 2]
   Data: Ne has less intermolecular forces [student 2]
   Warrant: So more heat would be going towards kinetic energy [Student 2]

3. *Alternate argument*
   Claim: Ne is hotter [Student 3]
   Data: Because Ne is lighter [student 3]
   Warrant: *It'll move faster with the same amount of heat added. And because there's a triple bond between nitrogen. The amount of heat given in the system, part of it will be considered to break the bonds* [Student 3]
   Rebuttal: *Oh, just because I heat it up nitrogen gas, does it dissociate bonds?* [Instructor]

4. *Alternate Argument*
   Claim: Ne is hotter [Student 4]

Data: N_2 also has vibrational, rotational energy, not just translational [Student 4]
Warrant: There are other ways to devote heat to it, it doesn't necessarily have the same increasing kinetic energy for that kind of input. You'd have more if your input of energy towards translational movement of the neon, so it was up for hotter average kinetic energy and hotter [Student 4].

5. *Initial Argument*
Claim: Nitrogen has a larger intermolecular force than neon [Student 5]
Data: It's more polarizable, because it's larger [Student 5]
Warrant: It's a molecule, it's more polarizable, it's larger, therefore it can have more van der Waals forces [Instructor]

In this specific example, the number of argument units and student contributions is greater in Case Study B2 compared to B1. These changes in argumentation point to a potential relationship between the instructor's moves and observed argumentation. Further, the primary change in argumentation was an increase in student participation in argument construction, reflecting increased student engagement. We believe this increase is, in part, due to changes in the instructor's moves; particularly, the increase in instructor prompting for justification. Additionally, in the B2 Case Study, group whiteboards were introduced as a means of presenting to the whole class. This likely explains some of the observed increase in number of student-generated warrants.

Our results show that instructor facilitation not only had an effect on the number of arguments and student contributions to those arguments, but also the content of the arguments. Of particular interest in chemistry as a whole, and especially physical chemistry, is the ability to connect macroscopic, submicroscopic, and symbolic forms of reasoning.[11] Because chemistry spans these three levels, it is expected that classroom discourse would span these levels. However, analysis revealed that translation across these levels depended upon instructor facilitation. This was evidenced by the abundance of only symbolic reasoning in small group discourse, where the instructor contribution was relatively absent, shown in Figure 11.2.[43] The figure is followed by an example of small group and whole class responses to the same question, illustrating which levels are addressed by each argument component (Tables 11.4 and 11.5).

A notable difference between these two argument sequences is the homogeneous level reasoning in the small group, contrary to the translation across levels present in the whole class argument. Incorporating the IODM analysis served to elucidate the impact of instructor facilitation on translation between levels. There was a sequence of instructor moves present in a majority of argument sequences that spanned at least two levels of reasoning (68%). The sequence of questioning—revoicing/repeating—revoicing/expanding (QRE) involved the instructor posing a question, repeating and

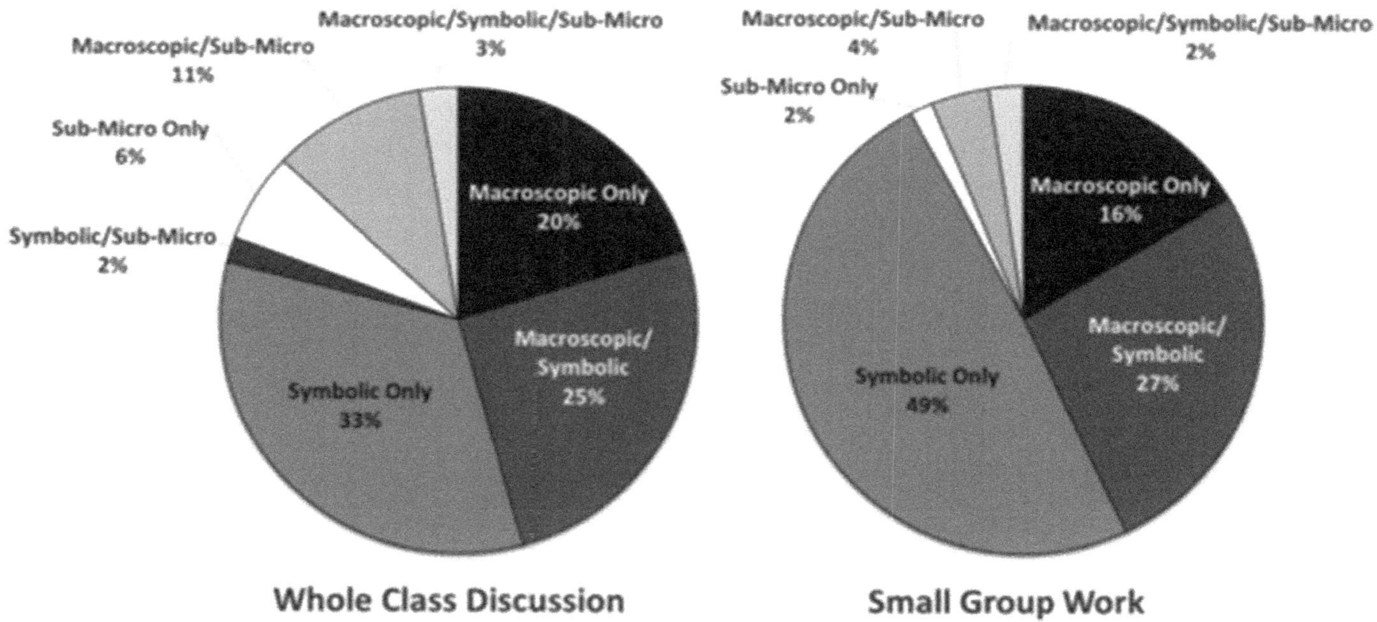

Figure 11.2 Comparison of macroscopic, submicroscopic, and symbolic forms of reasoning in whole class discussion and small group work.
Reproduced from ref. 43 with permission from the Royal Society of Chemistry.

Table 11.4 Small group argument about calculated value of $\Delta_{rxn}H$.

Claim: Delta $H_r = 44.01$ kJ mol^{-1} (Carrie)	Symbolic
Data: *Because you're ending with the H_2 gas here, if you flip this one you're starting with it. So this one's going to be positive* (Melissa describing the application of Hess's Law)	Symbolic
Warrant: *So it's positive 285 minus 241.84…* (Adam describing procedure for calculating $\Delta_{rxn}H$)	Symbolic

Table 11.5 Whole class argument about calculated value of $\Delta_{rxn}H$.

Claim: H_2O (l) $\rightarrow H_2O$ (g) is endothermic	Macroscopic
Data: $\Delta_{rxn}H = 44.01$ kJ mol^{-1} (Carrie)	Symbolic
Warrant: *It takes energy to go from water to water vapor* (Tom)	Macroscopic
Backing: *To go from liquid to the gas I have put energy in because the molecules are more excited* (Instructor)	Submicroscopic

Table 11.6 QRE sequence.

Instructor: How did you rank the substances in terms of their heat capacities?	Questioning: Evaluating
Beth: H_2O has the highest heat capacity, followed by CH_4, N_2, and Ne	Claim (Macroscopic)
Instructor: You basically said heat capacity was increasing this way (draws arrow pointing down on board)	Revoicing: Repeating
Instructor: What was your basis for that judgment?	Questioning: Justifying
Beth: Cause you put the same amount of energy into all of them but H_2O has the least temperature rise	Data (Symbolic)
Instructor: Right	Telling: Responding
Instructor: So if you put the same amount of energy, Whichever one has the least temperature change, will have the greatest heat capacity.	Revoicing: Rephrasing/ Warrant (Macroscopic)

confirming the student(s)' response, and eventually elaborating on the response. An example of this is illustrated in the argument sequence below (Table 11.6).

In this illustration, the instructor engaged in the QRE sequence, thereby moving the students from a macroscopic claim to symbolic data, and ultimately providing the warrant. We believe that this sequence is important for facilitating student translation across levels in their argumentation. Instructor facilitation impacts the quality of resulting argumentation. By increasing the number of justification moves and employing the QRE sequence of reasoning, instructors can promote an increase in student-generated arguments and can promote reasoning across representational levels.

11.6 Conclusions and Implications

Our work in argumentation in physical chemistry established a methodology for analyzing classroom discourse that allowed researchers to capture the dynamic nature of the classroom.[31] Across the arc of our research we combined argumentation analysis with methodological frameworks from chemistry, the chemical thinking learning progression,[44] and Johnstone's triangle,[43] to provide further insight into the nature of argumentation in physical chemistry classrooms.[31,42,46,51–53] Additionally, we adapted a framework from undergraduate mathematics education research known as IODM[37] to reveal which discursive moves faculty use to facilitate argumentation. Combining these methodological approaches was a novel way to generate greater depth in our findings, specifically implications for the classroom. We believe that such implementation of more than one framework can serve as a model for other researchers and inspire new research vistas.

Our findings provided new directions for research in terms of supporting student reasoning through interactions with other students and the professor. The results from this research program have identified three main factors that impact the quality of scientific argumentation in physical chemistry: (1) instructor facilitation; (2) curriculum materials (prompts and implementation); and (3) classroom reasoning norms. Using numerous quality criteria for argumentation, our investigation elucidated the relationships between these three factors and the quality of resulting argumentation.

Based upon the analysis of chemical thinking in student arguments, we developed implications for the classroom and for further research. We recognized that the instructor plays a role in facilitating the argumentation in whole class discussion and thus influences the type of discussion that emerges. We encourage instructors to prompt for student reasoning and build a classroom culture where argumentation is normative. While students construct arguments, instructors need to monitor the type of reasoning. We propose that faculty push for linear and multicomponent reasoning that takes into account more than one variable and thus is co-variational. Faculty can model this reasoning for students and they can form arguments collaboratively by providing warrants or data to support co-variational reasoning.

While facilitation played a big role in promoting argumentation, the nature of question prompts in the materials was still influential. Questions that prompt students to predict or explain generally result in more complex arguments with stronger evidence of student reasoning. Questions that simply focus on mathematical processes are not likely to result in meaningful understanding of the equations as models of chemical processes unless students are explicitly prompted to explore what is being represented by the equation and when and why it is useful.

The most significant implication for teaching is that instructors must take an active role in helping students develop the ability to generate scientific arguments. While course materials play a role in scaffolding student argumentation, instructor facilitation is still critical in promoting a classroom culture that encourages student discourse.

Practical Digest

Argumentation is a key scientific practice that should be developed in the undergraduate curriculum. We have identified three factors that affect the quality of classroom argumentation: (1) instructor facilitation strategies; (2) nature of curricular materials; and (3) classroom reasoning norms. Each of these factors offers unique implications for practice. Instructors can effectively support argumentation, particularly in physical chemistry, by attending to features of students' arguments and modeling desired features in their own arguments. Faculty play a key role in supporting translation across representational levels. Most physical chemistry textbooks focus almost exclusively on the symbolic level, and students rely primarily on symbolic and macroscopic reasoning when interacting with the curriculum. Given the critical nature of submicroscopic data and arguments, instructors should recognize that students require explicit support to construct such arguments. This includes connecting submicroscopic evidence and warrants to the symbolic and macroscopic claims, evidence, and warrants with which students seem to be more comfortable. In our data, the facilitation strategies of asking students to justify responses in terms of submicroscopic descriptors or phenomena and elaborating on student responses at the submicroscopic level were effective in supporting students in translating across levels and encouraging this type of reasoning as normative practice in the classroom. Further, students showed a tendency to rely on relational reasoning in which they used one variable to entirely explain an outcome. This is problematic given the complexity of thermodynamic explanations of chemical phenomena, which depend on dynamic relationships between multiple variables. Instructors can support this more complex reasoning by prompting students to consider more variables and engage in co-variational reasoning and also by modeling co-variational reasoning in their own arguments.

The selection (and design) of curricular materials employed in the classroom also influences student reasoning and understanding. In the case studies we investigated, curricular materials contributed to the quality of resulting arguments, in some ways similarly to the instructor facilitation. The information provided to students and the questions prompting them to use that information strongly influenced the kind of reasoning which emerged. While directed questions that help students explore macroscopic models or symbolic representations such as equations are not expected to generate arguments, they provide a means for students to extract the evidence they will use in later arguments. However, developing a shared interpretation of the information presented often results in argumentation when students work collaboratively. Tasks that support complex reasoning are more likely to result in argumentation and require students to generate something (*e.g.* a prediction or an experiment). These tasks should appear more frequently in curricular materials, but with an appropriate amount of scaffolding, as these are markedly more difficult for students. Questions are most likely to result in argumentation when students are explicitly prompted to justify answers (claims) or to explain their reasoning. This is true for mathematically intensive tasks as well as for more conceptually oriented questions and should be encouraged as instructors develop or modify activities for the classroom.

The materials also direct the representational levels students use in their reasoning. For example, our analysis revealed that the information provided in the POGIL activities tended to be mostly at the symbolic or macroscopic level and questions tended to prompt students to use primarily mathematical reasoning. This is reflected in student arguments, 75% of which incorporate symbolic representations or reasoning in some part of the argument. To support a greater variety of reasoning within arguments, curricular materials can provide a greater variety of representational levels in information and tasks. In addition, questions should explicitly call for explanations, even for mathematical tasks, that require students to translate across representational levels.

Classroom norms emerge from the interaction of instructor facilitation, curriculum materials, and student actions. The instructor can support the development of classroom reasoning norms necessary for argumentation through instructional decisions. One norm identified in this work was decentering—the explicit differentiation between differing perspectives—which made more students' reasoning visible and thus addressable by other students or the instructor. This norm can be established by directly eliciting perspectives from multiple students or groups of students. Having students report out in whole class discussion supported decentering in small group work as students, particularly the group spokesperson, ensured they understood the group reasoning for the consensus answer. Using strategies for simultaneous reporting out also encouraged students to consider the responses of other students and engage in argumentation as they resolved differences in reasoning. In line with eliciting more student ideas, establishing a norm of answer justification can support an increase in student-generated warrants. By establishing this expectation as a norm, students can begin to demand warrants of each other and co-construct arguments more effectively. The results from this investigation of argumentation in physical chemistry suggest that the implementation of these practices can support argumentation and understanding in physical chemistry classrooms.

Acknowledgements

This work was supported in part by the National Science Foundation under grants #0816792, #0817467, and #0816948. Any opinions, findings, and conclusions or recommendations expressed in this material are those of the authors and do not necessarily reflect the views of the National Science Foundation. We thank Chris Rasmussen, Megan Wawro, George Sweeney, and Nicole Becker, who played key roles in the development of the methodologies used across this research and in the research and publications that led to the findings discussed in this chapter. We also thank the instructors and students who allow us to document and analyze their classroom experiences.

References

1. Undergraduate Professional Education in Chemistry, American Chemical Society, https://www.acs.org/content/dam/acsorg/about/governance/committees/training/2015-acs-guidelines-for-bachelors-degree-programs.pdf (accessed August 2017).
2. K. Bain, A. Moon, M. Mack and M. H. Towns, *Chem. Educ. Res. Pract.*, 2014, **15**, 320.
3. G. Tsaparlis, *Chem. Educ. Res. Pract.*, 2016, **17**, 320.
4. G. Tsaparlis and O. E. Finlayson, *Chem. Educ. Res. Pract.*, 2014, **15**, 257.
5. D. Wren and J. Barbara, *J. Chem. Educ.*, 2013, **90**, 1590.
6. M. E. Derrick and F. W. Derrick, *J. Chem. Educ.*, 2002, **79**(8), 1013.
7. K. E. Hahn and W. F. Polik, *J. Chem. Educ.*, 2004, **81**(4), 567.
8. G. Nicoll and J. S. Francisco, *J. Chem. Educ.*, 2001, **78**(1), 99.
9. R. Kozma, E. Chin, J. Russell and N. Marx, *J. Learn. Sci.*, 2000, **9**, 105.
10. A. H. Johnstone, *Sch. Sci. Rev.*, 1982, **64**, 377.
11. A. H. Johnstone, *J. Comput. Assist, Learn.*, 1991, **7**, 75.
12. G. E. Hernández, B. A. Criswell, N. J. Kirk, D. G. Sauder and G. T. Rushton, *Chem. Educ. Res. Pract.*, 2014, **15**, 354.
13. H.-K. Wu, J. S. Krajcik and E. Soloway, *J. Res. Sci. Teach.*, 2001, **38**(7), 821.
14. M. Cooper, N. Grove, S. M. Underwood and M. W. Klymkowsky, *J. Chem. Educ.*, 2010, **87**, 869.
15. M. R. Abraham, V. M. Williamson and S. L. Westbrook, *J. Res. Sci. Teach.*, 1994, **31**, 147.
16. M. Stieff, M. Ryu and J. C. Yip, Speaking across levels – generating and addressing levels confusion in discourse, *Chem. Educ. Res. Pract.*, 2013, **14**, 376.
17. N. Becker, C. Stanford, M. Towns and R. Cole, *Chem. Educ. Res. Pract.*, 2015, **16**, 769–785.
18. J. N. Nyachwaya and N. B. Wood, *Chem. Educ. Res. Pract.*, 2014, **15**, 720–728.
19. M. M. Cooper, M. D. Caballero, D. Ebert-May, C. L. Fata-Hartley, S. E. Jardeleza, J. S. Krajcik, J. T. Laverty, R. L. Matz, L. A. Posey and S. M. Underwood, *Science*, 2015, **350**(6258), 281–282.
20. National Research Council, (2012), *A Framework for K-12 Science Education: Practices, Crosscutting Concepts, and Core Ideas*, National Academies Press, Washington, DC.

21. National Research Council, (2012), *Discipline-Based Education Research: Understanding and Improving Learning in Undergraduate Science and Engineering*, National Academies Press, Washington, DC.
22. Appendix F – Science and Engineering Practices in the NGSS, http://www.nextgenscience.org/sites/default/files/Appendix%20F%20%20Science%20and%20Engineering%20Practices%20in%20the%20NGSS%20-%20FINAL%20060513.pdf, (Accessed August 2017).
23. L. A. Bricker and P. Bell, *Sci. Educ.*, 2008, **92**, 473–498.
24. E. Forman, J. Larreamendy-Joerns, M. Stein and C. Brown, *Learn. Instr.*, 1998, **8**, 527–548.
25. J. Osborne, *Science*, 2010, **328**(5984), 1354.
26. V. Sampson and D. Clark, *Sci. Educ.*, 2009, **93**(3), 448–484.
27. A. Zohar and F. Nemet, *J. Res. Sci. Teach.*, 2002, **39**(1), 35–62.
28. S. Erduran, S. Simon and J. Osborne, *Sci. Educ.*, 2004, **88**(6), 915–933.
29. M. P. Jiménez-Aleixandre, A. Bugallo Rodríguez and R. A. Duschl, *Sci. Educ.*, 2000, **84**(6), 757–792.
30. G. Eberlein, J. Kampmeier, V. Minderhout, R. S. Moog, T. Platt, P. Varma-Nelson and H. B. White, *Biochem. Mol. Biol. Educ.*, 2008, **36**(4), 262–273.
31. R. Cole, N. Becker, M. Towns, G. Sweeney, M. Wawro and C. Rasmussen, *Int. J. Sci. Math. Educ.*, 2012, **10**, 193.
32. S. Toulmin, *The Uses of Argument*, Cambridge University Press, Cambridge, 1958.
33. R. S. Moog, F. J. Creegan, D. M. Hanson, J. N. Spencer and A. R. Straumanis, *Met. Uni.*, 2006, **17**(4), 41–52.
34. R. S. Moog and J. N. Spencer, *Process-Oriented Guided Inquiry Learning*, ed. R. S. Moog and J. N. Spencer, American Chemical Society, Washington DC, 2008.
35. J. N. Spencer, R. S. Moog and J. J. Farrell, *Physical Chemistry: A Guided Inquiry Thermodynamics*, Houghton Mifflin, Boston, 2004.
36. D. Kuhn, *Sci. Educ.*, 2010, **94**, 810–824.
37. C. Rasmussen and M. Stephan, in *Design Research in Education*, ed. A. E. Kelly and R. Lesh, Erlbaum, Mahwah, N. J., 2008, pp. 195–215.
38. K. S. Taber, *Chem. Educ. Res. Pract.*, 2013, **14**, 156–168.
39. V. Talanquer, *Int. J. Sci. Educ.*, 2011, **33**(2), 179–195.
40. H. Sevian and V. Talanquer, *Chem. Educ. Res. Pract.*, 2014, **15**(1), 10–23.
41. M. Tekkumru-Kisa, M. K. Stein and C. Schunn, *J. Res. Sci. Teach.*, 2015, **52**(5), 659–685.
42. N. Becker, M. Towns, R. Cole, G. Sweeney, M. Wawro and C. Rasmussen, *Chem. Educ. Res. Pract.*, 2013, **14**, 81.
43. N. Becker, C. Stanford, M. Towns and R. Cole, *Chem. Educ. Res. Pract.*, 2015, **16**, 769.
44. A. Moon, C. Stanford, R. Cole and M. Towns, *Chem. Educ. Res. Pract.*, 2016, **17**(2), 353.
45. J. Piaget, *The Language and Thought of the Child*, Meridian Books, Cleveland, OH, 1955.

46. A. Moon, C. Stanford, R. Cole and M. Towns, *J. Chem. Educ.*, 2017, **94**(7), 829–836.
47. J. J. Farrell, R. S. Moog and J. N. Spencer, *J. Chem. Educ.*, 1999, **76**, 570–574.
48. C. Kussmaul and M. Sullivan, Activity Selection and Writing, in *POGIL: An Introduction to Process Oriented Guided Inquiry Learning for Those Who Wish to Empower Learners*, ed. Shawn Simonson, Stylus Publishing, Sterling, VA. 2019, (in press).
49. C. Stanford, A. Moon, M. Towns and R. Cole, The Impact of Guided Inquiry Materials on Student Representational Level Understanding of Thermodynamics, in *Engaging Students in Physical Chemistry*, American Chemical Society, 2018, vol. **1279**, pp. 141–168.
50. U. Kulatunga, R. S. Moog and J. E. Lewis, *J. Coll. Sci. Teach.*, 2014, **43**, 78.
51. C. Stanford, A. Moon, M. Towns and R. Cole, *J. Chem. Educ.*, 2016, **93**(9), 1501.
52. R. Cole, N. Becker, and C. Stanford, *Discourse analysis as a tool to examine teaching and learning in the classroom*, in *Tools of Chemistry Education Research*, ed. D. Bunce and R. Cole, ACS Symposium Series, Oxford University Press, 2014, pp. 61–81.
53. A. Moon, C. Stanford, R. Cole and M. Towns, *J. Res. Sci. Teach.*, 2017, **54**(10), 1322–1346.

CHAPTER 12

Supporting Argumentation in Chemistry Education in Low-income Contexts

AUDREY MSIMANGA* AND BRIGHTON MUDADIGWA

University of the Witwatersrand, School of Education, 27 St Andrews Road, Parktown, Johannesburg 2193, South Africa
*Email: audrey.msimanga@wits.ac.za

12.1 Introduction

Argumentation is a form of high level engagement which is difficult to establish, support or sustain in constrained teaching and learning contexts such as prevail in many South African science classrooms – contexts of poor school resourcing, inadequate infrastructure (large classes and crowded classrooms), inadequate teacher provision and/or preparation, learner under preparedness for the grade level, poverty and hunger as well as teaching and learning science in a language that learners and some teachers are not proficient in. In order for argumentation to happen learners need to be talking and all the factors listed here militate against meaningful classroom talk.

However, our research has revealed some teacher pedagogical practices that are specific to the context. We have watched teacher efforts to get learners who previously would not talk at all to engage in argumentation in ways that have not been reported elsewhere. We recorded collaborative teacher–learner argument construction in whole class discourse and argumentation in small groups in learners' home languages. We concluded that

Advances in Chemistry Education Series No. 2
Argumentation in Chemistry Education: Research, Policy and Practice
Edited by Sibel Erduran

Published by the Royal Society of Chemistry, www.rsc.org

the following are some of the strategies that work in these classrooms and we illustrate these through vignettes from selected classrooms:

- Teacher questioning and teacher response to learner ideas.[1] Similar to Traver (1998) we observed that learner engagement was dependent on the kinds of questions the teacher asked. Also, the teachers' F-move was important in opening up or shutting down learner argumentation depending on whether it was elaborative or evaluative according to Virri and Saari's categorisation of teacher feedback.[2]
- Enabling a shared understanding of goals of the task or activity including mobilising the language resources of the classroom. Here we illustrate how learner engagement in chemistry classrooms can go beyond Jimenez-Aleixandre "doing the lesson" towards "doing the science"[3] if they have a shared understanding of the goal of the activity. In South Africa, English is the preferred language of instruction, yet most learners (and some teachers) are not proficient in the English language. Successful chemistry teachers therefore, tend to pay attention to language matters in their teaching, explicitly mobilising the language resources of the classroom.

We begin with a brief recap on the role of argumentation in science teaching and learning and then go on to discuss low-income teaching and learning environments and the South African context, leading to a discussion of the challenges of working with argumentation in these atypical low-income contexts. Our main argument is that the combination of historical, political, cultural, linguistic and socio-economic state of the South African education system makes it unique, complex and atypical. Thus supporting argumentation in these contexts should begin with supporting teacher and learner orientation to classroom talk and forms of interaction that open up the classroom as a space for a genuine give and take with shared understandings of the goals of classroom interaction. The challenges of creating the requisite dialogic discourse in such classrooms are discussed with reference to vignettes from some South African classroom to illustrate how teachers take up and adapt argumentation for their contexts.

12.2 The South African Context

South Africa is hardly a typical example of a low-income context. According to the June 2017 World Bank list of economies, South Africa is in the upper middle-income category. Low-income economies were defined as those with a gross national income per capita of \$1005 or less in 2016; lower-middle income economies were between \$1006 and \$3955 while upper middle-income economies like South Africa were at \$3956–\$12 235 per capita.[4] However, research shows that South Africa's upper middle-income status does not translate into social and economic benefit for all its citizens. For instance, Van der Berg argued that South Africa's social indicators like life expectancy, infant mortality and quality of education are closer to those of lower-middle income and even low-income countries and that in fact South Africa's social inequality is on the rise.[5] Education research further indicates that these inequalities are mirrored in education. According to Spaull[6] "South Africa has the worst education system of all middle-income countries ... (and) perform(s) worse than many low-income African countries." Spaull[7] further argues that "The strong legacy of apartheid and the consequent correlation between education and wealth have meant that, generally speaking, poorer South African students perform worse academically" (p. 436). According to Spaull the South African primary school comprises "two different education systems... one which is functional, wealthy, and able to educate students; with the other being poor, dysfunctional, and unable to equip students with the necessary numeracy and literacy skills they should be acquiring." Thus, any thinking and/or planning for teaching and learning and resourcing should take these inequalities into account.

This diversity and complexity of the South African education system creates a similarly diverse and complex range of classroom contexts, from the well-resourced schools in affluent communities to poor schools in townships and informal settlements and down to some rural schools with almost non-existent infrastructure. South African schools are categorised according to the quintile system, a system of ranking schools according to their infrastructure in order to systematically improve equity in education.[8] Quintile 1, 2 and 3 schools cater for learners in the poorest neighbourhoods while Quintile 4 and 5 schools are located among the more affluent communities. The quintile system was conceived as a way of addressing the diversity and inequality of schooling and classroom environments. Government funding is apportioned according to Quintiles. Thus, Quintile 1 schools are the poorest 20% of schools and receive the highest allocation per learner, while Quintile 2 caters for the next poorest 20% of schools and Quintile 5 schools represent the most affluent schools and receives the lowest government funding. Unfortunately, in science as in other subjects, it is also a reflection of the vast differences in learning experiences. Learner performance and achievement at the exit level national examinations still mirror the quintile categories and tend to follow racial and socio-economic trends.[3] Learners in low socio-economic contexts still perform the worst and many are unable to proceed with any form of education beyond the 12th Grade.

The vignettes cited in this chapter are from classrooms in Quintile 1–3 schools serving township communities. There is considerable internal variation among this group of schools in terms of resourcing, teacher and learner attributes as well as learner preparation for the grade level. The teaching and learning of science is similarly context dependent and varies among the Quintile 1–3 schools. There is a much higher demand on science teachers in these contexts to draw on a variety of teaching strategies in order to be able to create appropriate learning opportunities for the diversity of learners in their classrooms. All of this is against a back drop of generally low performance in science in South Africa. For instance for the years 2016 and 2017 only 39.5% and 42.2% of the Grade 12 candidates performed above 40% level in the National Senior Certificate examinations, respectively.

12.3 The South African Chemistry Curriculum Context

In South Africa science is compulsory for all learners up to the 9th Year of schooling. At the Senior Phase, Grades 7–9, the curriculum provision is Natural Sciences (NS) which is seen as serving "a dual purpose: it must enable learners to make sense of the world in scientific terms and prepare learners for continuing with a science(s) into the Further Education and Training (FET) phase... and beyond"[9] (p. 7). The NS curriculum covers the Physical Sciences, Life Sciences and Earth Sciences as a broad introduction to the specialisations later in the FET Phase at Grades 10–12. The NS curriculum specifies four knowledge areas: Life and Living (which leads to Life Sciences in the FET); Matter and Materials (which leads to Physical Sciences in the FET); Energy and Change (which leads to aspects of both Life Sciences and Physical Sciences); and Earth and Beyond (which leads to both Life Sciences and Geography). At this level NS teaching should promote and develop scientific literacy in all learners with the recognition that some may not continue with science subjects beyond Grade 9.

At the FET level the science curriculum is organised in the subjects, Life Sciences and Physical Sciences. The Physical Sciences covers both physics and chemistry content in six main knowledge areas, three in chemistry: matter and materials; chemical change; and chemical systems as well as three in physics: mechanics; waves, sound and light; and electricity and magnetism. The South African science curriculum emphasises the inclusion of Indigenous Knowledge Systems (IKS) in the teaching of science with a view to develop in learners the ability to draw critically from multiple worldviews in understand the physical environment.[10,11]

The Physical Sciences curriculum document advocates a spiral approach in the teaching of the science content, in which concepts are introduced in the lower grades and gradually built upon and interlinked progressively with other concepts in successive higher grades.[11] However, in spite of this envisaged connectedness in concept coverage, the Department of Basic Education's annual diagnostic reports continue to show poor learner performance at Grade 12 in questions that require an understanding of basic concepts and linking to concepts covered in higher grades.[12] In another publication we examine this curricular claim for the spiral approach in coverage of chemistry concepts at the FET level. We look at the extent to which the curriculum is comprehensible and progressional in the organisation of concepts and the teaching of electrolytic cells at the FET phase. We argue that there is a mismatch between the curricular aim for a spiral approach and how the concepts are arranged in the curriculum document according to grades which makes it difficult for teaching to take the envisaged spiral approach.[13] In this chapter however, we assume the curricular objective of progressive coverage of chemistry concepts and only focus on how teachers employ argumentation in teaching the chemistry content in the classrooms observed. In particular we focus on contextual issues such as how teachers work with the diversity of learners in South African science classrooms or how IKS might be related to the chemistry curriculum.

In an earlier study Erduran and Msimanga[14] pointed out the unique positioning of argumentation in the South African curriculum context alongside other themes such as IKS. According to Ogunniyi[15] "IKS reflects the wisdom about the environment developed over centuries by the inhabitants of South Africa, and much of this valuable wisdom believed to have been lost in the past 300 years of colonization now needs to be rediscovered and utilized to improve the quality of life of all South Africans." Erduran and Msimanga (2014) noted that other curricula around the world did not conjointly promote the acquisition of IKS within the science curriculum. Thus, the South African curriculum aims to incorporate IKS in science teaching, while also promoting evidence-based reasoning through Nature of Science (NoS) and argumentation. The authors also noted the tensions arising from this curricular decision in terms of the professional development of teachers. Teachers are expected to be able to decide on the appropriate content and pedagogical approaches to use to meet these curriculum objectives. Erduran and Msimanga argue that this requires a high level of teacher subject matter knowledge and confidence, a task which not all South African science teachers are adequately prepared to perform. Argumentation in itself is viewed as a higher-order thinking skill that places significant demands on teachers.

12.4 The Role of Argumentation

The role and importance of argumentation in the teaching and learning of science has been researched extensively internationally and recently in South Africa as well. Arguments are critical for development of epistemological knowledge of the discipline, they facilitate knowledge construction and participation in a dialectical process of critiquing and warranting of claims, whether one's own or others' claims. Thus, argumentation is implicated in the development of learner scientific reasoning and learner conceptual understanding.[16–18] Argumentation is also important in enculturation of learners into the practices of scientists.[19,20] Argumentation is seen to be at the core of science teaching and learning:

> "Argument—justification of claims with evidence—is not a peripheral aspect of science; it operates at the heart of science... Without a chance for pupils to talk to each other, without the space to debate and communicate their ideas, it is difficult to see how they can learn any concepts, let alone the discourse of science" (p. 29)[21]

Argumentation has therefore been the subject of much research as a teaching strategy for stimulating learner participation in structured discussion and as a tool for knowledge construction in science classrooms around the world. Hence argumentation can be regarded as an important form of knowledge to be taught to science learners (the skills, norms and values in the use of evidence to support claims) as well as an important part of the teaching processes.[22] In this chapter we discuss both the role of argumentation as a tool to foster learner participation and engagement and its role in the meaning-making process. Other studies have looked at curricula and pedagogical implications of argumentation.[23,24] Berland and colleagues demonstrated the importance of understanding the goals of argumentation; thus understanding the differences in argumentation for sense-making, for articulating understanding and for persuading others of these understandings. An understanding of these goals of argumentation had implications for teacher decisions in the design and implementation of instructional interventions to support student argumentation and conceptual understanding.

Also important is the work of Sampson and Clark on the value of collaborative argumentation in small groups, the impact of collaboration on scientific argumentation.[25] High school chemistry students were observed to be able to construct arguments in groups more easily than individually. Collaboration during argumentation in small groups enhanced individual student learning. Furthermore, Sampson and Clark (2008) explored the efficacy of argumentation as a research tool and argued that it should be viewed as a tool that can be adapted for different contexts and/or goals.[26] They argued for research beyond micro-analysis of structure of student arguments to research to understand the interaction of "structural, conceptual, epistemic, and social aspects of argument generation" and the "underlying reasons for these patterns" (pp. 469–470). Recent argumentation research in South Africa is exploring these intricate interactions.

12.5 Argumentation in the South African Context

Argumentation research in South Africa is still emergent, having established here only in the last decade. Most of the research has focused on the potential for argumentation in South Africa's curriculum change context. The South African curriculum has undergone several iterations of change and revision, from an outcomes focus with loose framing of the content knowledge to be covered to the current strongly framed content-heavy Curriculum and Assessment Policy Statement (CAPS).[27] As already noted in the previous section, in its outcomes-based era the science curriculum required the inclusion of indigenous knowledge in science teaching. In that phase, argumentation research then linked closely to research on the NoS and the Nature of Indigenous Knowledge Systems (NoIKS). For example, Ogunniyi proposed the Contiguity Argumentation Theory or CAT, an adaptation of Toulmin's model TAP for exploring the interface between different thought systems – in this case science and IKS.[28] Ogunniyi argues that when confronted with counter-intuitive school science learners are exposed to five possible types of adaptive co-existence: the dominant, suppressed, assimilated, emergent and equipollent situations. In the dominant state the underlying assumptions of one thought system are amenable to the teaching and learning context while in the other states the one system is either suppressed by or assimilated into the other in the face of compelling evidence. Then in the emergent state the learner is confronted by a concept for which s/he does not have pre-existing schema and in that case new ones are formed. Finally, in the equipollent state two competing schema exist as parallel notions of the concept. Ogunniyi's argument is that a variety of possible interactions can be expected when argumentation activities are used in South African classrooms particularly in conjunction with IKS.

Other argumentation work in South Africa has focused on development of practicing science teachers' argumentation skills. Scholtz and her colleagues for instance were concerned with understanding the nature of interaction in small group discussions using the TAP model in the context of continuous professional development programmes.[29,30] They reported a unique style of argumentation which precluded rebuttals, which they named inclusive argumentation. Teacher disagreements were either in the form of an affirmative statement followed by a counter claim or alternative warrant, or they were phrased as a question. This form of engagement was explained in terms of the concept of *Ubuntu*. *Ubuntu* is a way of talking in the African cultural context based on a world-view that emphasises consideration for the other person, "instead of comparing and judging individual contributions on their intrinsic merit, contributions are judged in terms of the extent to which they promote harmony and reciprocity"[31] (p. 31). Rebuttals were used differently from the Toulmin's argument structure as participants softened the challenge, as it were, in an effort to promote this harmony and reciprocity. Both the Ogunniyi and Scholtz research involved adult participants. Recent work on argumentation in South Africa has focused on more adult–child and child–child or teacher–learner and learner–learner classroom interactions.[32]

Previous argumentation research in South Africa focused on small group discussions by largely bilingual teachers and student teachers in the Cape Province of South Africa. Argumentation research with younger learners in the more linguistically diverse Gauteng Province, particularly in township schools, has come up with forms of argumentation unique to multi-lingual contexts. For instance, hybrid forms of engagement were observed as teachers adopted pedagogical strategies that required previously silent learners to engage in classroom talk. Instead of using the traditional argumentation activities that require learners to engage with each other and support their thinking among themselves, teachers were observed modelling argumentation in whole class interaction. As teachers changed the kinds of questions they asked and the forms of feedback that they gave in response to learner answers, learners started talking and engaging each other about the science content. Interactions transformed from teacher–learner 1, teacher–learner 2 to triadic teacher–learner 1–learner 2 forms. Eventually, the teacher and learners were engaging in collaborative argumentation. Ultimately as learner talk became entrenched and teachers started using tasks designed to promote small group discussions another unique form of argumentation was observed. Learners were observed to immediately switch to their home language, using the more familiar language to make sense of the science and to challenge each other in small group discussions.[33]

The emergence of these contextual forms of engagement; inclusive argumentation, contiguity argumentation, collaborative argumentation, argumentation in home languages all attest to the potential for adaptations of argumentation in South African contexts.[34] The study of argumentation in South African classrooms is dependent on transformation of classroom discourse from interaction dominated by teacher talk to classrooms where learners are actually talking in class. Thus, for the most part, supporting argumentation in these contexts is about developing dialogic discourses in the science classrooms – helping teachers to get their learners talking.

12.6 Development of Dialogic Discourse – How to Get Learners Talking and Arguing

Literature emphasises the importance of whole class discussion to develop learner ability to dialogue.[35,36] According to Roth, whole class discussions are just as important as small group discussions in providing opportunities for developing interactive classroom discourse. Teacher and peer questions, comments, and suggestions help to sustain the conversations. These kinds of conversations have the potential to break the barrier between formal classroom discourse, formal science and learner personal experience, providing for Aikenhead and Jegede's idea of learner navigation of borders between the different sub-cultures that they belong to.[37] In the South African context this would include border crossing between formal science and IK and between the language domains.

According to Yackel dialogic discourse results from the negotiation of social roles which have to be taken-as-shared, in other words intersubjectivity between the teacher and learners and also among the learners.[38] This happens as the teacher validates, suppresses, recognises, reiterates, clarifies, acknowledges and emphasises learner ideas and involving as many learners as possible in the co-construction of arguments (explaining and justifying own thinking). Among learners this intersubjectivity happens through inclusiveness, asking questions, and challenging each other's ideas. According to Driver and Newton (2000) it is also important for participants to be awareness of the norms of engagement.

However, establishing the desired dialogic discourse in the science classroom is not easy, particularly where learning happens in a foreign language. Learners are not able to engage confidently in classroom talk, much less construct scientific arguments. Rojas-Drummond and Zapata[39] describe a dual demand on learners to build the requisite registers for the language of instruction as well as engage in the meaning-making processes of the discussion. Learner engagement in their natural languages seems to mediate the dual demand on learners, facilitating easier engagement in the meaning-making process. However, following small group discussions learners are usually required to report back to the rest of the class or present their work in writing in English. This introduces different demands on learners to make the transition from meaning-making in their own language to articulation of their understanding in English. Supporting argumentation in these classrooms can sometimes mean a choice between investing in the emerging learner talk and focusing on developing learner argumentation skills. For many South African learners acquainted with traditional teaching methods this shift to interactive methods that requires them to participate in classroom discussions is not easy. The vignettes discussed in the next section illustrate some of the interaction observed in chemistry lessons and how the teachers supported development of learner argumentation skills.

12.7 Illustrating Argumentation in South African Chemistry Classrooms

We observed teacher efforts to get learners talking in these classrooms and learners ultimately began to engage in argumentation in various ways. Two key forms of teacher support merged; creating an environment conducive to learner talk, and mediating language transitions. The first vignette illustrates how one teacher's questions and the kind of feedback that she gave in response to learners' contributions created opportunities for learner engagement in classroom discussion and argumentation.

12.7.1 Teacher Questioning and Responses to Learner Contributions

The following excerpts are taken from my other published work.[33,40] The first is from Msimanga and Lelliott (2014) (pp. 195–200) and the second is from Msimanga (2013) (pp. 192–206).

The vignettes in this section illustrate co-constructed argumentation between a teacher and her chemistry class talking about chemical bonding. The events presented here happened in the middle of the lesson around an error that became apparent when a learner responded to a teacher question. The teacher asked what would happen when two atoms involved in formation of a covalent bond drew closer together as their potential energy decreased. The learner Tahari answered that "*the atoms are now going to be negatively charged*". Instead of simply ignoring the incorrect answer and asking another learner as she would normally do, the teacher actually engaged the class in a question and answer session to address the error. We focus here on the questions that she asked and how she worked with the learners' responses to her questions, resulting in a previously unplanned for collaborative argumentation episode. The utterances are transcribed verbatim in the language that they were made in (in italics) and translated into English (Table 12.1).

The transcripts open with the teacher posing a thinking question '*What do you think is going to happen?*' to which Tahari answers '*…they are now going to be negatively charged.*' The teacher's '*Why?*' seemed to be a genuine reaction to an unexpected or even surprising learner answer. This is borne out in the teacher's follow up question when it becomes clear from Tahari's answer in turn 117 that she is articulating a deep misunderstanding, perhaps even misconception. Tahari supported her claim that the two atoms will become negatively charged as a result of moving closer to each other with the evidence that '*they are about to come together*' warranted by the assertion that, '*they are attracting each other.*' Normally, this teacher would have simply ignored the error, directed the original question to another learner and kept going until she got the correct answer or provided the answer herself. This she said "saved time" and got the class back on task. In this instance however, she decided to pursue the error, inviting other learners to, as it were, interrogate their peer's thinking. By the end of the episode eleven learners had spoken and participated in resolving the error. It could be argued that the teacher facilitated and mediated shared meaning making by more than just the one learner.

It also became clear as the teacher continued to question them that Tahari's error was shared by other learners as well. According to Berland and Reiser (2009) in sense-making argumentation the aim is to get learners to open up to their peers' ideas and in the process enable them to modify their own understanding. We see this in the next excerpt as the teacher continued her line of questioning with another learner, Vuma who joined the argument to articulate his own (erroneous) understanding (Table 12.2).

Vuma seemed to agree with Tahari's claim but on different grounds. Tahari warranted her claim on the nature of forces operating between the atoms as the distance between them decreases; Vuma, on the other hand, premised his argument on evidence about covalent bonds (turn 124). Of course, this new argument did not bring the class any closer to resolving the problem but it helped reveal more learner erroneous thinking. The teacher's response at the beginning of the next excerpt below '*When does an atom*

Table 12.1 Turns 114–118, transcript of bond energy lesson.

114	Teacher	If they move closer to one another this means that the distance between them decreases. What do you think is going to happen?
115	Tahari	Er if they decrease they are now going to be negatively charged
116	Teacher	Why?
117	Tahari	Ma'am because now they are attracting each other Ma'am *Angithi manje sezahlangana sezifunukuhlangana* (they are now just about to come together)
118	Teacher	(*addressing the rest of the class*) What do you think of that statement?

Table 12.2 Turns 121–126, transcript of bond energy lesson.

121	Teacher	She says that if they come closer the hydrogen is going to become negative because of the attraction force. What do you think about that?
122	Vuma	Er Ma'am I much agree
123	Teacher	Oh you agree?
124	Vuma	Yes Ma'am coz Ma'am when they bond covalently they are bonding to achieve the nearest noble gas
125	Class	Yes
126	Vuma	Ok fine then so soon as they reach the noble the nearest noble gas they share and then when they are sharing, er, they become negatively charged

Table 12.3 Turns 151–164, transcript of bond energy lesson.

151	Teacher	When does an atom become negatively charged?
152	Teacher	Let me give you a chance
153	Thinta	Ma'am I disagree with the statement coz Ma'am I think when the two atoms (inaudible) the chemical potential energy will increase
154	Teacher	Why do you disagree with the statement?
155	Thinta	It's because Ma'am when the two atoms interact it's impossible for them to be negatively charged
156	Tahari	Ma'am didn't you say...?
157	Teacher	Why?
158	Thinta	*Azikathintani* Ma'am (they are not yet touching)
159	Class	Yes yes yes
160	Tahari	Ma'am didn't you say when they er when they get closer to each other when they attract each other the potential energy it will decrease
161	Thinta	It's like this... (*holding pen and set square I each hand and moving them towards each other*)
162	Tahari	It will decrease...
163	Thinta	*Azikathintani* (they are not yet touching)
164	Tahari	Ma'am you said you said the potential energy will decrease and therefore those atoms are going to be negatively charged

become negatively charged?' Seems to suggest that she had identified the root of the learners' misunderstanding – the circumstances under which an atom becomes negatively charged.

Of particular interest to us is how the teacher maintains the cognitive demand of the task and does not simplify the question as the discussion unfolds. Research in mathematics classrooms in South Africa indicates that in many cases as the teacher senses that her learners are struggling to get to the correct answer there is a tendency to then ask simpler questions thus reducing the cognitive demand of the task at hand.[41] In this case the teacher asked the question differently but without reducing its difficulty and so maintained the cognitive demand of the task.

What was also interesting was how as the discussion continued, learners started responding more directly to each other's ideas rather than focusing on answering or responding to the teacher's question, thus rebutting each other's claims directly – turns 160–164 (Table 12.3). The teacher had

Table 12.4 Turns 184–185 (transcript of bond energy lesson).

184	Bonga	I don't think of any er er negative charge forming coz we are dealing with covalent bonding here (inaudible) they share their electrons with each other so I think negatively charged electrons only form when you are dealing with ionic bonding
185	Class	Yes yes

managed to work with learner contributions in ways that opened up the classroom space for learner talk and argumentation. The difficulty for the teacher in this case is finding the balance between allowing the engagement to continue sufficiently to facilitate learner conceptual understanding and moving the lesson ahead, thus shutting down and opening up dialogic engagement in the lesson.

The teacher did in fact shut down the dialogic engagement at this point and proceeded to ask some guiding questions that led the class to the correct answer. At the end of the episode one learner who had been silent throughout the discussion raised his hand to make an unsolicited contribution (Table 12.4).

This raises the question of how participation is defined in this context. What can one conclude about of the level of participation when only eleven out of more than fifty learners in the class engage in the 15-minute episode of teacher–learner talk? Are the other forty participating? If so what is the nature of their participation? Are they engaging in silent personal argumentation and like Bonga are they able to ultimately construct and support correct arguments in the end? How would a teacher know if she is supporting argumentation for all?

In the next set of vignettes we demonstrate how argumentation played out in another chemistry class during small group discussions in ways that speak to the current debates on the role of language in multi-lingual science classrooms. In South Africa the debate is mainly about the challenge of teaching and learning science in English, a language that most learners are not proficient in while attempting to prepare learners for final assessment (and national examinations) in the English language.

The vignettes are used to illustrate how learners created for themselves safe spaces to draw on their linguistic capital, using their own home languages to engage with the science concepts for meaning-making and conceptual understanding. They then have to negotiate the border crossing back to the language of instruction as they articulated these understandings in their reports in English to the whole class.

12.7.2 Shared Understanding of Goals and Working with Learner Language

Some of the data cited here is published in Msimanga and Lelliott 2014, pp. 1159–1183.

We present vignettes from a chemistry lesson on reactions of acids and bases. This teacher used small group tasks quite frequently in her classroom with learners in self-selected groups of four and five. We noted that learners always deliberately sought out the same members each time to form the groups. We learnt later that these were not only friendship groups, but that learners selected those who share similar home languages. The utterances are in the language in which they were made (in italics) with English translation. We present data from one of the small groups.

The task was to determine the products of the reaction of sulphuric acid with magnesium oxide, identify the salt formed and balance the equation. At the end of their discussion, they put up on the board the equation $H_2SO_4 + MgO = MgSO_4 + H_2O$. The group comprised one female (Lg1) and three male learners (Lb1, Lb2 and Lb3). This is only part of the discussion to illustrate how they drew on their different home languages (Table 12.5).

The excerpt begins with the group discussion the procedure to follow in resolving the problem. The learners debated which of the tasks to start with; determine the products first to identify the salt or construct and balance the

Table 12.5 Group 1 discussion in isiZulu and seSotho.

113	Lg1	*Nanku bhekani bhekani ukuthi nje kufanele* ***sibhalanse i-equation*** *ngendlela esijayele* ***ukubhalansa*** *ngayo* ***i-equation*** *and then ke sesizo* ***fomuleyitha*** *i-gama laleyo* ***salt and because of ama results*** *esowathola out of* ***i-equation*** *niya* ***andastenda****?* (here here look this is it *(pointing to an equation in her book)* it's just that we have to balance the equation in the way we always do and then we can formulate the name of the salt from the results we get out of the equation, do you understand?)
114	Lb2	*Phinda futhi* (say that again)
115	Lg1	*Sibhalansa i-equation masiqeda ukubhalansa i-equation i-end result esizoyithola sizo neyima i salt leyo because out of i-equation let's say we get sodium* (We balance the equation and then after balancing the equation the end result that we get we name the salt based on that equation let's say we get sodium...)
116	Lb3	*Lamoutlwisisa?* (do you understand her?)
117	Lb1	*Ushukuthi njengoba iso soqala sibhalanse le-equation?* (do you mean that since it is like this we first need to balance this equation?)
118	Lg1	*Hhmm masiqeda lentole le answer le siyitholile sineyima isalt* (yes and thereafter from the answer that we get we name the salt)
119	Lb2	*Sibhalansa yona kuqala?* (so we balance this one first?)
120	Lg1	*Kuzoba yigama let's say kuzoba yi nitric...* (it is going to be the name of the salt let's say it is nitric)
121	Lb1	*Oh sibhalansa le kuqala?* (so we balance this one first?)
122	Lg1	*Ja* (yes)
123	Lb3	*Kanti a reetsi reaction pele? Keona reyi balanse?* (but don't we need to do the reaction first? And then balance it?)
126	Lg1	*Maubhalansile sesibhalansile sesizoneyima ...* (we'll name it after balancing)
125	Lb3	Ja e ya-*naming* I do understand mara you know what I'm saying? *mara ha riyibhalansa retswantsi rebhalanse* reaction this side *pele* (I do understand the naming but what I'm saying is that we need to balance the reaction this side first)

equation first and then name the salt. Lg1 argues that they should start by balancing the equation "...in the way we always do and then we can formulate the name of the salt from the results we get out of the equation." A lengthy exchange ensues with all four learners debating Lg1's suggestion until they reach consensus by turn 122, all in their home languages. The group went on to construct the correct equation and balance it, after which they successfully named the salt.

We believe that using their home languages made it easier for these learners to engage both procedurally, going beyond the routine process of the mathematical manipulation of the chemical equation and conceptually in joint meaning making. They challenged and pushed each other to explain and defend their thinking. They were able to support their arguments without being distracted by the language shifts, moving effortlessly between transliteration and code switching.

12.8 Discussion and Conclusions

Research on chemistry education has established that because of its complexity, chemistry is a difficult subject to teach and to learn. Over four decades ago Johnstone[42] proposed a model that recognises three domains of chemical knowledge; the macroscopic, the sub-microscopic and the symbolic to guide the teaching of chemistry. Johnstone argued that some of the difficulties in learning chemistry involved what he termed multilevel thought.[43] Thus, chemistry learning is about understanding abstract theoretical explanations of chemical phenomena and chemistry teaching must take into consideration this need for learners to think and understand the multiple domains of chemistry all at once.

The need to teach chemistry in ways that mediate learning amid this complexity imposes huge demands not only on teacher subject matter knowledge but also the use of appropriate pedagogical approaches. The implications for teacher preparation and resourcing of chemistry education cannot be over emphasised. However, in low socio-economic environments, investment in education is very low including teacher education.

In this chapter we have demonstrated the potential for argumentation as a teaching strategy and a learning tool in these low socio-economic contexts in South Africa. The vignettes illustrate possible pedagogical approaches that can be adopted to maximise the learning of chemistry in difficult contexts. With adequate preparation and support teachers can and do adapt argumentation strategies to facilitate the learner multiple thinking required to understand chemistry. We concur with Aydeniz and Gurcay's[44] argument to develop in teachers "epistemic and pedagogical knowledge of argumentation" and that the science teacher educators must "provide epistemic and professional development experiences to future science teachers so that they can help their students appropriate the epistemic and social norms of science in the context of argumentation" (p. 270).

Sound teacher epistemic and pedagogical knowledge of argumentation has potential to shift interactions in science classrooms towards a more dialogic discourse that creates opportunities for learner involvement and meaningful engagement with both the teacher and their peers.

However, in many low-income countries the challenges of teaching and learning chemistry are compounded by the teaching and learning of chemistry in a language that the learner is not proficient in. Thus, science teacher education has the added responsibility to model for future teachers strategies to mobilise and maximise all the language resources of the classroom.

References

1. R. Traver, What is a good guiding question?, *Educ. Leadership*, 1998, **55**(6), 70–74.
2. J. Viiri and H. Saari, Teacher talk patterns in science lessons: Use in teacher education, *J. Sci. Teach. Educ.*, 2006, **17**, 347–365.
3. M. P. Jimenez-Aleixandre, A. B. Rodrigues and R. Duschl, "Doing the lesson" or "doing science": Argument in High School Genetics, *Sci. Educ.*, 2000, **84**, 757–792.
4. Group TWB. World Bank list of economies: The World Bank; 2017 [Available from: https://muse.jhu.edu/about_files/uploaded/wdi2017.pdf.].
5. S. van der Berg, Inequality, poverty and prospects for redistribution, *Dev. Southern Africa*, 2014, **31**(2), 197–218.
6. N. Spaull, *South Africa's Education Crisis: The Quality of Education in South Africa 1994–2011*, CDE, Johannesburg, 2013.
7. N. Spaull, Poverty & privilege: Primary school inequality in South Africa, *Int. J. Educ. Dev.*, 2013, **33**(5), 436–447.
8. South Africa D. South African Schools Act, 1996 (Act No 84 of 1996), *Amended National Norms and Standards for School Funding*, Department of Education, Pretoria, 2006.
9. Education DoB, Curriculum and Assessment Policyy Statement (CAPS): Natural Sciences, in *Education*, DBE, Pretoria, 2011.
10. Department of Education, National Curriculum Statement Grades R – 12: Curriculum and Assessment Policy (CAPS) Life Sciences, in *Education*, Department of Education, Pretoria, 2011.
11. Department of Education, National Curriculum Statement Grades R – 12: Curriculum and Assessment Policy (CAPS) Physical Sciences, in *Education*, Department of Education, Pretoria, 2011.
12. Education DoB, National Senior Certificate Diagnostic Report Part 1, in *Education*, DBE, Pretoria, 2017.
13. B. Mudadigwa and A. Msimanga, How spiral is the Chemistry FET curriculum, presented at XVIII IOSTE conference, Malmo, Sweden, August 13–17, 2018.
14. S. Erduran and A. Msimanga, Science curriculum reform in SOuth Africa: Lessons for professional development from research on argumentation in science education, *Educ. Change*, 2014, **18**(Sup 1), S33–S46.

15. M. B. Ogunniyi Contiguity Argumentation Theory and Educators' Conceptions of the Nature of Science and Indigenous Knowledge Systems. SAARMSTE Proceedings. 2007.
16. *Research in Argumentation in Science Education: Perspectives from Classroom-based Research*, ed. S. Erduran and M. P. Jimenez-Aleixandre, Dordrecht, Springer, 2007.
17. K. L. McNeill and D. S. Pimentel, Scientific discourse in three urban classrooms: The role of the teacher in engaging high school students in argumentation, *Sci. Educ.*, 2010, **94**, 203–229.
18. A. Zohar and F. Nemet, Fostering students' knowledge and argumentation skills through dilemmas in human genetics, *J. Res. Sci. Teach.*, 2002, **39**(1), 35–62.
19. R. Driver, P. Newton and J. Osborne, Establishing the norms of argumentation in classrooms, *Sci. Educ.*, 2000, **84**(3), 287–312.
20. M. P. Jimenez-Aleixandre and S. Erduran, Argumentation in science education: An overview, in *Argumentation in Science Education: Perspectives from Classroom-based Research Dordrecht*, ed. S. Erduran and M. P. Jimenez-Aleixandre, Springer, The Netherlands, 2008, pp. 3–28.
21. S. Erduran, Special editorial: Argument, discourse and interactivity, *Sch. Sci. Rev.*, 2007, **88**(324), 29–30.
22. A. Tiberghien, Foreword, in *Argumentation in Science Education: Perspectives from Classroom-based Research*, ed. S. Erduran and M. Jimenez-Aleixandre, *Science & Technology Education Library. 35*, Springer, Tallahasse, USA, 2008, pp. ix–xv.
23. L. K. Berland and D. Hammer, Framing for scientific argumentation, *J. Res. Sci. Teach.*, 2012, **49**(1), 68–94.
24. L. K. Berland and B. R. Reiser, Making sense of argumentation and explanation, *Sci. Educ.*, 2009, **93**, 26–55.
25. V. Sampson and B. C. Clark, A Comparison of the Collaborative Scientific Argumentation Practices of Two High and Two Low Performing Groups, *Res. Sci. Educ.*, 2009, **41**(1), 63–97.
26. V. Sampson and D. Clark, Assessment of the ways students generate arguments in science education: Current perspectives and recommendations for future directions, *Sci. Educ.*, 2008, **92**, 447–472.
27. Department of Education, *The National Curriculum Statement Grade 10–12 (General) Physical Science*, Department of Education, Pretoria, 2003.
28. S. Kwofie and M. B. Ogunniyi, Effects of dialogical argumentation-based workshops on primary and secondary school teachers' ability to co-construct the concept of solubility, in *Proceedings of the 19th Annual Meeting of the Southern African Association for Research in Mathematics*, ed. T. Mamiala and F. Kwaisi, *Science and Technology*, SAARMSTE, Mafeking, 2011, pp. 277–286.
29. M. Braund, F. Lubben, Z. Scholtz, M. Sadeck and M. Hodges, Comparing the effect of scientific and socio-scientific argumentation tasks: lessons from South Africa, *Sch. Sci. Rev.*, 2007, **88**(324), 67–76.

30. Z. Scholtz, R. Watson and O. Amosum, Investigating teachers' responses to curriculum innovation, *Afr. J. Res. Sci. Math. Technol. Educ.*, 2004, **8**(1), 41–52.
31. Z. Scholtz, M. Braund, M. Hodges, R. Koopman and F. Lubben, South African teachers' ability to argue: The emergence of inclusive argumentation, *Int. J. Educ. Dev.*, 2008, **28**(1), 21–34.
32. A. Msimanga and A. Lelliott, Making sense of science: Argumentation for meaning-making in a teacher-led whole class discussion, *Afr. J. Res. Math. Sci. Technol. Educ.*, 2012, **16**(2), 192–206.
33. A. Msimanga and A. Lelliott, Talking Science in Multilingual Contexts in South Africa: Possibilities and challenges for engagement in learners home languages in high school classrooms, *Int. J. Sci. Educ.*, 2014, **36**(7), 1159–1183.
34. F. S. Òtúlàjà, A. Cameron and A. Msimanga, Rethinking argumentation-teaching strategies and indigenous knowledge in South African science classrooms, *Cultural Stud. Sci. Educ.*, 2011, **6**(3), 693–703.
35. J. L. Lemke, *Talking Science: Language, Learning and Values*, Ablex Publishing, Norwood, NJ, 1990.
36. W.-M. Roth, Learning to Talk Engineering Design: Results from an Interpretive Study in a Grade 4/5 Classroom, *Int. J. Technol. Design Educ.*, 1996, **6**, 107–135.
37. G. S. Aikenhead and O. J. Jegede, Cross-cultural science education: A cognitive explanation of a cultural phenomenon, *J. Res. Sci. Teach.*, 1999, **36**, 269–287.
38. E. Yackel, What we can learn from analysing the teacher's role in collective argumentation, *J. Math. Behav.*, 2002, **21**, 423–440.
39. S. Rojas-Drummond and M. P. Zapata, Exploratory talk, argumentation and reasoning in Mexican school children, *Lang. Educ.*, 2004, **18**(6), 539–557.
40. A. Msimanga *Talking Science in South Africa: Case Studies of Grade 10–12 Science Classrooms in Soweto*, University of the Witwatersrand, Johannesburg, 2013.
41. K. Brodie, Dialogue in mathematics classrooms: beyond question-and-answer methods, *Pythagoras*, 2007, **66**, 3–13.
42. H. Ja, Macro- and microchemsitry, *Sch. Sci. Rev.*, 1982, **64**, 377–379.
43. A. H. Johnstone, Why is science difficult to learn? Things are seldom what they seem, *J. Comp. Assisted Learn.*, 1991, **7**, 75–83.
44. A. Aydeniz and D. Gurcay, Assessing quality of pre-service physics teachers' written arguments, *Res. Sci. Technol. Educ.*, 2013, **31**(3), 269–287.

Subject Index